Lecture Notes in Control and Information Sciences

Volume 499

This series reports new developments in the fields of control and information sciences—quickly, informally and at a high level. The type of material considered for publication includes:

1. Preliminary drafts of monographs and advanced textbooks
2. Lectures on a new field, or presenting a new angle on a classical field
3. Research reports
4. Reports of meetings, provided they are

(a) of exceptional interest and
(b) devoted to a specific topic. The timeliness of subject material is very important.

Indexed by EI-Compendex, SCOPUS, Ulrich's, MathSciNet, Current Index to Statistics, Current Mathematical Publications, Mathematical Reviews, IngentaConnect, MetaPress and Springerlink.

Sebastiaan van den Eijnden · Marcel Heertjes ·
Maurice Heemels · Henk Nijmeijer

Hybrid Integrator-Gain Systems

Analysis, Design and Applications

Springer

Sebastiaan van den Eijnden
Department of Mechanical Engineering
Eindhoven University of Technology
Eindhoven, Noord-Brabant
The Netherlands

Marcel Heertjes
Department of Mechanical Engineering
Eindhoven University of Technology
Eindhoven, Noord-Brabant
The Netherlands

Maurice Heemels
Department of Mechanical Engineering
Eindhoven University of Technology
Eindhoven, Noord-Brabant
The Netherlands

Henk Nijmeijer
Department of Mechanical Engineering
Eindhoven University of Technology
Eindhoven, Noord-Brabant
The Netherlands

ISSN 0170-8643 ISSN 1610-7411 (electronic)
Lecture Notes in Control and Information Sciences
ISBN 978-3-032-18100-8 ISBN 978-3-032-18101-5 (eBook)
https://doi.org/10.1007/978-3-032-18101-5

Mathematics Subject Classification: 93

This Springer imprint is published by the registered company Springer Nature Switzerland AG
The registered company address is: Gewerbestrasse 11, 6330 Cham, Switzerland

Preface

The desire of researchers and practitioners to overcome fundamental limitations of linear time-invariant control systems at least dates back to the 1950s when J. C. Clegg introduced his integrator with a resetting mechanism. Since its introduction, the Clegg integrator has inspired researchers and practitioners alike, which resulted in nonlinear control strategies and methods that enable the go beyond the classical limitations of linear control. Today, nonlinear control is firmly embedded in Systems and Control programs at universities, conferences, and enjoys a large amount of journal and textbook publications. It has a strong foundation in stability, signals, and system analysis, although it still offers challenges, for example, in the direction of performance. Addressing the problem of performance and its limitations is especially relevant in the high-tech systems industry, where performance demands are ever increasing and where linear control strategies still form the industrial standard.

This technical monograph is the result of collecting ideas for more than two decades at the Department of Mechanical Engineering from the Eindhoven University of Technology in the context of performance of nonlinear control systems, with special interest in the frequency-domain perspective. These ideas ended up in the development of variable-gain controllers with the aim of addressing Bode's sensitivity integral (the waterbed effect) as well as the development of reset and switching control schemes for dealing with Bode's gain–phase relationship. The latter in the form of nonlinear filters that mimic the gain characteristics of linear filters such as integrators, lead–lag filters, and low-pass filters, but improve on the phase lag. These developments were nurtured from within the high-tech ecosystem in the Eindhoven "Brainport" region of the Netherlands, with its unique motion control industry that offered, and still offers, an unparalleled environment for applying nonlinear control. This ecosystem specifically provided the conditions and working environment needed to develop hybrid integrator–gain systems (HIGS) to the state that the reader will find reflected in this monograph.

We are grateful to many people who influenced our thinking and helped us create our views on HIGS. However, two persons need to be mentioned. Daniel Deenen who introduced the basic design and analysis of HIGS in his Master's thesis and Bardia Sharif who formalized HIGS and brought it within the sampled-data context in his Ph.D. thesis. Also, we are grateful for the financial support of the European Research Council (ERC) under the Advanced ERC grant PROACTHIS, no. 101055384.

Eindhoven, The Netherlands
July 2025

Sebastiaan van den Eijnden
Marcel Heertjes
Maurice Heemels
Henk Nijmeijer

Competing Interests The authors have no competing interests to declare that are relevant to the content of this manuscript.

Notation

- $\mathbb{R}$: the set of real numbers; $\mathbb{C}$: the set of complex numbers; $\mathbb{N}$: the set of natural numbers, i.e., $\mathbb{N} = \{0, 1, 2, \dots\}$; $\mathbb{Z}$: the set of integers. Let X denote one of these sets. For $a \in X$, denote $X_{\geq a} = \{x \in X | \, x \geq a\}$ to indicate subsets of X consisting of elements greater than or equal to a. Similar convention is used for $X_{\leq a}$, $X_{>a}$, and $X_{<a}$. In particular, $\mathbb{R}_{\geq 0}$ is the set of nonnegative real numbers.
- For a complex number $z \in \mathbb{C}$, $\mathrm{Re}\{z\}$ and $\mathrm{Im}\{z\}$ denote its real and imaginary parts, respectively. Its complex conjugate is denoted by $\overline{z}$.
- The transpose of an $n \times m$ matrix $M \in \mathbb{R}^{n \times m}$ is denoted by $M^\top \in \mathbb{R}^{m \times n}$.
- $\mathbb{S}^{n \times n} := \left\{M \in \mathbb{R}^{n \times n} | M = M^\top\right\}$: the set of real symmetric $n \times n$ matrices.
- $\mathbb{S}^{n \times n}_{\geq 0}$: the set of real symmetric $n \times n$ matrices with nonnegative elements.
- A matrix $M \in \mathbb{S}^{n \times n}$ is:
  - *positive definite, denoted* $M \succ 0$, if $x^\top M x > 0$ for all $x \in \mathbb{R}^n \backslash \{0\}$;
  - *positive semi-definite, denoted* $M \succeq 0$, if $x^\top M x \geq 0$ for all $x \in \mathbb{R}^n$;
  - *negative definite, denoted* $M \prec 0$, if $x^\top M x < 0$ for all $x \in \mathbb{R}^n \backslash \{0\}$;
  - *negative semi-definite, denoted* $M \preceq 0$, if $x^\top M x \leq 0$ for all $x \in \mathbb{R}^n$.
- For any matrix M, the notation $\mathrm{He}(M) := M + M^\top$ and $\mathrm{Pr}(M) := M^\top M$ is used.
- $\ker M$: the kernel (also denoted nullspace) of a matrix $M \in \mathbb{R}^{n \times m}$, i.e., $\ker M = \{x \in \mathbb{R}^m \mid Mx = 0\}$; $M_\perp$: a matrix whose columns form a basis of $\ker M$, so that $MM_\perp = 0$; $\mathrm{im}\, M$: the image of M, i.e., $\mathrm{im}\, M = \{Mv \mid \, v \in \mathbb{R}^m\}$.
- Let $A_1, A_2, \dots, A_n$ be square matrices (not necessarily of the same size). Then $\mathrm{diag}(A_1, A_2, \dots, A_n)$ denotes the block diagonal matrix with $A_1, A_2, \dots, A_n$ on the main diagonal and all off-diagonal blocks equal to zero matrices of appropriate dimension.
- The Euclidean inner product between two vectors $v \in \mathbb{R}^n$ and $w \in \mathbb{R}^n$, denoted by $\langle v, w \rangle$, is defined as $\langle v, w \rangle = v^\top w$.
- Inequalities $>, \geq, <, \leq$ applied to vectors are understood componentwise. For instance, $v \geq 0$ means each component of $v \in \mathbb{R}^n$ is nonnegative.
- The eigenvalues of a matrix $A \in \mathbb{R}^{n \times n}$ are denoted by $\lambda(A)$; individual eigenvalues are referred to as $\lambda_i(A)$.

- A matrix $A \in \mathbb{R}^{n \times n}$ is called Hurwitz (or stable) if all its eigenvalues have strictly negative real parts, i.e., $\mathrm{Re}\{\lambda_i(A)\} < 0$ for all $i = 1, 2, \ldots, n$, or stated otherwise, when all eigenvalues lie in the open left-half complex plane.
- A multi-input multi-output (MIMO) transfer function matrix $G(s)$, $s \in \mathbb{C}$, with elements $G_{ij}(s)$ being rational functions of s with real coefficients, is said to be:

  - *stable* if all poles of each $G_{ij}(s)$ lie in the open left-half complex plane;
  - *proper* if each $G_{ij}(s)$ satisfies $\lim_{s \to \infty} G_{ij}(s)$ is finite;
  - *strictly proper* if each $G_{ij}(s)$ satisfies $\lim_{s \to \infty} G_{ij}(s) = 0$.

  Real, rational, proper, and stable transfer function matrices of size $m \times n$ are denoted by $G(s) \in \mathcal{RH}_{\infty}^{m \times n}$.
- For linear time-invariant (LTI) systems, $G(j\omega)$, $\omega \in \mathbb{R}$, denotes the frequency response of $G(s)$. The Hermitian transpose is denoted by $G^*(j\omega) := G^\top(-j\omega)$. The H_∞-norm is given by $\|G(j\omega)\|_\infty := \sup_{\omega \in \mathbb{R}} \sigma(G(j\omega))$, where σ denotes the maximum singular value of a complex matrix. In the single-input single-output (SISO) case, $\|G(j\omega)\|$ and $\angle G(j\omega)$ denote the magnitude and phase of $G(j\omega) \in \mathbb{C}$.
- $\overline{\mathcal{S}}$: the closure of a set $\mathcal{S} \subset \mathbb{R}^n$.
- $\mathrm{int}(\mathcal{S})$: the interior of a set $\mathcal{S} \subset \mathbb{R}^n$.
- $\|\cdot\|$: the standard Euclidean norm on $\mathbb{R}^n$, i.e., $\|x\| = \sqrt{x^\top x}$ for $x \in \mathbb{R}^n$.
- $\|f\|_p$: the $\mathcal{L}_p$-norm of a measurable function $f : \mathbb{R}_{\geq 0} \to \mathbb{R}^n$, that is, $\|f\|_p = \left(\int_0^\infty \|f(t)\|^p dt\right)^{1/p}$ for $1 \leq p < \infty$ provided the integral exists and is finite. The notation $f \in \mathcal{L}_p$ indicates that f has a finite $\mathcal{L}_p$-norm. Functions with $\|f\|_2 < \infty$ are called square integrable, denoted $f \in \mathcal{L}_2$.
- $\mathcal{K}$-function: A continuous function α: $[0, a) \to \mathbb{R}_{\geq 0}$, with $a > 0$ or $a = \infty$, is said to be a $\mathcal{K}$-function if it is strictly increasing and satisfies $\alpha(0) = 0$.
- $\mathcal{K}_\infty$-function: A continuous function $\alpha : [0, a) \to \mathbb{R}_{\geq 0}$ is said to be a $\mathcal{K}_\infty$-function if $a = \infty$ and $\lim_{r \to \infty} \alpha(r) = \infty$; that is, it is unbounded on $\mathbb{R}_{\geq 0}$.
- $\mathcal{KL}$-function: A continuous function $\beta : [0, a) \times \mathbb{R}_{\geq 0} \to \mathbb{R}_{\geq 0}$, with $a > 0$ or $a = \infty$, is said to be a $\mathcal{KL}$-function if for each fixed $s \geq 0$, the mapping $r \mapsto \beta(r, s)$ is a $\mathcal{K}$-function, and for each fixed $r \geq 0$, the mapping $s \mapsto \beta(r, s)$ is decreasing and satisfies $\lim_{s \to \infty} \beta(r, s) = 0$.
- C^1: the set of once continuously differentiable functions; i.e., functions with continuous first derivatives.
- PC: the set of piecewise continuous functions on $\mathbb{R}_{\geq t_0}$ meaning that for every $f \in PC$ there exists a sequence $\{t_k\}_{k \in \mathbb{N}}$ with $t_{k+1} > t_k > t_0$ for all $k \in \mathbb{N}$ and $t_k \to \infty$ when $k \to \infty$ such that f is continuous on (t_k, t_{k+1}), where $\lim_{t \uparrow t_k} f(t)$ exists for all $k \in \mathbb{N}_{>0}$ and $\lim_{t \downarrow t_k} f(t)$ exists for all $k \in \mathbb{N}$ with $\lim_{t \downarrow t_k} f(t) = f(t_k)$. It is assumed that $t_0 \in \mathbb{R}$ is clear from the context.
- AC: the set of absolutely continuous functions.
- a.e.: almost everywhere (usually with respect to the Lebesgue measure).

Contents

Chapter 1
Introduction

1.1 Control of High-Precision Positioning Systems

Over the last decades, progress in science and technology has been a major determining factor for economic growth and social welfare [26, 64, 79]. A fundamental mechanism for driving scientific and technological innovations is an increase in precision with which it is possible to observe and interact with the physical world. A prominent example is found in the semiconductor industry, where increased precision in complex lithography machines such as wafer scanners enables a continued exponential growth in available computing power, greatly impacting all aspects of science and technology as well as everyday life [65, 89]. Other examples illustrating the impact of increased precision include electron microscopy for molecular research [21], pick-and-place equipment in the manufacturing industry [54], and laser interferometry used, for example, in the detection of gravitational waves [3].

To meet the extreme requirements on precision, as well as the increasing demands on operating speed, stability, and reliability, high-precision systems heavily depend on the control solutions being used. Control of high-precision systems is often simplified by a stiff mechanical design combined with strategically placed actuators and sensors. The electro-mechanical components such as motors, bearings, and transmissions are optimized as to achieve a functional behaviour that is highly deterministic, predictable, and reproducible [66]. As such, the plant dynamics can be modelled accurately according to the linear paradigm. A decoupling strategy is typically applied for reducing multivariable design aspects to multi-loop single-input single-output (SISO) design [20]. Given this practice, the class of SISO linear time-invariant (LTI) plants is of particular relevance in the high-precision industry, and will be the primary scope in this book. The specific class of (to-be-controlled) SISO LTI plants allows for the application of well-developed LTI feedback control methods such as proportional-integral-derivative (PID) control [9] that can be designed using intuitive

S. van den Eijnden et al., *Hybrid Integrator-Gain Systems*, Lecture Notes in Control and Information Sciences 499, https://doi.org/10.1007/978-3-032-18101-5_1

techniques such as loop-shaping [33, 80], and more advanced methods including H_∞-optimal design and μ-synthesis [78, 84]. The sense and simplicity of an LTI control design in meeting performance requirements makes it a widely adopted method in practice that often sets the industrial standard [73].

1.2 Trade-Offs in Linear Control Design

Despite the clear advantages, restricting the applied control solutions to the linear time-invariant domain only is not without drawbacks. No matter the variety of LTI control structures or the method used to design the controllers, all LTI designs are bound to fundamental limitations in both time- and frequency-domain, leading to inevitable design trade-offs [35, 63, 74]. These trade-offs are generally encountered in high-precision positioning systems where an increase in precision and speed often requires enhanced low-frequency disturbance suppression and increased controller bandwidths [44, 58]. However, improving low-frequency disturbance rejection properties of linear systems, e.g., by increasing the bandwidth or adding integral action to the controller, directly comes at the cost of increased sensitivity to high-frequency noise. This is an inevitable consequence of the Bode sensitivity integral, often referred to as the "waterbed effect" [34, 35]. Simply stated: a change in controller design to achieve sensitivity reduction in some frequency interval necessarily comes at the expense of a sensitivity increase in another frequency interval.

Even if a desired shape on the sensitivity is determined that yields an acceptable trade-off between low-frequency disturbance suppression and high-frequency noise amplification, it may still be possible that no robustly stabilizing LTI controller exists that achieves such a shape. This is due to a limitation known as Bode's gain-phase relationship, which states that the phase of the frequency response of a linear system is completely determined by its magnitude characteristics and vice versa. Hence, the two parameters—gain and phase—that describe a complex function yield only one degree of freedom in design [35]. As a consequence, attempts to specify certain design requirements in one frequency range may compromise those in other frequency ranges. An example is given by the use of low-pass filters in the controller design with the aim to provide high-frequency roll-off. As prescribed by Bode's gain-phase relationship, this roll-off comes at the expense of phase lag that may compromise robust stability margins and limits the achievable bandwidth, see [8].

Besides suffering from limitations in the frequency-domain, time-domain limitations also pose restrictions on linear time-invariant feedback designs. For example, it is well-known [74] that a stable closed-loop controlled LTI system having at least two open-loop integrators must necessarily exhibit overshoot in its step-response. Typical high-precision positioning systems often contain a free-moving mass (or inertia) and, therefore, generally satisfy this limitation. Moreover, adding integral action in the controller for the purpose of counteracting static disturbances and improving low-frequency disturbance rejection properties may come at the expense of overshoot, which can compromise settling time, resulting in loss of system throughput.

The fact that modern linear techniques such as H_∞-design and μ-synthesis [78] cannot overcome the fundamental design trade-offs in LTI control systems poses a serious challenge for meeting the ever tightening performance requirements for high-precision positioning systems.

1.3 Hybrid Control as a Possible Solution?

Part of the limitations discussed above are a consequence of the LTI structure imposed on the feedback controller. This naturally raises the question whether these limitations can be overcome by using nonlinear/hybrid feedback control strategies. Being aware of the complexity normally associated with nonlinear systems, at first sight it may appear counter-intuitive to intentionally de-linearise an otherwise linear system. However, in some cases there may be good reason for doing so. Nonlinear/hybrid systems generally do not obey the same rules as LTI systems, and thus may potentially offer the necessary design freedom to break away from the fundamental limitations imposed by classical LTI control. Over the years, many different nonlinear/hybrid control strategies that specifically aim at overcoming the limitations of LTI control have been proposed. Examples include variable-gain control [6, 7, 40–42, 55], sliding-mode control [4, 72, 83], switching control [28, 30, 43, 48, 59, 61, 75], and reset control [2, 16, 25, 45, 49, 68, 91]. Some of these strategies have, in fact, explicitly been shown to be able to outperform LTI feedback control for LTI plants, see, for instance, the works in [16, 55, 81, 94].

In this book, the focus is explicitly on dealing with the unavoidable phase lag involved with LTI filters that limits the increase of bandwidth and may compromise time-domain performance properties. Within this scope, several hybrid control concepts are of particular interest. Hybrid controllers consist of a dynamical system with interacting continuous and discrete (or non-smooth) components [37]. Three prominent "classical" examples of hybrid control strategies dedicated to overcoming fundamental limitations of LTI control are given by reset control, split-path nonlinear filters, and hybrid integrator-gain systems. For a comparative performance study of these hybrid strategies within a motion control context, the reader is referred to [46].

1.3.1 Reset Control

Reset control emerged with the introduction of the Clegg integrator [25], being an integrator that resets its state to zero upon a zero crossing of its input. Compared to an LTI integrator, the resetting mechanism in the Clegg integrator forces the sign of the integrator output to be equivalent to the sign of its input at all times. A typical response of the classical Clegg integrator when subjected to a sinusoidal input is shown in Fig. 1.1 in black, together with the response of an LTI integrator in grey.

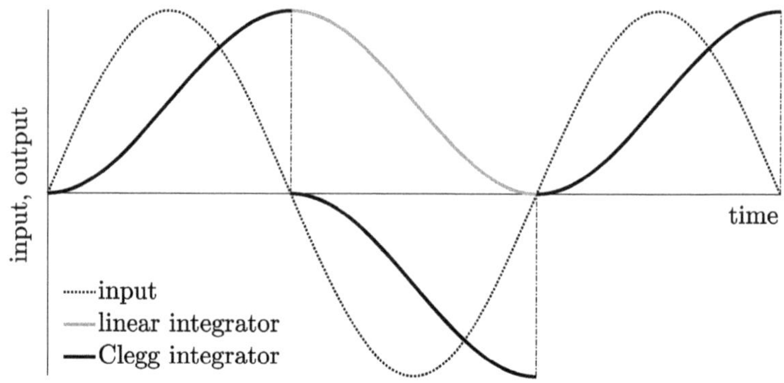

Fig. 1.1 Typical time response of a linear integrator (grey) and the Clegg integrator (black) when subject to a sinusoidal input signal (dotted)

A desirable feature of the resetting mechanism in the Clegg integrator is characterized by its describing function, which relates amplitude and phase properties of the driving sinusoidal input to amplitude and phase properties of the first harmonic in the corresponding steady-state response [36]. The describing function of the Clegg integrator exhibits a 20 dB/decade amplitude decay similar to that of a linear integrator, but with an induced phase lag of only 38.15° as opposed to 90° for the linear integrator [24]. Clearly, such favourable phase behaviour hints toward improved performance of LTI systems by designing a compensator capable of supplying the required bandwidth with less gain at high frequencies. The development of reset control involved various generalizations such as the first-order reset element [17, 24, 49, 94], second-order reset element [39], generalized fractional order reset elements [53], and conditional reset integrators [38] to name but a few. Extensive research on reset control systems has led to various fruitful results regarding stability and performance analysis [2, 11, 12, 17, 23, 62, 67, 86, 91–93], hybrid formulations [90], and experimental demonstrations of reset control systems achieving improved performance [13, 22, 38, 52, 69, 71, 95]. To date, reset control is one of the few nonlinear strategies that has been shown to overcome fundamental time-domain limitations of LTI control for LTI plants [16, 94].

1.3.2 Split-Path Nonlinear (SPAN) Filters

Related strategies that specifically challenge Bode's gain-phase relationship, but apparently have received less attention than reset control are given by the driven limiter [10] and split-path nonlinear (SPAN) filter [1, 31, 32, 57]. The main design philosophy of these strategies is to "split" the magnitude and phase characteristics of a filter as to design these separately, thereby facilitating an additional degree-of-freedom for controller design. Variants of the SPAN filter that include integral action

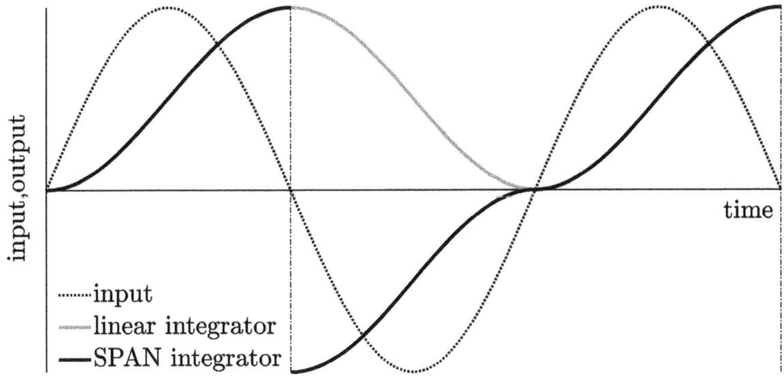

Fig. 1.2 Typical time response of a linear integrator (grey) and the SPAN integrator (black) when subject to a sinusoidal input signal (dotted)

have recently been proposed in [75, 85, 87, 88]. These so-called SPAN integrators (SPANI) aim for an improved integrator design, with all the benefits of a linear integrator in terms of achieving zero steady-state tracking error, but without introducing the usual 90° phase lag. For achieving this, the output of the SPAN integrator is composed as the product of the absolute value (magnitude) of the integrator state, and the sign (phase) of the input signal. A typical response of a SPAN integrator subject to a sinusoidal input is shown in Fig. 1.2 in black, along with the response of an LTI integrator in grey. In terms of its describing function, the SPAN integrator exhibits similar magnitude characteristics as an LTI integrator, but with zero phase lag [75]. The absence of phase lag is attributed to the fact that the output of the SPAN filter shows a certain symmetry after the switching instance.

Regarding its design philosophy, the SPAN(I) filter exhibits some interesting analogies with the previously discussed reset control strategy. Both strategies adopt "close-to-linear" thinking in their design. That is, the designs adopt the structure of basic LTI filters such as an integrator to which certain modifications are made in order to provide additional functionality. In particular, SPAN(I) filters and reset strategies use a switching surface to trigger changes in the control signal, leading to the injection of *discontinuous* control signals into an otherwise smooth (linear) feedback system.

Although promising, the aforementioned control structures were only partially successful. The reasons vary from lack of appealing tools for broad adoption in industrial applications or complications in establishing essential system-theoretic results. For instance, the hard resets in the Clegg integrator as in Fig. 1.1, and the discontinuous control signals of the SPAN filter as shown in Fig. 1.2 increase the risk of exciting higher-order modes in the controlled plant [7]. In particular, such signals may distribute low-frequency input energy over a broad frequency interval, including the high-frequency range in which the controlled system may contain multiple lightly-damped resonances. As a result, the controlled system is usually more sensitive in the high-frequency range. This may lead to unwanted behaviour that generally deteriorates performance and may even lead to system failure or damage. Moreover, the need for taking into account discontinuities in the state trajectories may hamper the formal derivation of key properties such as incremental stability [5, 19] and convergence [70], as well as the development of easy-to-use (frequency-domain) tools for controller design and analysis [56]. These aspects are instrumental to set up comprehensive design and analysis frameworks that adhere to the "close-to-linear" way of thinking, something which is highly valued by control engineers in industry and which could expedite the acceptance in industry.

1.3.3 Hybrid Integrator-Gain Systems (HIGS)

As an alternative to discontinuous control strategies, a nonlinear integrator referred to as the hybrid integrator-gain system (HIGS) has been proposed recently in [27], see also [28]. The underlying philosophy of HIGS is inspired by the Clegg integrator, i.e., keeping the sign of the integrator's output equal to that of the input at all times, thereby inheriting the hinted phase advantages of reset control. However, HIGS avoids hard resets of the integrator state, and instead exploits mode switching between an integrator mode and a gain mode (hence the terminology hybrid integrator-gain system) to enforce equal sign. Switching is done in a manner that results in a *continuous* output signal as opposed to the *discontinuous* signals generated by reset control or as generated by SPAN(I) filters. A typical response of HIGS subject to a sinusoidal input is shown in Fig. 1.3. The describing function characteristics of HIGS demonstrate an amplitude decay comparable to that of an LTI integrator, but with an induced phase lag of at most 38.15°.

Given the previous discussion on the possible difficulties arising from the use of discontinuous control signals with reset control and SPAN(I) filters, continuous control with HIGS offers comparable performance benefits as the previous strategies,

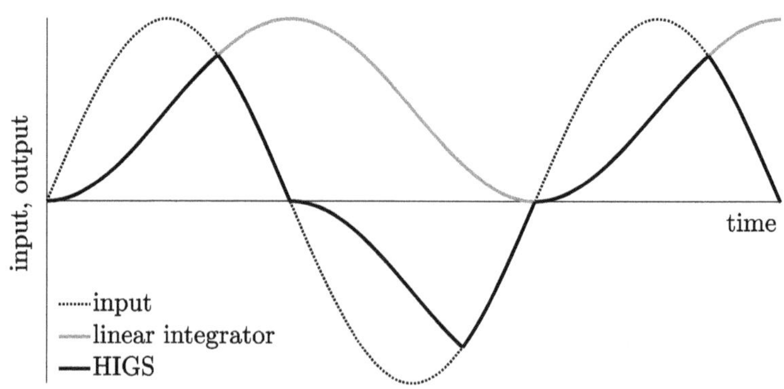

Fig. 1.3 Typical time response of a linear integrator (grey) and HIGS (black) when subject to a sinusoidal input signal (dotted)

for example in terms of reduced phase lag, but may provide different opportunities in both application-based as well as system-theoretic directions. In particular, due to their continuous nature, HIGS-based controllers may be easier to construct, analyse, employ, and adopt than their discontinuous counterparts. Recent successful application of HIGS for control of industrial mechatronic systems such as wafer scanners, active vibration isolation systems, wire bonders, micro-electromechanical nanopositioning systems, and medical devices is demonstrated in [14, 15, 29, 43, 47, 76, 77]. Emerging research directions that extend the definition of HIGS considered in this book to include fractional-order elements can be found in [50, 51].

Finally, within the context of continuous control with HIGS, the related line of research on so-called "soft-reset" control systems that has emerged recently in [18, 60, 82] should be mentioned. This strategy aims at approximating hard reset controllers via a differential inclusion, resulting in continuous states/outputs, which is also the objective with HIGS. As discussed in [82], however, within a closed-loop setting this approach may not always bear much resemblance to reset control, thereby potentially loosing the intuitive and desirable features of a reset control strategy.

1.4 Outline of the Book

This book offers a comprehensive overview of the theory and practice of hybrid integrator-gain systems (HIGS). In order for any nonlinear/hybrid control strategy to be accepted as a viable extension to the well-developed class of linear control methods, three key aspects should be addressed. First, performance advantages in comparison to linear control should be evident. Second, techniques for design and analysis should be accurate, systematic, and easy-to-use, the latter possibly conflicting with the former objectives. Third, performance benefits should be demonstrated by experiment in representative real-world (industrial) settings.

In line with the above, this book is divided into three parts. The first part discusses fundamentals of HIGS. In particular, Chap. 2 presents a mathematical introduction to HIGS as well as several of its open-loop properties, and introduces the formal closed-loop setting with HIGS that is considered throughout this book. In Chap. 3 the possibility to overcome a fundamental time-domain limitation of LTI control for LTI systems by carefully embedding HIGS into an otherwise linear feedback control structure is established. The fundamental limitation that is shown to be overcome is the unavoidable presence of overshoot in the step-response of a closed-loop controlled single-input single-output (SISO) LTI plant having a real unstable open-loop pole. It is shown that with a HIGS-based controller design, overshoot can be completely avoided. Key for alleviating the overshoot limitations is the sequence in which the linear filters in the controller appear with respect to the HIGS component(s). This insight provides a valuable guideline for HIGS-based controller design to enable genuine performance advantages over LTI controller designs.

The second part of this book covers stability and performance analysis of HIGS-based feedback control systems. Chapters 4 and 5 address the development of tools for input-to-state stability (ISS) analysis. Two directions are pursued: a direction developing frequency-domain based conditions for stability in Chap. 4, and a direction that presents time-domain conditions in the form of numerically tractable linear matrix inequalities (LMIs) in Chap. 5. Underlying the frequency-domain conditions is the connection via the Kalman–Yakubovich–Popov (KYP) lemma to a quadratic Lyapunov function that is only positive in a subset of the state space where trajectories of the controlled system can exist. The time-domain conditions developed in Chap. 5 are based on the use of piecewise quadratic (PWQ) Lyapunov functions. For constructing such functions, a novel partitioning of the subset determining the active modes of HIGS is considered. Rigorous conditions for stability are formulated as sets of LMIs, which are extended toward performance analysis in terms of the $\mathcal{L}_2$-gain and the $\mathcal{H}_2$-norm, providing quantitative measures for disturbance rejection and transient performance. Both the frequency-domain approach and time-domain approach have benefits and limitations, and may complement each other to some extent. The frequency-domain approach allows for easy-to-verify conditions based on measured plant data, but possibly at the cost of increased conservatism, whereas the time-domain approach allows for very tight stability estimates, but possibly at the cost of increased (numerical) complexity in verifying the conditions.

In addition to the aforementioned tools for ISS and performance analysis, tools for incremental stability and convergence analysis of HIGS-controlled systems are presented in Chap. 6. Convergent systems enjoy the property of possessing for each bounded (time-varying) input, a unique and bounded steady-state response. This property opens up the possibility for characterizing steady-state performance in a more accurate manner as compared to, e.g., the $\mathcal{L}_2$-gain in Chap. 5. Namely, the generality of the $\mathcal{L}_2$-gain may not always reflect the actual performance objective of the system under study, concerning, for instance, specific steady-state response characteristics in the presence of specific inputs. Both time- and frequency-domain conditions for verifying incremental stability and convergence of HIGS-controlled systems are presented. Though specifically tailored to these systems, the approaches may open up fruitful directions for analysis of other classes of hybrid/switched systems as well.

The third part of the book presents approaches toward HIGS-based controller design, as well as an experimental case-study. Regarding design, two distinct approaches are presented in Chap. 7. The first approach reformulates the frequency-domain conditions presented in Chap. 4 as matrix inequalities. Combined with conditions for $\mathcal{L}_2$-performance as derived in Chap. 5, these matrix inequalities lead to appropriate (robust) synthesis conditions. The second design approach exploits a describing function characterization of HIGS within a robust loop-shaping framework. Key in this approach is to express the modelling error induced by the describing function as an uncertainty in the controller. Deriving such an uncertainty is highly specific to the problem at hand. Nevertheless, practically useful guidelines are presented which allow for effective frequency-domain based controller design. The approach strongly connects to the industrial practice and may aid further (practical) development of HIGS as a nonlinear control solution. Chapter 8 presents an experimental validation of a HIGS-based PID control strategy on an industrial wafer scanner.

Controller design is based on frequency-domain loop-shaping such as outlined in Chap. 7, and which is demonstrated to be valuable in an industrial setting. Compared to the state-of-practice linear control solution, the HIGS-based strategy allows for a significant increase in both controller bandwidth and low-frequency disturbance rejection properties as hinted by a describing function analysis. This is supported by measurements which show a significant improvement in tracking accuracy without strong penalties on high-frequency noise amplification, thereby underlining the potential of HIGS-based control in the industrial practice.

Finally, Chap. 9 provides a reflection of the results in this book by summarizing the main conclusions together with a vision by the authors on future research directions.

References

1. Aangenent, W., van de Molengraft, R., Steinbuch, M.: Nonlinear control of a linear motion system. IFAC Proc. Vol. **38**(1), 446–451 (2005)
2. Aangenent, W.H.T.M., Witvoet, G., Heemels, W.P.M.H., Molengraft, M.J.G., Steinbuch, M.: Performance analysis of reset control systems. Int. J. Robust Nonlinear Control **20**, 1213–1233 (2010)
3. Abbott, B., Abbott, R., Abbott, T., Abernathy, M., Acernese, F., Ackley, K., Adams, C., Adams, T., Addesso, P., Adya, V., Affeldt, C., Agathos, M., Agatsuma, K., Aggarwal, N., Aguiar, O., Ain, A., Ajith, P., Zweizig, J.: Observation of gravitational waves from a binary black hole merger. Phys. Rev. Lett. **116**(6), 1–16 (2016)
4. Abidi, K., Sabanovic, A.: Sliding-mode control for high-precision motion of a piezostage. IEEE Trans. Industr. Electron. **54**, 629–637 (2007)
5. Angeli, D.: A Lyapunov approach to incremental stability properties. IEEE Trans. Autom. Control **47**, 410–421 (2002)
6. Armstrong, B., McPherson, J., Li, Y.: A Lyapunov stability proof for nonlinear-stiffness PD control. In: Proceedings of IEEE International Conference on Robotics and Automation, vol. 1, pp. 945–950 (1996)
7. Armstrong, B., Gutierrez, J., Wade, B., Joseph, R.: Stability of phase-based gain modulation with designer-chosen switch functions. Int. J. Robot. Res. **25**, 781–796 (2006)
8. Åström, K.: Limitations on control system performance. Eur. J. Control **6**(1), 2–20 (2000)
9. Åström, K., Hägglund, T.: The future of PID control. Control. Eng. Pract. **9**(11), 1163–1175 (2001)
10. Bailey, A.: Stabilization of control systems by the use of driven limiters. Proc. IEE **113**(1), 169–174 (1963)
11. Baños, A., Barreiro, A.: Reset Control Systems. Springer, London (2012)
12. Beerens, R., Bisoffi, A., Zaccarian, L., Heemels, W., Nijmeijer, H., van de Wouw, N.: Reset integral control for improved settling of PID-based motion systems with friction. Automatica **107**, 483–492 (2019)
13. Beerens, R., Bisoffi, A., Zaccarian, L., Nijmeijer, H., Heemels, M., van de Wouw, N.: Reset PID design for motion systems with Stribeck friction. IEEE Trans. Control Syst. Technol. **30**(1), 294–310 (2022)
14. Beerens, R, Konings, P, Heertjes, M, van den Eijnden, S.: Stability and performance assessment of a mimo higs-based controller design. In: 2024 European Control Conference (ECC), pp. 1723–1728 (2024)
15. Beijen, M., op Smeijers, T.T., Boerrigter, G., van den Eijnden, S.: Hybrid control of a variable-speed peristaltic pump. In: 2024 American Control Conference (ACC), pp. 4296–4301 (2024)

16. Beker, O., Hollot, C., Chait, Y.: Plant with integrator: an example of reset control overcoming limitations of linear feedback. IEEE Trans. Autom. Control **46**, 1797–1799 (2001)
17. Beker, O., Hollot, C., Chait, Y., Han, H.: Fundamental properties of reset control systems. Automatica **40**(6), 905–915 (2004)
18. Bertollo, R., Teel, A.R., Zaccarian, L.: Soft-reset control with max-of-quadratics Lyapunov certificates. IEEE Trans. Autom. Control **68**(9), 5245–5257 (2023)
19. Biemond, J.J.B., Postoyan, R., Heemels, W.P.M.H., van de Wouw, N.: Incremental stability of hybrid dynamical systems. IEEE Trans. Autom. Control **63**(12), 4094–4109 (2018)
20. Butler, H.: Position control in lithographic equipment. Control. Syst., IEEE **31**, 28–47 (2011)
21. Callaway, E.: The revolution will not be crystallized: a new method sweeps through structural biology. Nature **525**, 172–174 (2015)
22. Caporale, D., van Eijk, L.F., Karbasizadeh, N., Beer, S., Kostić, D., HosseinNia, S.H.: Practical implementation of a reset controller to improve performance of an industrial motion stage. IEEE Trans. Control Syst. Technol. **32**(4), 1451–1462 (2024)
23. Carrasco, J., Baños, A., Schaft, A.: A passivity-based approach to reset control systems stability. Syst. Control Lett. **59**, 18–24 (2010)
24. Chait, Y., Hollot, C.V.: On Horowitz's contributions to reset control. Int. J. Robust Nonlinear Control **12**(4), 335–355 (2002)
25. Clegg, J.C.: A nonlinear integrator for servomechanisms. Trans. Am. Inst. Electr. Eng., Part II: Appl. Ind. **77**(1), 41–42 (1958)
26. Coccia, M.: Why do nations produce science advances and new technology? Technol. Soc. **59**, 1–9 (2019)
27. Deenen, D.A., Heertjes, M.F., Heemels, W.P.M.H., Nijmeijer, H.: Hybrid integrator design for enhanced tracking in motion control. In: Proceedings of the American Control Conference (ACC'17), pp. 2863–2868 (2017)
28. Deenen, D.A., Sharif, B., van den Eijnden, S., Nijmeijer, H., Heemels, M., Heertjes, M.: Projection-based integrators for improved motion control: formalization, well-posedness and stability of hybrid integrator-gain systems. Automatica **133**, 109830 (2021)
29. van Eijk, L.F., Beer, S., van Es, R.M.J., Kostić, D., Nijmeijer, H.: Frequency-domain properties of the hybrid integrator-gain system and its application as a nonlinear lag filter. IEEE Trans. Control Syst. Technol. **31**(2), 905–912 (2023)
30. Feuer, A., Goodwin, G.C., Salgado, M.: Potential benefits of hybrid control for linear time invariant plants. In: Proceedings of the 1997 American Control Conference (ACC'97), vol. 5, pp. 2790–2794 (1997)
31. Fong, M., Szeto, W.: Application of a nonlinear filter to a conditionally stable system. Int. J. Control **32**(6), 963–981 (1980)
32. Foster, W., Gieseking, D., Waymayer, W.: A nonlinear filter for independent gain and phase (with applications). Trans. ASME J. Basic Eng. **88**, 457–462 (1966)
33. Franklin, G., Powell, J., Emami-Naeini, A.: Feedback Control of Dynamic Systems, 5th edn. Prentice Hall (2005)
34. Freudenberg, J., Looze, D.: Right half plane poles and zeros and design tradeoffs in feedback systems. IEEE Trans. Autom. Control **30**(6), 555–565 (1985)
35. Freudenberg, J., Middleton, R., Stefanpoulou, A.: A survey of inherent design limitations. In: Proceedings of the 2000 American Control Conference (ACC), vol. 5, pp. 2987–3001 (2000)
36. Gelb, A., Vander Velde, W.: Multiple-input Describing Functions and Nonlinear Systems Design. McGraw Hill, New York (1968)
37. Goebel, R., Sanfelice, R., Teel, A.: Hybrid dynamical systems. IEEE Control. Syst. **29**, 28–93 (2009)
38. Gruntjens, K., Heertjes, M., Van Loon, S., Van De Wouw, N., Heemels, W.: Hybrid integral reset control with application to a lens motion system. In: 2019 American Control Conference (ACC), pp. 2408–2413 (2019)
39. Hazeleger, L., Heertjes, M., Nijmeijer, H.: Second-order reset elements for stage control design. In: 2016 American Control Conference (ACC), pp. 2643–2648 (2016)

40. Heertjes, M., Steinbuch, M.: Stability and performance of a variable gain controller with application to a DVD storage drive. Automatica **40**(4), 591–602 (2004)
41. Heertjes, M., Verstappen, R.: Self-tuning in integral sliding mode control with a Levenberg–Marquardt algorithm. Mechatronics **24**(4), 385–393 (2014)
42. Heertjes, M., Schuurbiers, X., Nijmeijer, H.: Performance-improved design of N-PID controlled motion systems with applications to wafer stages. IEEE Trans. Industr. Electron. **56**, 1347–1355 (2009)
43. Heertjes, M., van den Eijnden, S., Sharif, B., Heemels, M., Nijmeijer, H.: Hybrid integrator-gain system for active vibration isolation with improved transient response. IFAC-PapersOnLine **52**(15), 454–459, 8th IFAC Symposium on Mechatronic Systems MECHATRONICS 2019 (2019)
44. Heertjes, M., Butler, H., Dirkx, N., van der Meulen, S., Ahlawat, R., O'Brien, K., Simonelli, J., Teng, K.T., Zhao, Y.: Control of wafer scanners: methods and developments. In: 2020 American Control Conference (ACC), pp. 3686–3703 (2020)
45. Heertjes, M.F., Sahin, I.H., van de Wouw, N., Heemels, W.P.M.H.: Switching control in vibration isolation systems. IEEE Trans. Control Syst. Technol. **21**(3), 626–635 (2013)
46. Heertjes, M.F., Hoogeveen, T.P., van den Eijnden, S.J.: Stage motion control: nonlinear integrators revisited. Annu. Rev. Control., Robot., Auton. Syst. (2024)
47. Heertjes, M., van den Eijnden, S., Sharif, B.: An overview on hybrid integrator-gain systems with applications to wafer scanners. In: 2023 IEEE International Conference on Mechatronics (ICM), pp. 1–8 (2023)
48. Hespanha, J.P., Morse, A.: Switching between stabilizing controllers. Automatica **38**(11), 1905–1917 (2002)
49. Horowitz, I., Rosenbaum, P.: Non-linear design for cost of feedback reduction in systems with large parameter uncertainty. Int. J. Control **21**(6), 977–1001 (1975)
50. Hosseini, S.A., Tavazoei, M.S., Van Eijk, L.F., HosseinNia, S.H.: Generalizing hybrid integrator-gain systems using fractional calculus. In: 2022 IEEE Conference on Control Technology and Applications (CCTA), pp. 1050–1055 (2022)
51. Hosseini, S.A., van Eijk, L.F., Kaczmarek, M.B., HosseinNia, S.H.: Higher-order sinusoidal-input describing function analysis of fractional-order hybrid integrator-gain systems. IFAC-PapersOnLine **58**(7), 400–405, 4th IFAC Conference on Advances in Proportional-Integral-Derivate Control PID 2024 (2024)
52. HosseinNia, S., Tejado, I., Vinagre, B.: A method for the design of robust controllers ensuring the quadratic stability for switching systems. J. Vib. Control **20**, 1085–1098 (2014)
53. HosseinNia, S.H., Tejado, I., Vinagre, B.M.: Fractional-order reset control: application to a servomotor. Mechatronics **23**(7), 781–788 (2013)
54. Huang, H.H., Wey, J.: Research on the high-speed pick and place device for die bonders. IEEE ICCA **2010**, 1683–1687 (2010)
55. Hunnekens, B., van de Wouw, N., Nešić, D.: Overcoming a fundamental time-domain performance limitation by nonlinear control. Automatica **67**, 277–281 (2016)
56. Kamenetskiy, V.: Frequency-domain stability conditions for hybrid systems. Autom. Remote Control **78**, 2101–2119 (2017)
57. Karybakas, C.A.: Nonlinear integrator with zero phase shift. IEEE Trans. Ind. Electron. Control. Instrum. **IECI-24**(2), 150–152 (1977)
58. Lamnabhi-Lagarrigue, F., Annaswamy, A., Engell, S., Isaksson, A., Khargonekar, P., Murray, R.M., Nijmeijer, H., Samad, T., Tilbury, D., Van den Hof, P.: Systems & control for the future of humanity, research agenda: current and future roles, impact and grand challenges. Annu. Rev. Control. **43**, 1–64 (2017)
59. Lau, K., Middleton, R.: Switched integrator control schemes for integrating plants. Eur. J. Control **9**, 539–559 (2003)
60. Le, J.H., Teel, A.R.: Passive soft-reset control for nonlinear systems. In: 60th IEEE Conference on Decision and Control (CDC), pp. 5320–5325 (2021)
61. Liberzon, D.: Switching in Systems and Control. Birkhauser, Boston (2003)

62. Loquen, T., Nešić, D., Prieur, C., Tarbouriech, S., Teel, A., Zaccarian, L.: Piecewise quadratic Lyapunov functions for linear control systems with first order reset elements. IFAC Proc. Vol. **43**, 807–812 (2010)
63. Middleton, R.: Trade-offs in linear control system design. Automatica **27**(2), 281–292 (1991)
64. Mokyr, J.: The past and the future of innovation: some lessons from economic history. Explor. Econ. Hist. **69**, 13–26 (2018)
65. Moore, G.: Cramming more components onto integrated circuits. Proc. IEEE **86**(1), 82–85 (1998)
66. Munnig Schmidt, R., Schitter, G., Rankers, A., van Eijk, J.: The Design of High Performance Mechatronics: High-tech Functionality by Multidisciplinary System Integration (2nd revised edition). IOS Press (2014)
67. Nešić, D., Zaccarian, L., Teel, A.: Stability properties of reset systems. Automatica **44**(8), 2019–2026 (2008)
68. Nešić, D., Teel, A.R., Zaccarian, L.: Stability and performance of SISO control systems with first-order reset elements. IEEE Trans. Autom. Control **56**(11), 2567–2582 (2011)
69. Panni, F.S., Waschl, H., Alberer, D., Zaccarian, L.: Position regulation of an EGR valve using reset control with adaptive feedforward. IEEE Trans. Control Syst. Technol. **22**(6), 2424–2431 (2014)
70. Pavlov, A., Pogromsky, A., van de Wouw, N., Nijmeijer, H., Rooda, K.: Convergent piecewise affine systems: analysis and design part ii: discontinuous case. In: Proceedings of the 44th IEEE Conference on Decision and Control, pp. 5397–5402 (2005)
71. Saikumar, N., Sinha, R.K., HosseinNia, S.H.: "Constant in gain lead in phase" element– application in precision motion control. IEEE/ASME Trans. Mechatron. **24**(3), 1176–1185 (2019)
72. Sam, Y.M., Osman, J.H., Ghani, M.A.: A class of proportional-integral sliding mode control with application to active suspension system. Syst. Control Lett. **51**(3), 217–223 (2004)
73. Samad, T., Bauer, M., Bortoff, S., Di Cairano, S., Fagiano, L., Odgaard, P.F., Rhinehart, R.R., Sánchez-Peña, R., Serbezov, A., Ankersen, F., Goupil, P., Grosman, B., Heertjes, M., Mareels, I., Sosseh, R.: Industry engagement with control research: perspective and messages. Annu. Rev. Control. **49**, 1–14 (2020)
74. Seron, M.M., Braslavsky, J.H., Goodwin, G.C.: Fundamental Limitations in Filtering and Control. Springer, Berlin (1997)
75. Sharif, B., van der Maas, A., van de Wouw, N., Heemels, W.P.M.H.: Filtered split-path nonlinear integrator: a hybrid controller for transient performance improvement. IEEE Trans. Control. Syst. Technol. 1–13 (2021)
76. Shi, K., Nikooienejad, N., Petersen, I.R., Moheimani, S.O.R.: Negative imaginary control using hybrid integrator-gain systems: application to mems nanopositioner. IEEE Trans. Control Syst. Technol. **32**(3), 718–730 (2024)
77. Shi, K., Khodabakhshi, E., Biswas, P., Petersen, I.R., Moheimani, S.O.R.: Discrete-time higs based digital control of negative imaginary systems. Control. Eng. Pract. **163**, 106386 (2025)
78. Skogestad, S., Postlethwaite, I.: Multivariable Feedback Control: Analysis and Design, 2nd edn. Wiley and Sons (2010)
79. Solarin, S., Yuen, Y.: A global analysis of the impact of research output on economic growth. Scientometrics **108**, 855–874 (2016)
80. Steinbuch, M., Norg, M.: Advanced motion control: an industrial perspective. Eur. J. Control **4**(4), 278–293 (1998)
81. Stoorvogel, A.: Nonlinear l_1 optimal controllers for linear systems. IEEE Trans. Autom. Control **40**(4), 694–696 (1995)
82. Teel, A.R.: Continuous-Time Implementation of Reset Control Systems, pp. 27–41. Springer, Cham (2022)
83. Utkin, V.: Sliding mode control design principles and applications to electric drives. IEEE Trans. Industr. Electron. **40**(1), 23–36 (1993)
84. van de Wal, M., Baars, G., Sperling, F., Bosgra, O.: Multivariable h_∞/μ feedback control design for high-precision wafer stage motion. Control Eng. Pract. **10**, 739–755 (2002)

85. van den Eijnden, S., Sharif, B., Heertjes, M., Heemels, W.: Frequency-domain stability conditions for split-path nonlinear systems. IFAC-PapersOnLine **56**(1), 301–306 (2023)
86. van den Eijnden, S., Chaffey, T., Oomen, T., (Maurice) Heemels W.: Scaled graphs for reset control system analysis. Eur. J. Control. **80**, 101050 (2024)
87. van der Maas, A., van de Wouw, N., Heemels, W.: Filtered split-path nonlinear integrator (f-spani) for improved transient performance. In: 2017 American Control Conference (ACC), pp. 3500–3505 (2017)
88. van Loon, S., Hunnekens, B., Heemels, W., van de Wouw, N., Nijmeijer, H.: Split-path nonlinear integral control for transient performance improvement. Automatica **66**, 262–270 (2016)
89. van Schoot, J., Troost, K., Bornebroek, F., van Ballegoij, R., Lok, S., Krabbendam, P., Stoeldraijer, J., Loopstra, E., Benschop, J.P., Finders, J., et al.: High-NA EUV lithography enabling Moore's law in the next decade. In: International Conference on Extreme Ultraviolet Lithography 2017, International Society for Optics and Photonics, p. 10450 (2017)
90. Zaccarian, L., Nešić, D., Teel, A.: First order reset elements and the clegg integrator revisited. In: Proceedings of the 2005, American Control Conference, 2005, vol. 1, pp. 563–568 (2005)
91. Zaccarian, L., Nešić, D., Teel, A.R.: Analytical and numerical Lyapunov functions for SISO linear control systems with first-order reset elements. Int. J. Robust Nonlinear Control **21**, 1134–1158 (2011)
92. Zhao, G., Wang, J.: On l_2 gain performance improvement of linear systems with Lyapunov-based reset control. Nonlinear Anal. Hybrid Syst. **21**, 105–117 (2016)
93. Zhao, G., Nešić, D., Tan, Y., Wang, J.: Open problems in reset control. In: 52nd IEEE Conference on Decision and Control, pp. 3326–3331 (2013)
94. Zhao, G., Nešić, D., Tan, Y., Hua, C.: Overcoming overshoot performance limitations of linear systems with reset control. Automatica **101**, 27–35 (2019)
95. Zheng, Y., Chait, Y., Hollot, C., Steinbuch, M., Norg, M.: Experimental demonstration of reset control design. Control Eng. Pract. **8**, 113–120 (2000)

Part I
Fundamentals of HIGS

Chapter 2
Formalization and Basic Results

2.1 Introduction

The hybrid integrator–gain system, abbreviated as HIGS, first appeared in the literature in [12], where it was introduced for dealing with the classical trade-offs resulting from the use of linear time-invariant (LTI) integral control. The underlying principle of HIGS is inspired by reset control solutions, see, e.g., [1, 3, 4, 10, 28, 44], and can best be understood when considering the step response of the feedback interconnection of an LTI integrator and a plant. The integrator accumulates its input—the feedback error signal—over time for achieving zero steady-state error. When the error crosses zero for the first time, however, the accumulated past (positive) error signal is still stored in the integrator's state. As a consequence, the integrator produces a positive output and keeps on pushing the system in the direction away from zero error, causing overshoot. To improve upon this aspect, a simple idea is to keep the sign of the integrator's output the same as that of its input, thereby forcing the system toward zero error at all times. Reset integrators achieve this sign-equivalence by instantaneously resetting their buffer (state) to zero upon a sign change of the input. Discontinuities in the generated output signal resulting from the resetting action, however, may increase the risk of exciting structural modes of the plant dynamics, possibly leading to unwanted behaviour that potentially can damage the system. To avoid state resets, but still preserve equal sign of the integrator's input and output signals, HIGS exploits switching between integrator dynamics and gain characteristics. In particular, a HIGS element acts as a linear integrator as long as its input–output pair lies within (the interior of) a specified set, which is often referred to as the sector. This mode of operation is referred to as the *integrator mode*. At moments when the sector conditions tend to be violated, a switch is enforced to keep trajectories to stay on the boundary of the mentioned sector. This second mode of operation is referred to as the *gain mode*, explaining the terminology hybrid integrator–gain system. Switching between integrator and gain characteristics is orchestrated in a manner such that the generated output remains continuous. As a result, HIGS inherits the benefits of

S. van den Eijnden et al., *Hybrid Integrator-Gain Systems*, Lecture Notes in Control and Information Sciences 499, https://doi.org/10.1007/978-3-032-18101-5_2

reset controllers, often expressed in terms of phase advantages, but avoids explicit resets of controller states.

To introduce HIGS properly, this chapter presents relevant preliminaries. In particular, a mathematical description of HIGS in an open-loop setting is provided in Sect. 2.2. To obtain more insights in the working principle of HIGS, and better appreciate its benefits, time- and frequency-domain properties are discussed. In Sect. 2.3, the formal closed-loop setting with HIGS that will be considered throughout this book is presented. Important preliminaries regarding the upcoming stability and performance analysis are provided as well. All aforementioned aspects are treated in a continuous-time setting. However, discrete-time aspects play a crucial role in a digital implementation. Therefore, in Sect. 2.4 discrete-time versions of HIGS are discussed. The most important findings in this chapter are summarized in Sect. 2.5.

2.2 Open-Loop System Representation

In this section, the hybrid integrator–gain system is introduced. In order to develop insights into its working principle, and to appreciate the benefits it can offer for control, an open-loop analysis is conducted in both time- and frequency-domain; the latter is based on the describing function method.

2.2.1 *State Space Description of HIGS*

The hybrid integrator–gain system (HIGS) can mathematically be formulated as a piecewise linear (PWL) system with discontinuous right-hand side, given by

$$\mathcal{H}:\begin{cases}\dot{x}_h(t)=\omega_h z(t), & \text{if } (z(t),u(t),\dot{z}(t))\in\mathcal{F}_1, \qquad (2.1a)\\ x_h(t)=k_h z(t), & \text{if } (z(t),u(t),\dot{z}(t))\in\mathcal{F}_2, \qquad (2.1b)\\ u(t)=x_h(t), & \qquad (2.1c)\end{cases}$$

where $x_h(t)\in\mathbb{R}$ denotes the state of the integrator, $z(t)\in\mathbb{R}$ is the input, and $u(t)\in\mathbb{R}$ is the generated output, all at time $t\in\mathbb{R}_{\geq 0}$. Note that (2.1b) can alternatively be described by the differential equation $\dot{x}_h(t)=k_h\dot{z}(t)$, which in combination with (2.1a) indicates the discontinuous nature of the vector field of (2.1). To have well-defined behaviour, it is assumed for now that the input z is a continuously differentiable function, such that its time-derivative $\dot{z}(t)$ exists for all times $t\in\mathbb{R}_{\geq 0}$ and $\dot{z}$ is continuous, i.e., $z\in C^1$. Though in general this assumption is restrictive, for the purpose of illustrating the working principle of HIGS and restricting the amount of mathematical details needed here, it is convenient at this point. In Sect. 2.3, regularity conditions on the input signals will be discussed and motivated in detail within a closed-loop system setting with HIGS.

The parameters $\omega_h \in \mathbb{R}_{\geq 0}$ and $k_h \in \mathbb{R}_{>0}$ in (2.1) represent the integrator frequency and the proportional gain, respectively. The active mode of HIGS is dictated by the flow sets $\mathcal{F}_1$ and $\mathcal{F}_2$. When (2.1a) is active, HIGS is said to operate in *integrator mode*, whereas if (2.1b) is active, it is said to be in *gain mode*. Before specifying the sets $\mathcal{F}_1$ and $\mathcal{F}_2$ in more detail, the design philosophy of HIGS is discussed first.

In its primary mode of operation, HIGS exhibits integrator dynamics, i.e.,

$$\begin{aligned} \dot{x}_h(t) &= \omega_h z(t), \\ u(t) &= x_h(t). \end{aligned} \tag{2.2}$$

To maintain the idea of keeping the sign of the integrator's output u in (2.2) equal to that of its input z at all times, operating in integrator mode is allowed as long as the input–output pair (z, u) remains inside the $[0, k_h]$-sector defined by

$$\mathcal{F} := \left\{ (z, u, \dot{z}) \in \mathbb{R}^3 \mid zu \geq \frac{1}{k_h} u^2 \right\}. \tag{2.3}$$

Using integrator dynamics on the boundary $u = k_h z$ of the sector $\mathcal{F}$, may result in the vector field $(\dot{x}_h, \dot{z}) = (\omega_h z, \dot{z})$ to point outside of the sector $\mathcal{F}$. As a consequence, the corresponding trajectory can escape the sector when sticking to the integrator mode. Specifically, provided $u = k_h z$, this occurs when $\dot{x}_h > k_h \dot{z}$ for $z > 0$, or when $\dot{x}_h < k_h \dot{z}$ for $z < 0$. Since in integrator mode one has $\dot{x}_h = \omega_h z$, the inequalities can equivalently be expressed in a quadratic form by $\omega_h z^2 > k_h \dot{z} z$. To prevent a trajectory from escaping the sector, a switch to gain mode (2.1b) is initiated, thereby enforcing the trajectory to move over the sector boundary according to the dynamics $\dot{x}_h = k_h \dot{z}$. A transition from gain mode back to integrator mode is allowed whenever $\omega_h z^2 \leq k_h z \dot{z}$, i.e., when the vector field $(\dot{x}_h, \dot{z}) = (\omega_h z, \dot{z})$ that results from applying the integrator dynamics, points to the interior of the sector $\mathcal{F}$. Note that on the boundary of $\mathcal{F}$ defined by $u = 0$, enforcing the integrator dynamics $\dot{x}_h = \omega_h z$ will ensure the vector field $(\dot{x}_h, \dot{z}) = (\omega_h z, \dot{z})$ to point toward the interior of the sector.

On the basis of the above reasoning, the sets governing the active dynamics of HIGS are constructed as

$$\mathcal{F}_1 := \left\{ (z, u, \dot{z}) \in \mathbb{R}^3 \mid (z, u, \dot{z}) \in \mathcal{F} \setminus \mathcal{F}_2 \right\}, \tag{2.4a}$$

$$\mathcal{F}_2 := \left\{ (z, u, \dot{z}) \in \mathbb{R}^3 \mid (z, u, \dot{z}) \in \mathcal{F} \vee u = k_h z \wedge \omega_h z^2 > k_h \dot{z} z \right\}. \tag{2.4b}$$

By construction, it follows that $\mathcal{F} = \mathcal{F}_1 \cup \mathcal{F}_2$. A visualization of the sets in (2.4) is provided in Fig. 2.1. Note that given that z is continuous, the switching strategy renders the set $\mathcal{F}$ forward invariant, i.e., trajectories starting in $\mathcal{F}$ remain in $\mathcal{F}$ for all times. All initial state values $x_h(0) = x_{h,0} \in \mathbb{R}$ that satisfy $(z(0), x_h(0), \dot{z}(0)) \in \mathcal{F}$ are hereafter called *consistent* initial states, given $z \in C^1$.

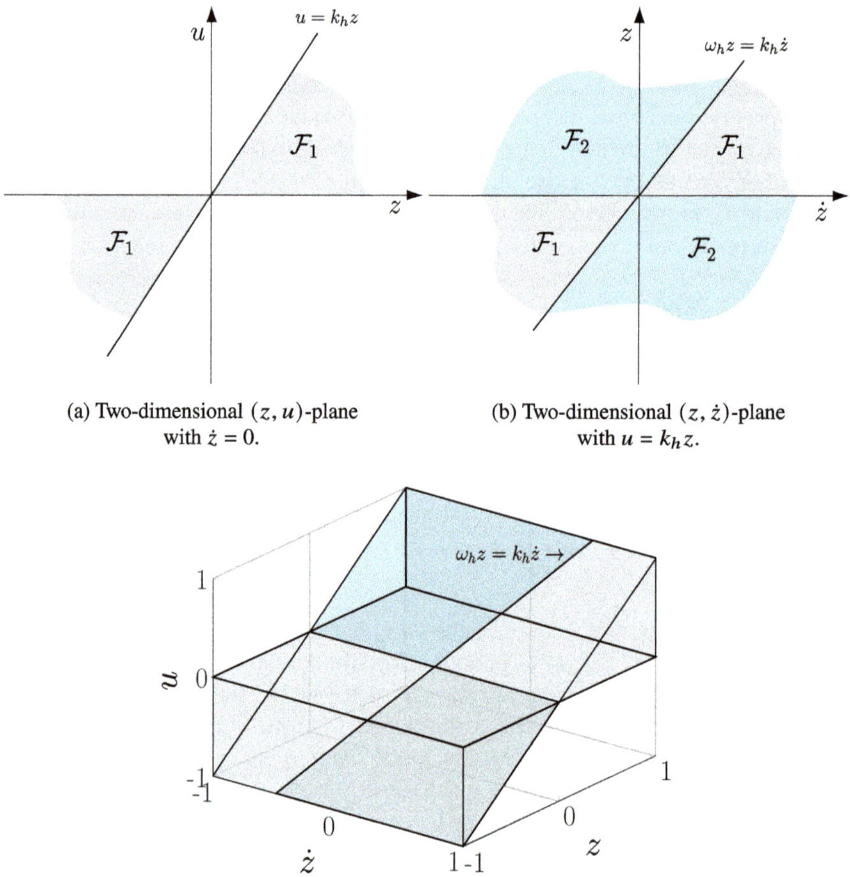

(a) Two-dimensional (z, u)-plane with $\dot{z} = 0$.

(b) Two-dimensional $(z, \dot{z})$-plane with $u = k_h z$.

(c) Three-dimensional region $\mathcal{F} = \mathcal{F}_1 \cup \mathcal{F}_2$.

Fig. 2.1 Flow sets $\mathcal{F}_1$ and $\mathcal{F}_2$ as described in (2.4), which determine the active dynamics of HIGS. The integrator mode region $\mathcal{F}_1$ is depicted in grey, and the gain mode region $\mathcal{F}_2$ is depicted in blue

Remark 2.1 The dynamics in (2.1) are defined for initial values $(z(0), u(0), \dot{z}(0)) \in \mathcal{F}$, but can be extended such that these are also defined for $(z(0), u(0), \dot{z}(0)) \notin \mathcal{F}$. To account for the latter, additional jump dynamics can be defined as

$$x_h(t^+) = \rho z(t), \quad \text{if } (z, u, \dot{z}) \in \mathcal{J}, \tag{2.5}$$

where $\rho \in [0, k_h]$, and in which the jump set is defined by

$$\mathcal{J} := \left\{ (z, u, \dot{z}) \in \mathbb{R}^2 \mid zu < \frac{1}{k_h} u^2 \right\}. \tag{2.6}$$

These dynamics describe an instantaneous reset which, by the choice of $\rho \in [0, k_h]$, results in $(z(t^+), u(t^+), \dot{z}(t^+)) \in \mathcal{F}$. Note that such a reset occurs only at the initial time $t = 0$ and not at times $t > 0$, as from the previous reasoning (and the fact that z is continuous) the specific switching strategy of HIGS will enforce $(z(t), u(t), \dot{z}(t))$ to remain within the sector $\mathcal{F}$ for all times $t \in \mathbb{R}_{>0}$.

To illustrate the switching strategy in (2.4), Fig. 2.2 depicts a simulated response of HIGS in (2.1) to a multi-sine input $z(t) = \sin(t) + 0.5\sin(3t)$, with the parameters set to $k_h = 1.5$, and $\omega_h = 1$ rad/s. When starting in integrator mode, the system generates the output u, depicted by the red curve in Fig. 2.2. Upon reaching the sector boundary $u = k_h z$, HIGS switches to gain mode, resulting in the output u as depicted by the blue curve which slides along the sector boundary until the integrator output no longer leads to a violation of the sector constraint. At that point, the gain mode terminates and HIGS re-enters the integrator mode. The system continues in integrator mode, again followed by a switch to gain mode when necessary, and

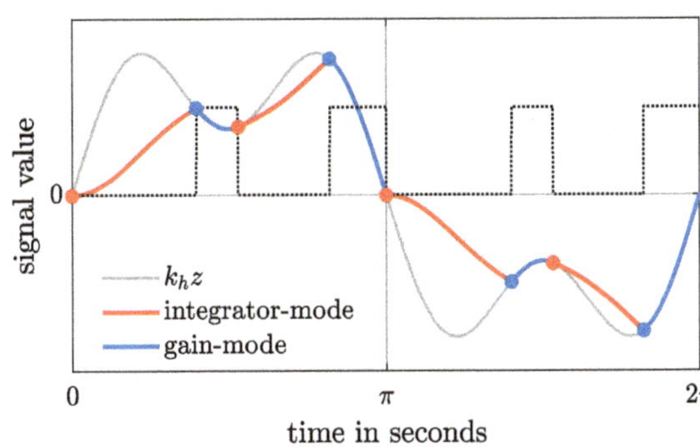

(a) Time-series response of HIGS to a multi-sine input.

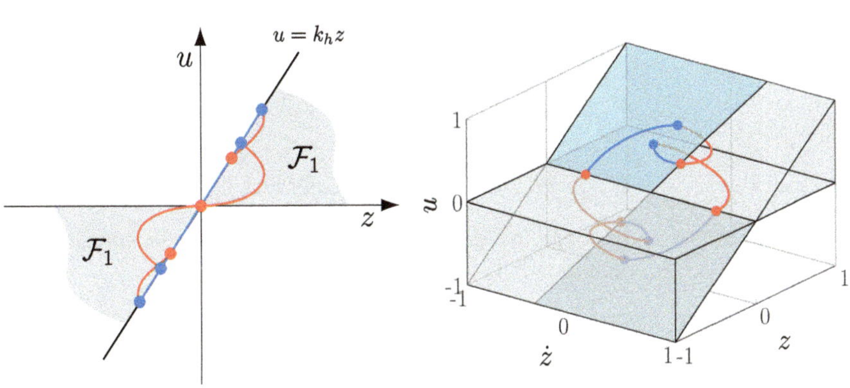

(b) Trajectory in (z, u)-plane. (c) Trajectory in $(z, u, \dot{z})$-space.

Fig. 2.2 Time response of HIGS when subjected to a multi-sine input $z(t) = \sin(t) + 0.5\sin(3t)$, $t \in \mathbb{R}_{\geq 0}$ (top figure) and corresponding spatial plots of the input–output trajectories (bottom figures) so on. In this manner, HIGS generates a *continuous* (but non-smooth) output that satisfies the sector constraint in (2.3) at all times. Figure 2.2 additionally shows the two-valued switching sequence (dotted), where the zero values correspond to HIGS being in integrator mode, and the non-zero values indicate HIGS being in gain mode.

2.2.2 Time-Domain Properties

In this section, useful time-domain properties of HIGS in (2.1) are discussed. Before going into more detail, however, the adopted solution concept for the discontinuous differential equation in (2.1) is formalized first.

Definition 2.1 Let $\mathbb{T} \subset \mathbb{R}_{\geq 0}$ be an interval of the form $[0, T]$ or $[0, T)$ with $T \in \mathbb{R}_{\geq 0}$ a finite number, or $\mathbb{T} = \mathbb{R}_{\geq 0}$. A locally absolutely continuous (AC) function $x_h : \mathbb{T} \to \mathbb{R}$ is called a solution to the discontinuous differential equation in (2.1) on $\mathbb{T}$ with consistent initial state $x_{h,0} \in \mathbb{R}$ and input $z \in C^1$, if $(z(t), x_h(t), \dot{z}(t)) \in \mathcal{F}$ for almost all $t \in \mathbb{T}$, and (2.1) holds almost everywhere in $\mathbb{T}$.

The solutions in Definition 2.1 are of *Carathéodory* type, see, e.g., [11] for more details regarding solution concepts for discontinuous dynamical systems. Under the assumption that the input z is a bounded continuously differentiable function (or even bounded AC function) defined on $\mathbb{T}$, there exists a (Carathéodory) solution $x_h : \mathbb{T} \to \mathbb{R}$ to (2.1). This can be established based on arguments as summarized in [7, Sect. 2.1.1]. The result also holds for $\mathbb{T} = \mathbb{R}_{\geq 0}$ (requiring locally AC and boundedness of the input z on $\mathbb{T}$). Having established a meaningful solution concept, the focus will now be on deriving useful (input–output) properties of HIGS.

Theorem 2.1 *The hybrid integrator–gain system in* (2.1) *is strictly passive from input z to output u in the sense that there exists a positive definite storage function $V(x_h) = kx_h^2$ with $0 < k < \frac{1}{2\omega_h}$ that satisfies*

$$\dot{V} \leq -cx_h^2 + zu, \quad \text{with} \quad 0 < c < \frac{1}{k_h}, \tag{2.7}$$

along all solutions of (2.1) *with $z \in C^1$ for almost all times $t \in \mathbb{R}_{\geq 0}$.*

Proof The proof can be found in Sect. A.2.1 and builds on some technical results that deal with non-smooth functions introduced in Sect. 2.3. □

Theorem 2.2 *The hybrid integrator–gain system in* (2.1) *is negative imaginary from input z to output u in the sense that there exists a positive definite storage function $V(x_h) = \frac{1}{2k_h}x_h^2$ that satisfies*

$$\dot{V} \leq z\dot{u}, \tag{2.8}$$

along all solutions of (2.1) *with $z \in C^1$ for almost all times $t \in \mathbb{R}_{\geq 0}$.*

Proof The proof is provided in Sect. A.2.2. □

Strict passivity [25, 43] is a particularly useful property for stability analysis as it implies that a well-posed feedback interconnection of HIGS with any strictly passive nonlinear plant is passive, see also [25, Theorem 6.1]. Under mild detectability conditions, this, in turn, can imply asymptotic stability of the closed-loop system, see also some recent results in [8]. The negative imaginary property in Theorem 2.2 relates to passivity and was first established in [40]. This property is useful when studying positive feedback interconnections and in the context of, for example, resonant control [39]. For further results on feedback stability theory for (nonlinear) negative imaginary systems, see, e.g., [5, 30].

The next result establishes useful *incremental stability* properties of HIGS.

Theorem 2.3 *Consider the hybrid integrator–gain system in* (2.1). *For all bounded inputs* $z \in C^1$, *and consistent initial states* $x'_{h,0}, x''_{h,0} \in \mathbb{R}$ *it holds that*

(1) $\|x'_h(t, x'_{h,0}, z) - x''_h(t, x''_{h,0}, z)\| \leq \|x'_{h,0} - x''_{h,0}\|$ *for all* $t \in \mathbb{R}_{\geq 0}$;
(2) $\|x'_h(t, x'_{h,0}, z) - x''_h(t, x''_{h,0}, z)\| \to 0$ as $t \to \infty$.

Proof The proof can be found in Sect. A.2.3. □

Theorem 2.4 *The hybrid integrator–gain system in* (2.1) *is homogeneous of degree one in the sense that for any* $\alpha \in \mathbb{R}$, *inputs* $z \in C^1$, *and consistent initial state* $x_{h,0} \in \mathbb{R}$, *the solutions satisfy*

$$\alpha x_h(t, x_{h,0}, z(t)) = x_h(t, \alpha x_{h,0}, \alpha z(t)), \tag{2.9}$$

for all times $t \in \mathbb{R}_{\geq 0}$.

Proof Denote $x_h(t) := x_h(t, x_h(0), z(t))$ and $x_h^\alpha(t) := x_h(t, \alpha x_{h,0}, \alpha z(t))$. Then, one can deduce in a similar manner as in the proof of Theorem 2.3 that a function of the form $V(x_h, x_h^\alpha) = \frac{1}{2}(\alpha x_h - x_h^\alpha)^2$ satisfies $\dot{V} \leq 0$ for almost all $t \in \mathbb{R}_{\geq 0}$. Since $\alpha x_{h,0} = x_{h,0}^\alpha$ one finds $V(0) = 0$, which implies $V(t) = 0$ for all $t \in \mathbb{R}_{\geq 0}$. □

Several important properties of HIGS follow from Theorems 2.3 and 2.4. First, from the result in Theorem 2.4, it immediately follows (with $\alpha = 1$) that for a consistent initial state $x_{h,0}$ and $z \in C^1$ the solution to the discontinuous differential equation in (2.1) is unique. From (2.9), it is also immediate that the output of HIGS scales linearly with input amplitude, hence demonstrating amplitude invariance. Moreover, it is an odd function in the sense that $-x_h(t, x_{h,0}, z(t)) = x_h(t, -x_{h,0}, -z(t))$. The above properties are useful for the describing function analysis as will be done in the next section.

2.2.3 Describing Function Analysis

A cornerstone of the success of LTI control is the fact that industry relies heavily on frequency-domain design techniques. A reason for the latter is that frequency-domain characterizations such as magnitude and phase provide transparent insights in the behaviour of a feedback loop in terms of robust stability and performance. However, the behaviour of nonlinear systems in general, and HIGS specifically, cannot be evaluated in the frequency-domain in the same manner as is done for LTI systems, the main reasons being: (i) the absence of the superposition principle, (ii) the fact that a nonlinear system may possess multiple steady-state solutions to a single (periodic) excitation, and (iii) the fact that a single-frequency excitation signal typically induces a response that contains harmonic content at multiple frequencies. To deal with these shortcomings, the describing function method is adopted [16]. In describing function analysis, an approximate frequency-domain characterization of nonlinear systems is determined by relating a sinusoidal input signal to the fundamental harmonic of the system's steady-state response. Higher harmonics in the output signal are usually neglected in the analysis.

In the previous section, it was shown in Theorems 2.3 and 2.4 that for a given input signal, HIGS in (2.1) has unique solutions, and that these solutions are input amplitude-invariant. This helps the use of the describing function approach to study frequency-domain properties of HIGS. For this purpose, consider a sinusoidal input (which is continuously differentiable) of the form

$$z(t) = A\sin(\omega t), \tag{2.10}$$

with amplitude $A \in \mathbb{R}$, and frequency $\omega \in \mathbb{R}_{\geq 0}$. Remark that by virtue of Theorem 2.4 the steady-state response of HIGS scales linearly with the input amplitude, such that without loss of generality the amplitude of (2.10) can be set to one, i.e., $A = 1$. Moreover, the steady-state response of HIGS to the sinusoidal input in (2.10) is unique, and has a period of $2\pi/\omega$ seconds. The response is found analytically by solving (2.1) for $x_h(0) = 0$ within the time-interval $[0, 2\pi/\omega]$, resulting in

$$u(\tau) = \begin{cases} \frac{\omega_h}{\omega}\left(1-\cos(\tau)\right), & \text{if } 0 \leq \tau \leq \gamma(\omega), \\ k_h \sin(\tau), & \text{if } \gamma(\omega) \leq \tau \leq \pi, \\ \frac{\omega_h}{\omega}\left(-1-\cos(\tau)\right), & \text{if } \pi \leq \tau \leq \pi + \gamma(\omega), \\ k_h \sin(\tau), & \text{if } \pi + \gamma(\omega) \leq \tau \leq 2\pi, \end{cases} \tag{2.11}$$

where $\tau = \omega t$ and $\gamma(\omega)$ represents the switching instant, which is specified hereafter. The steady-state response in (2.11) is depicted in Fig. 2.3.

An explicit expression for the switching instance $\gamma(\omega)$ is found from equating the generated output in integrator mode to the output in gain mode, that is,

$$\frac{\omega_h}{\omega}(1-\cos(\gamma(\omega))) = k_h \sin(\gamma(\omega)). \tag{2.12}$$

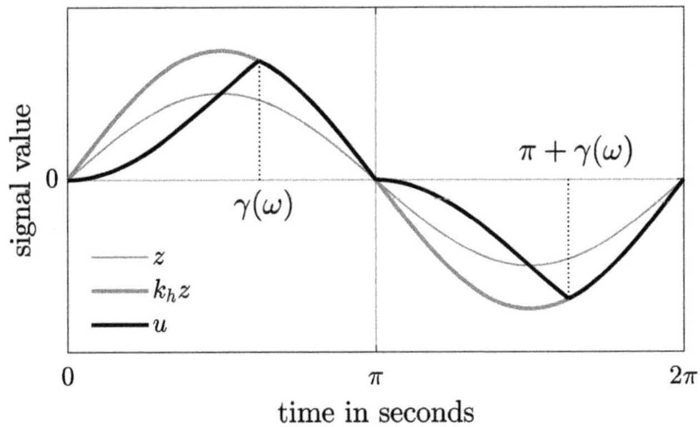

Fig. 2.3 Steady-state response of HIGS (black) to a sinusoidal input (thin grey) for $\omega_h = \omega$ and $k_h = 1.5$, alongside the maximum allowable output as given by the sector $[0, k_h]$ (bold grey)

Using the standard trigonometric identities $1 - \cos(a) = 2\sin^2(\frac{a}{2})$ and $\sin(a) = 2\sin(\frac{a}{2})\cos(\frac{a}{2})$ yields

$$\frac{\omega_h}{\omega}\sin\left(\frac{\gamma(\omega)}{2}\right) = k_h\cos\left(\frac{\gamma(\omega)}{2}\right), \tag{2.13}$$

which by $\tan(a) = \frac{\sin(a)}{\cos(a)}$ gives

$$\gamma(\omega) = 2\arctan\left(\frac{k_h\omega}{\omega_h}\right). \tag{2.14}$$

Given periodicity of the response in (2.11), it is possible to write $u(\tau)$ in the form of the Fourier series expansion

$$u(\tau) = \frac{a_0}{2} + \sum_{k=1}^{\infty}(a_k(\omega)\cos(k\tau) + b_k(\omega)\sin(k\tau)), \tag{2.15}$$

where the Fourier coefficients are given by

$$a_k(\omega) = \frac{\omega}{\pi}\int_0^{2\pi} u(\tau)\cos(k\tau)d\tau, \tag{2.16a}$$

$$b_k(\omega) = \frac{\omega}{\pi}\int_0^{2\pi} u(\tau)\sin(k\tau)d\tau, \tag{2.16b}$$

with $k \in \mathbb{Z}_{\geq 0}$ a multitude of the input frequency. Since $u(\tau)$ in (2.11) is an odd function, i.e., $u(-\tau) = -u(\tau)$, it follows that $a_0 = 0$ in (2.15). The coefficients for

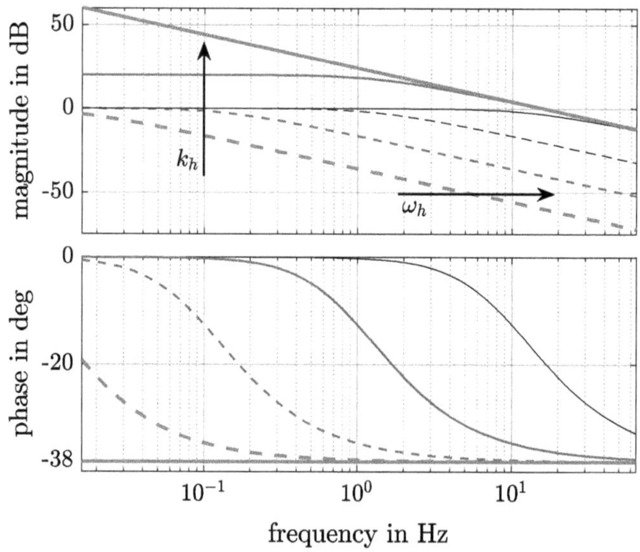

Fig. 2.4 Bode-like diagram of the describing function in (2.18) for different values of k_h and ω_h. The cross-over frequency is given by $\omega_c = |1 + 4j/\pi|\frac{\omega_h}{k_h}$ rad/s

$k = 1$ obtained after some algebra are given by (where for brevity dependency of γ on ω is omitted)

$$a_1(\omega) = \frac{1}{2\pi}\left(\frac{\omega_h}{\omega}\left(4\sin(\gamma) - \sin(2\gamma) - 2\gamma\right) - k_h\left(1 - \cos(2\gamma)\right)\right), \tag{2.17a}$$

$$b_1(\omega) = \frac{1}{2\pi}\left(\frac{\omega_h}{\omega}\left(3 - 4\cos(\gamma) + \cos(2\gamma)\right) + k_h\left(2(\pi - \gamma) + \sin(2\gamma)\right)\right). \tag{2.17b}$$

By neglecting all contributions in (2.15) at $k > 1$, and using Euler's formula, the describing function of HIGS is derived as

$$\begin{aligned} \mathcal{D}(j\omega) &= b_1(\omega) + ja_1(\omega) \\ &= \frac{\omega_h}{j\omega}\left(\frac{\gamma(\omega)}{\pi} + j\frac{e^{-2j\gamma(\omega)} - 4e^{-j\gamma(\omega)} + 3}{2\pi}\right) + k_h\left(\frac{\pi - \gamma(\omega)}{\pi} + j\frac{e^{-2j\gamma(\omega)} - 1}{2\pi}\right). \end{aligned} \tag{2.18}$$

The direct contribution of the integrator dynamics and gain dynamics can be recognized by the combination of the terms $\omega_h/j\omega$ and k_h in the respective ratios $\gamma(\omega)/\pi$ and $(\pi - \gamma(\omega))/\pi$, which correspond to the relative time-intervals in which these dynamics are active, see also Fig. 2.3.

A Bode-like plot of the describing function in (2.18) for different values of ω_h and k_h is shown in Fig. 2.4. Two asymptotes appear in this figure:

- For $\omega \to 0$, the describing function (2.18) tends to a static gain, i.e.,

$$\lim_{\omega \to 0} \mathcal{D}(j\omega) = k_h, \tag{2.19}$$

as a result of $\lim_{\omega \to 0} \gamma(\omega) = 0$. This can be explained in the time-domain as for harmonic inputs with $\omega \ll \omega_h/k_h$, the integrator buffer reaches its sector-admissible value after a time-interval negligible when compared to the input frequency, meaning HIGS effectively reduces to a static gain. In the same spirit, for $\omega_h/k_h \to \infty$ the above asymptote holds for all $\omega \in \mathbb{R}_{\geq 0}$.
- In contrast, for $\omega \to \infty$, the describing function characteristics tend to that of an integrator in terms of the magnitude, but with significantly less phase lag. Given (2.14), (2.18) it is found that $\lim_{\omega \to \infty} \gamma(\omega) = \pi$, yielding

$$\lim_{\omega \to \infty} \mathcal{D}(j\omega) = \frac{\omega_h}{j\omega}\left(1 + \frac{4j}{\pi}\right), \tag{2.20}$$

for which the magnitude and phase characteristics can be computed as

$$\lim_{\omega \to \infty} |\mathcal{D}(j\omega)| = \left|1 + \frac{4j}{\pi}\right| \frac{\omega_h}{\omega} \approx 1.62\frac{\omega_h}{\omega}, \tag{2.21a}$$

$$\lim_{\omega \to \infty} \angle\mathcal{D}(j\omega) = \arctan\left(-\frac{\pi}{4}\right) \approx -38.15^\circ. \tag{2.21b}$$

Interestingly, for $\omega \to \infty$, (2.20) tends to the describing function characteristics of the Clegg (reset) integrator, see [10, 24]. A time-domain interpretation is given as follows. For $\omega \gg \omega_h/k_h$, the integrator buffer builds up very slowly compared to the time scale of excitation, making the time-interval in which the gain mode is active negligible compared to the interval in which integrator dynamics are active. As sign-equivalence is enforced due to switching, this behaviour somewhat resembles that of the Clegg integrator, though without hard resets. The asymptote is also found for $\omega_h/k_h \to 0$.

From the above reasoning, one may observe the ratio ω_h/k_h as the leading factor that determines whether integrator or gain characteristics dominate the overall behaviour of HIGS both in time- and frequency-domain. This is also seen from the cross-over frequency of the describing function in (2.18), which is given by $\omega_c := |1 + 4j/\pi|\frac{\omega_h}{k_h}$ rad/s. Loosely speaking, for $\omega_h/k_h \to \infty$, HIGS behaves as a linear gain k_h, whereas for $\omega_h/k_h \to 0$, HIGS mimics the behaviour of a Clegg (reset) integrator without hard resets. This behaviour can be recognized in time-domain from the simulated time responses as depicted in Fig. 2.5.

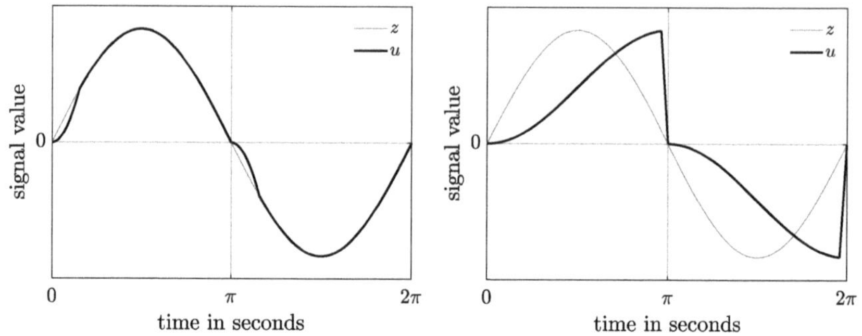

(a) Time-response for $\omega_h/k_h = 6\omega$.

(b) Time-response for $\omega_h/k_h = \omega/16$.

Fig. 2.5 Time response of HIGS (black) for different ratios of ω_h/k_h. The response in **a** resembles that of a linear gain, whereas the response in **b** resembles that of the Clegg integrator. Note that in both cases the responses are continuous, but not smooth

2.2.4 HIGS and Projected Dynamical Systems

A strongly related, but seemingly different approach for formalizing the working principle and closed-loop description of HIGS comes from considering a class of systems referred to as projected dynamical systems (PDS), see e.g., [15, 20, 23, 27]. Such systems are described by differential equations of which the solutions are restricted to be contained in a constraint set. At moments when the solutions tend to leave this set, the vector field of the system is changed by means of *projection* to ensure that the solutions remain inside the constraint set. Although the PDS philosophy naturally resembles that of HIGS, there are key differences that prevent a direct description of HIGS as a PDS. For instance, the constraint set of HIGS (the sector), does not satisfy certain regularity requirements necessary for the PDS framework to be applicable, see [23]. Furthermore, in the case of PDS, the complete vector field can be changed by projection on the constraint set. In a control context, however, when considering HIGS in feedback with a physical plant as done in the next section, it is only possible to project the dynamics of the controller and not the full dynamics including the plant. For that reason, in [13, 33] a generalization of PDS, referred to as *extended* projected dynamical systems (ePDS) has been proposed, which deals with these issues in a natural manner, thereby allowing for an alternative and intuitive representation of HIGS. One of the main advantages of formulating a HIGS-controlled system in the ePDS framework is that it may link to other classes of dynamical systems such that system theoretical properties directly transfer from one class to the other, see for instance the results in [18, 22, 34]. In [13], it was shown that in a closed-loop setting, the ePDS formulation can be written equivalently as a piecewise linear (PWL) model. As such, properties established for the former representation are directly inherited by the latter. Moreover, as for classical PDS stability and performance properties are well-established, these may naturally extend to systems with HIGS. In this book, however, the PWL formulation that will be introduced in the next section is adopted as it is more convenient for the stability and performance analysis approaches adopted in the subsequent parts. Recent connections in [14] between PDS and control barrier functions suggest an interesting direction for implementing discontinuous (e)PDS-based controllers in a continuous fashion, which can lead to alternative approaches toward stability and performance analysis of feedback control systems with HIGS. For a thorough exposition of the new ePDS framework in general, and the link with HIGS in particular, the reader is referred to [13, 21, 33].

2.3 Closed-Loop System Representation

Having developed more insight into the working principle and key properties of HIGS, this section formally discusses the closed-loop setting as considered for stability and performance analysis purposes throughout the remainder of this book. The generic set-up is depicted in Fig. 2.6, and consists of a linear time-invariant (LTI), multi-input multi-output (MIMO) generalized plant G interconnected with a single-input single-output (SISO) HIGS element $\mathcal{H}$. The generalized plant G contains the linear parts of the system, including the physical plant to be controlled and filters related to the LTI part of the feedback controller, possibly augmented with LTI input–output weighting filters. The latter weighting filters are added for the purpose of including problem-specific input knowledge into the system description, e.g., through known spectra of exogenous signals, and penalizing regulated output variables (see also [42]). In state space formulation, the plant G is given by

$$G : \begin{cases} \dot{x}_g(t) &= A_g x_g(t) + B_g u(t) + B_w w(t), \\ y(t) &= C_g x_g(t) + D_g u(t) + D_w w(t), \\ q(t) &= C_q x_g(t) + D_q w(t) + D_u u(t), \end{cases} \tag{2.22}$$

with states $x_g(t) \in \mathbb{R}^m$, regulated outputs $q(t) \in \mathbb{R}^r$ which contains performance variables such as servo errors and control actions, and external inputs $w(t) \in \mathbb{R}^p$ including disturbances, noise, and reference profiles, all at time $t \in \mathbb{R}_{\geq 0}$. The input and output of HIGS are denoted by $z(t) = -y(t) \in \mathbb{R}$, and $u(t) \in \mathbb{R}$, respectively.

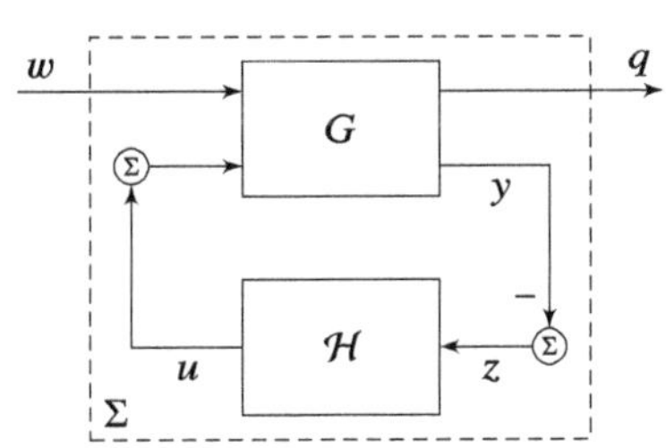

Fig. 2.6 Generic feedback interconnection of a MIMO LTI system G and HIGS $\mathcal{H}$ with exogenous inputs w and regulated outputs q

The realization (A_g, B_g, C_g, D_g) is assumed to be a minimal realization. For ensuring the closed-loop system in Fig. 2.6 to have well-behaved dynamics, the following standing assumption is made throughout this book.

Assumption 2.5 The LTI system G given in (2.22) and depicted in Fig. 2.6 satisfies $D_g = 0$, $D_w = 0$, $C_g B_g = 0$, and $C_g B_w = 0$.

Assumption 2.5 implies that the transfer functions from u to y and w to y have a relative degree of at least two. It removes the occurrence of an algebraic loop when HIGS is in gain mode, ensures the absence of a direct feedthrough from the signals w to z and $\dot{z}$, and guarantees the input to HIGS to be continuous and at least one time continuously differentiable. Note that within the relevant and broad class of motion systems, where plants are typically described by double integrators with additional structural dynamics, the relative degree assumption is often naturally satisfied.

2.3.1 *Closed-Loop System Description*

Due to the piecewise linear (PWL) nature of HIGS in (2.1), and under Assumption 2.5, the closed-loop system in Fig. 2.6 naturally admits the PWL formulation

$$\Sigma : \begin{cases} \dot{x}(t) &= A_i x(t) + B w(t), \quad \text{if } x(t) \in \mathcal{X}_i,\ i \in \{1, 2\} \\ q(t) &= C x(t) + D w(t), \end{cases} \tag{2.23}$$

with state $x(t) = \left[x_g(t)^\top x_h(t)\right]^\top \in \mathbb{R}^n$, in which x_g and x_h are the states of G and $\mathcal{H}$, respectively, so that $n = m + 1$. The sets $\mathcal{X}_i$, $i \in \{1, 2\}$, are defined as

$$\mathcal{X}_i := \left\{x \in \mathbb{R}^n \mid Ex \in \mathcal{F}_i\right\}, \tag{2.24}$$

with $\mathcal{F}_i$, $i = \{1, 2\}$, the sets defined in (2.4), and where the matrix $E \in \mathbb{R}^{3\times n}$ extracts those signals from x that determine mode switching of HIGS, i.e., E is such that $Ex := (z\ u\ \dot{z})^\top$, and is, therefore, given by

$$E^\top = \begin{bmatrix} -C_g^\top & 0 & -(C_g A_g)^\top \\ 0 & 1 & 0 \end{bmatrix}. \tag{2.25}$$

Remark that $\mathcal{X}_1 \cup \mathcal{X}_2 = \mathcal{X} := \{x \in \mathbb{R}^n \mid Ex \in \mathcal{F}_1 \cup \mathcal{F}_2\}$. The mode-dependent system matrices in (2.23) are given by

$$A_1 = \begin{bmatrix} A_g & B_g \\ -\omega_h C_g & 0 \end{bmatrix}, \quad \text{and} \quad A_2 = \begin{bmatrix} A_g & B_g \\ -k_h C_g A_g & 0 \end{bmatrix}, \tag{2.26}$$

and furthermore $B = [B_w^\top, 0]^\top$, $C = [C_q, D_u]$, and $D = D_q$. Note that the state space formulation in (2.23) results from Assumption 2.5 and explicit differentiation of the algebraic constraint $x_h = k_h z$ in (2.1) in gain mode.

Remark 2.2 The system matrix A_2 in (2.26) results from explicit differentiation of the algebraic constraint $x_h - k_h z = 0$ in gain mode. Consequently, A_2 is rank deficient and has at least one zero eigenvalue. In the upcoming parts, it will sometimes be useful to remove this rank deficiency. This can be accomplished by adding the algebraic constraint in gain mode to A_2, giving $\mathcal{A}_2 := A_2 + \mu b H$ with $\mu \in \mathbb{R}$, $b = [0 \;\; 1]^\top$ and $H = [-k_h C_g \; - 1]$. Since $A_2 x = \mathcal{A}_2 x$ for all $x \in \mathcal{X}_2$, this extension is not changing the dynamics in (2.23). Using the matrix $\mathcal{A}_2$ rather than A_2 is useful in Chaps. 4 and 5 where specific results require dynamics in gain mode to be stable, and thus the matrix $\mathcal{A}_2$ must be Hurwitz, i.e., all its eigenvalues have strictly negative real part.

It can be seen that (2.23) describes a *discontinuous* differential equation. This makes proving (global) existence of solutions, given an initial state x_0 and external signal w a difficult problem, since typical continuity properties used for studying differential equations/inclusions (such as upper-semicontinuity of the right-hand side conform [2]) are not given, see also [11]. The next section briefly revisits some of the recent results presented in [21] regarding well-posedness in the sense of global existence and forward completeness of solutions for the class of extended projected dynamical systems. Since the HIGS-controlled system in (2.23) fits this class—see [13] for an equivalent representation of the discontinuous PWL system formulation in (2.23) as an extended projected dynamical system—these results are of relevance here.

2.3.2 Solution Concept and Well-Posedness

The results in [21] are applicable to systems of the form (2.23) with w piecewise continuous functions (denoted $w \in PC$). The following solution concept, related to Definition 2.1 is adopted.

Definition 2.2 Let $\mathbb{T} \subset \mathbb{R}_{\geq 0}$ be an interval of the form $[0, T]$ or $[0, T)$ with $T \in \mathbb{R}_{\geq 0}$ a finite number, or $\mathbb{T} = \mathbb{R}_{\geq 0}$. A locally absolutely continuous (AC) function $x : \mathbb{T} \to \mathbb{R}^n$ is called a solution to the HIGS-controlled system in (2.23) on $\mathbb{T}$ with initial state $x_0 \in \mathcal{X}$ and $w \in PC$, if $x(0) = x_0$, $x(t) \in \mathcal{X}$ for all $t \in \mathbb{T}$, and (2.23) holds almost everywhere in $\mathbb{T}$.

Theorem 2.6 *For each initial state $x_0 \in \mathcal{X}$ and bounded inputs $w \in PC$, there exists a solution $x : \mathbb{R}_{\geq 0} \to \mathbb{R}^n$ with $x(0) = x_0$ and input w to the HIGS-controlled system in* (2.23) *on* $\mathbb{R}_{\geq 0}$.

Theorem 2.6 is an immediate consequence of [21, Corollary 2], and the fact that the plant G in (2.22) is linear (and, therefore, the linear growth condition in [21, Corollary 2] is satisfied).

The above result in Theorem 2.6 shows well-posedness of a HIGS-controlled system in terms of global existence of solutions. However, it is also true that these solutions are *unique*, in the sense that given a consistent initial condition $x(0) \in \mathcal{X}$, and input $w \in PC$, there is exactly one corresponding solution. The formal proof of this statement is postponed to Sect. 6.4.3, where incremental stability properties of closed-loop HIGS-controlled systems are further discussed in detail.

Remark 2.3 Since the response of HIGS is amplitude-invariant and an odd function (Theorem 2.4), it immediately follows that the solutions of the closed-loop interconnection of HIGS and an LTI system as in (2.23) scale linearly with the input amplitude as well, i.e., $x(t, \alpha x(0), \alpha w(t)) = \alpha x(t, x(0), w(t))$, $t \in \mathbb{R}_{\geq 0}$ for $\alpha \in \mathbb{R}$. This property is particularly useful in, e.g., performance analysis through nonlinear Bode plots as will be discussed in more detail in Chap. 6.

2.3.3 Stability and Performance: Preliminaries

One of the primary objectives in this book is to develop tools for verifying (robust) stability and performance of the HIGS-controlled system in (2.23). To make the discussions precise, preliminary definitions, notations, and results heavily used throughout the remainder of this book are provided in this section.

To study stability properties of the closed-loop system in (2.23), the notion of input-to-state stability (ISS) is adopted.

Definition 2.3 ([41]) The closed-loop system in (2.23) is said to be input-to-state stable (ISS), if there exist a $\mathcal{KL}$-function α and a $\mathcal{K}$-function β such that for any initial condition $x(0) = x_0 \in \mathcal{X}$ and any bounded input signal $w \in PC$, the corresponding solution to (2.23) satisfies

$$\|x(t)\| \leq \alpha(\|x(0)\|, t) + \beta\left(\sup_{0 \leq \tau \leq t} \|w(\tau)\|\right), \tag{2.27}$$

for all $t \in \mathbb{R}_{\geq 0}$.

In the study of ISS for discontinuous dynamical systems such as in (2.23), non-smooth Lyapunov functions are often employed, see e.g., [6, 19, 31, 37]. In this context, it is useful to revisit some of the machinery used in non-smooth analysis.

Definition 2.4 ([9, 37]) For a locally Lipschitz function $V : \mathbb{R}^n \to \mathbb{R}$, the generalized gradient of V at x is defined as

$$\partial V(x) = \overline{\text{co}}\left\{\lim_{i \to \infty} \nabla V(x_i) \mid x_i \to x, x_i \notin \Omega_V\right\}, \tag{2.28}$$

where $\overline{\text{co}}$ denotes the closed convex hull, ∇V denotes the gradient of V (at states where it is defined), and Ω_V is the set of measure zero where the gradient of V is not defined.

Theorem 2.7 *Let $V : \mathbb{R}^n \to \mathbb{R}_{\geq 0}$ be a locally Lipschitz continuous function and x a locally absolutely continuous solution to $\dot{x}(t) \in F(x(t), w(t))$ for some function $w \in PC$. Then $t \mapsto V(x(t))$ is locally Lipschitz continuous and*

$$\frac{d}{dt}V(x(t)) \leq \max_{p \in \partial V(x(t))} \max_{f \in F(x(t),w(t))} \langle p, f \rangle \tag{2.29}$$

holds for almost all times $t \in \mathbb{R}_{\geq 0}$.

Proof The proof largely proceeds as the proof of Theorem 3.3 in [19]. First, observe that since V is locally Lipschitz continuous, and x is locally absolutely continuous, the composite function $t \mapsto V(x(t))$ is locally absolutely continuous and consequently, $t \mapsto V(x(t))$ is differentiable almost everywhere with respect to time t. Suppose that at time t both $\dot{x}(t)$ and $dV(x(t))/dt$ exist. Note that $\dot{x}(t) = f \in F(x(t), w(t))$. Using the same arguments as in the proof of Theorem 3.3 in [19], it can be concluded that

$$\begin{aligned}\frac{d}{dt}V(x(t)) &= \lim_{h \to 0} \frac{V(x(t+h)) - V(x(t))}{h} \\ &= \lim_{h \downarrow 0} \frac{V(x(t) + hf) - V(x(t))}{h}.\end{aligned} \tag{2.30}$$

To evaluate the right-hand side of (2.30) use is made of Lebourg's Lipschitz mean-value theorem [9, Theorem 2.3.7], which states that for any V that is Lipschitz on an open set containing the closed line segment $[a, b]$, there is a point c in the open line segment (a, b) such that

$$V(a) - V(b) \in \langle a - b, \partial V(c) \rangle. \tag{2.31}$$

Applying this to $V(x(t) + h_i f) - V(x(t))$ for $h_i \downarrow 0$ when $i \to \infty$, one obtains

$$\frac{d}{dt}V(x(t) = \lim_{i \to \infty} \langle p_i, f \rangle, \tag{2.32}$$

where $p_i \in \partial V(x(t) + \alpha_i f)$ with $\alpha_i \in (0, h_i)$. Since V is a locally Lipschitz continuous function, it follows by application of [9, Proposition 2.1.2] and [9, Proposition 2.1.5] that for α_i converging to zero, there is a subsequence $\{p_i\}_{i \in \mathbb{N}}$ converging to p, and one has $p \in \partial V(x)$. It then follows that the right-hand side in (2.30) is equal to

$$\frac{d}{dt}V(x(t)) = \langle p, f \rangle \tag{2.33}$$

for some $p \in \partial V(x)$. Then, an upper-bound on the time-derivative of V follows as

$$\frac{d}{dt}V(x(t)) \leq \max_{p \in \partial V(x(t))} \max_{f \in F(x(t),w(t))} \langle p, f \rangle, \tag{2.34}$$

which holds for almost all times $t \in \mathbb{R}_{\geq 0}$. □

The above result provides a basis for the definition of an ISS-Lyapunov function for the HIGS-controlled system in (2.23), in line with [19, 41].

Definition 2.5 A locally Lipschitz continuous function $V : \mathbb{R}^n \to \mathbb{R}_{\geq 0}$ is said to be an ISS-Lyapunov function for the system in (2.23), if

- there exist $\mathcal{K}_\infty$-functions α_1, α_2 such that for all $x \in \mathbb{R}^n$

$$\alpha_1(|x|) \leq V(x) \leq \alpha_2(|x|); \tag{2.35}$$

- there exist $\mathcal{K}$-functions α and β such that for all $x \in \mathbb{R}^n$ and $w \in \mathbb{R}^p$

$$\max_{p \in \partial V(x)} \langle p, F(x, w) \rangle \leq -\alpha(|x|) + \beta(|w|), \tag{2.36}$$

  where

$$F(x, w) = \begin{cases} A_1 x + Bw & \text{if } x \in \mathcal{X}_1, \\ A_2 x + Bw & \text{if } x \in \mathcal{X}_2, \end{cases} \tag{2.37}$$

  with $\mathcal{X}_i, i = \{1, 2\}$ given in (2.24), and matrices A_1, A_2, B given in (2.26).

Theorem 2.8 *If there exists an ISS-Lyapunov function for* (2.23) *in the sense of Definition 2.5, then system* (2.23) *is ISS.*

Proof The proof follows similar steps as the proof of [19, Theorem 3.5]. □

In studying different (input–output) performance properties of the closed-loop system in (2.23) the concept of dissipativity will be particularly useful.

Definition 2.6 ([45]) The closed-loop system in (2.23) is said to be dissipative with respect to a supply function $s : \mathbb{R}^n \times \mathbb{R}^r \times \mathbb{R}^m \to \mathbb{R}$, if there exist a positive (semi-)definite continuous function $V : \mathbb{R}^n \to \mathbb{R}_{\geq 0}$, called a storage function, and $\mathcal{K}_\infty$-functions α_1, α_2 such that $\alpha_1(|x|) \leq V(x) \leq \alpha_2(|x|)$ for all $x \in \mathcal{X}$, and

$$V(x(t)) - V(x(0)) \leq \int_0^t s(w(\tau), q(\tau), x(t)) d\tau, \tag{2.38}$$

for all $t \in \mathbb{R}_{\geq 0}$ and all (x, q, w) satisfying (2.23) for almost all times t.

The definition of an ISS-Lyapunov function as given in Definition 2.5 can be extended toward a storage function by relaxing condition (2.36) to

$$\max_{p \in \partial V(x)} \langle p, F(x, w) \rangle \leq s(w, q, x). \tag{2.39}$$

Supply functions that are often used to assert performance in terms of, for example, the $\mathcal{L}_2$-gain and $\mathcal{H}_2$-norm are given by $s(w, q, x) = \gamma^2 \|w\|^2 - \|q\|^2$, with $\gamma > 0$, and $s(w, q, x) = -\|q\|^2$, see, e.g., [43]. Recall also in this context the passivity and negative imaginary properties of HIGS in Theorems 2.1 and 2.2.

2.3.4 Robustness

Some comments are in order regarding robustness issues that can arise for discontinuous differential equations of the form

$$\dot{x}(t) = f(x(t), w(t)), \quad \text{when } x(t) \in \mathcal{X}, \tag{2.40}$$

with $x(t) \in \mathbb{R}^n$, $w(t) \in \mathbb{R}^p$, and $\mathcal{X} \subseteq \mathbb{R}^n$, just as the closed-loop system in (2.23). As discussed in [17], in order to obtain robust stability/performance guarantees (with respect to arbitrary small state perturbations), it is important to consider the so-called Krasovskii regularization [26] of (2.40), which is defined as

$$\dot{x}(t) \in \bigcap_{\delta > 0} \overline{\text{co}}(f(\mathbb{B}(x(t), \delta) \cap \mathcal{X}, w(t)), \quad \text{when } x(t) \in \overline{\mathcal{X}}, \tag{2.41}$$

where $\mathbb{B}(x(t), \delta)$ is the open ball of radius δ around $x(t)$ and for a set $\Omega \subseteq \mathbb{R}^n$, $\overline{\text{co}}(\Omega)$ denotes its closed convex hull and $\overline{\Omega}$ denotes the closure. The Krasovskii regularization of (2.23) can be calculated explicitly, giving

$$\dot{x}(t) \in \begin{cases} A_1 x(t) + B w(t) & \text{if } x(t) \in \mathcal{X} \setminus \overline{\mathcal{X}}_2, \\ \overline{\text{co}}(A_1 x(t), A_2 x(t)) + B w(t) & \text{if } x(t) \in \overline{\mathcal{X}}_2. \end{cases} \tag{2.42}$$

Solutions to (2.40) in the sense of Krasovskii are locally AC solutions to the differential inclusion in (2.42), see, e.g., [32]. In the remainder, it will be shown that for specific choices of V, the conditions in Definition 2.5 guarantee ISS/dissipativity of the Krasovskii regularization in (2.42) as well, and thus robust versions of stability and performance guarantees for (2.23) can be obtained including state perturbations.

2.4 Discrete-Time Aspects

Although the main focus of this book is on continuous-time aspects of HIGS-based control systems, discrete-time aspects undoubtedly play an important role when implementing HIGS on digital platforms. This section revisits some discrete-time aspects, recently developed in [36].

2.4.1 Discrete-Time Formulation of HIGS

Discrete-time versions of HIGS that preserve the main characteristics and the working principle of the continuous-time version in (2.1) are introduced in [36]. A particular representation, the so-called trimodal version, is given by the discrete-time PWL system

$$\overline{\mathcal{H}}:\begin{cases} x_h[k] = x_h[k-1] + \omega_h T_s z[k] & \text{if } \xi[k], \in \overline{\mathcal{F}}_1, \quad (2.43a)\\ x_h[k] = k_h z[k], & \text{if } \xi[k] \in \overline{\mathcal{F}}_2, \quad (2.43b)\\ x_h[k] = 0, & \text{if } \xi[k] \in \overline{\mathcal{F}}_3, \quad (2.43c)\\ u[k] = x_h[k], & \quad (2.43d)\end{cases}$$

where $x_h[k] \in \mathbb{R}$ denotes the state, $z[k] \in \mathbb{R}$ is the input, and $u[k] \in \mathbb{R}$ is the output, all at time instant $t = kT_s$, with $k \in \mathbb{N}$ the discrete-time step, and $T_s \in \mathbb{R}_{>0}$ the sampling period. Note that the discrete-time integrator mode dynamics (2.43a) are obtained by backward Euler discretization of the dynamics in (2.1a). The active mode of operation is based on the decision variable $\xi[k] = (z[k], u^-[k], z^-[k]) := (z[k], u[k-1], z[k-1])$. The regions $\overline{\mathcal{F}}_i$, $i = \{1, 2, 3\}$, where different subsystems are active, are given by

$$\overline{\mathcal{F}}_1 := \left\{\xi \in \mathbb{R}^3 \mid (z^-, u^-, z) \in \mathcal{F} \wedge (u^- + \omega_h T_s z)z \geq \frac{1}{k_h}(u^- + \omega_h T_s z)^2\right\}, \quad (2.44a)$$

$$\overline{\mathcal{F}}_2 := \left\{\xi \in \mathbb{R}^3 \mid (z^-, u^-, z) \in \mathcal{F} \wedge (u^- + \omega_h T_s z)z < \frac{1}{k_h}(u^- + \omega_h T_s z)^2 \wedge zz^- \geq 0\right\}, \quad (2.44b)$$

$$\overline{\mathcal{F}}_3 := \left\{\xi \in \mathbb{R}^3 \mid (z^-, u^-, z) \in \mathcal{F} \wedge (u^- + \omega_h T_s z)z < \frac{1}{k_h}(u^- + \omega_h T_s z)^2 \wedge zz^- < 0\right\}, \quad (2.44c)$$

where $\mathcal{F} = \left\{(z^-, u^-, z) \in \mathbb{R}^3 \mid z^- u^- \geq \frac{1}{k_h}(u^-)^2\right\}$ resembles the $[0, k_h]$-sector in (2.3). The discrete-time representation in (2.43) is accompanied by the following assumption.

Assumption 2.9 The parameters ω_h, k_h, and T_s satisfy $0 < T_s \leq \frac{k_h}{\omega_h}$.

Assumption 2.9 can be satisfied by design, and ensures that (2.43) always continues in integrator mode from zero (initial) state value, i.e., when $x_h[k] = u[k] = 0$ with $k \in \mathbb{N}$. This is an important property as will be clarified later.

The trimodal representation in (2.43) and the regions (2.44) deserve some further explanation. The second condition defining the set $\overline{\mathcal{F}}_1$ in (2.44a) is "testing" if operation according to the integrator dynamics in (2.43a) would lead to trajectories that satisfy the sector constraint $uz \geq \frac{1}{k_h}u^2$, and, therefore, at the next timestep would remain in $\mathcal{F}$. When this is not the case, and trajectories resulting from the integrator mode (2.43a) would violate the sector constraint, the gain mode must be activated to keep trajectories in the set $\mathcal{F}$, adhering to the philosophy of HIGS in continuous-time. Next, observe from the continuous-time response of HIGS (see Fig. 2.2) that a transition from gain mode back to integrator mode, starting at zero, is enforced upon

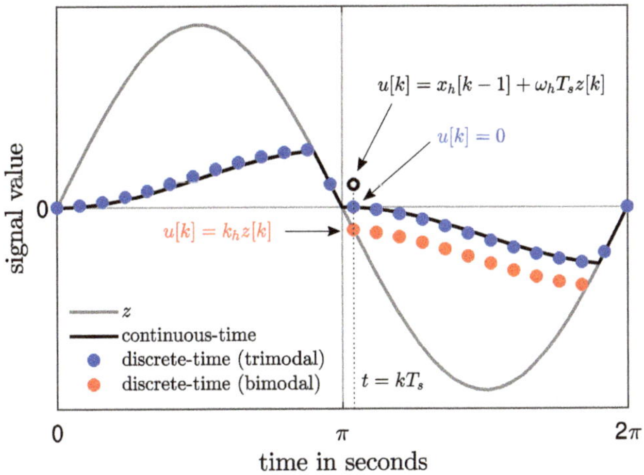

Fig. 2.7 Response of HIGS to a sinusoidal input z (thin grey) in continuous-time (black) and discrete-time with zeroing mode (trimodel, blue dots) and without zeroing mode (bimodal, red dots) for $k_h = 1$, $\omega_h = 2\pi$ rad/s and $T_s = 0.08\pi$ s. Reprinted from Automatica, Vol 167, Sharif B, Alferink D, Heertjes M, Nijmeijer H, Heemels M, Analysis of sampled-data hybrid integrator-gain systems: A discrete-time approach, 111765, Copyright 2024, with permission from Elsevier.

a zero-crossing of the input z. However, due to the sampled-data nature of the input in discrete-time, gain mode may still be enforced after a zero-crossing if no additional measures are taken, which goes against the idea that the primary mode of operation should be the integrator mode. This leads to the third condition in the set $\overline{\mathcal{F}}_2$ in (2.44b), which provides a zero-crossing detection mechanism. If no zero-crossing of the input is detected, i.e., $z[k]z[k-1] \geq 0$, operation in gain mode can continue. This motivates the set $\overline{\mathcal{F}}_2$ in which the gain mode dynamics (2.43b) are active. Upon a zero-crossing of the input, i.e., when $z[k]z[k-1] < 0$, the state of HIGS is reset to zero to ensure that after a zero-crossing of the input, $x_h[k] = 0$ and HIGS continues in integrator mode, the latter by virtue of Assumption 2.9. This motivates the third (zeroing) mode in (2.43c) as well as the set $\overline{\mathcal{F}}_3$ in (2.44c) where this mode is active.

The time responses of both the continuous-time HIGS in (2.1) and the discrete-time HIGS in (2.43) to a sinusoidal input are shown in Fig. 2.7. In addition, the response in absence of the zeroing mode (2.43c) is shown (this case corresponds to the so-called bimodal version considered in, e.g., [36, 38]). It can be seen that initially all three responses are similar to each other. Upon a zero-crossing of the input z, however, the trimodal HIGS in (2.43) generates an output that is closer to the output of the continuous-time HIGS in (2.1) as compared to the response without zeroing mode, indicating the importance of including the zeroing mode. In both cases, it can be seen that the sector condition in $\mathcal{F}$ is respected.

In summary, the trimodal version of HIGS in (2.43) preserves the main characteristics of HIGS in continuous-time, in the sense of (i) having the integrator mode as its primary mode of operation, and (ii) ensuring that the input–output trajectories remain within the sector $\mathcal{F}$ at all discrete times, see also Fig. 2.7.

2.4.2 Remarks on Implementation

In the remainder of this book, the discrete-time version of HIGS in (2.43) will be adopted for implementation purposes in a digital environment. Consequently, the numerical and experimental closed-loop control systems discussed are of a digital (sampled-data) nature, rather than continuous time. To justify the use of the continuous-time tools for analysis and design of HIGS-controlled systems presented in this book, the sampling frequency $F_s = 1/T_s$ in the examples is set significantly higher than the bandwidth of the system (over a factor of 20). This ensures that the sampling interval is much smaller than the time scale of the system dynamics, resulting in behaviour that closely approximates that of a continuous-time system, see also [29, Chap. 2]. For a formal analysis of discrete-time and sampled-data HIGS-based control systems, the reader is referred to the works in [35, 36, 39].

2.5 Summary

In this chapter, the hybrid integrator–gain system, HIGS for short, is formally introduced. It is shown that specific switching between integrator and gain characteristics ensures the generated output of the hybrid integrator to have a sign that is equal to the sign of the input at all times, thereby adhering to the traditional philosophy of the Clegg (reset) integrator. Different from reset integrators, however, HIGS avoids hard resets, and generates continuous (non-smooth) outputs, which may provide benefits. Both time- and frequency-domain properties of HIGS are studied, the latter in an approximate sense through its describing function, which reveals magnitude characteristics that correspond to a (weak) integrator, but with a phase lag of only $38.15°$.

In the next chapter, the properties of HIGS are exploited for constructing nonlinear controllers that are able to overcome some of the fundamental performance limitations of linear time-invariant control.

References

1. Aangenent, W.H.T.M., Witvoet, G., Heemels, W.P.M.H., Molengraft, M.J.G., Steinbuch, M.: Performance analysis of reset control systems. Int. J. Robust Nonlinear Control **20**, 1213–1233 (2010)
2. Aubin, J.P., Cellina, A.: Differential Inclusions. Springer, Berlin Heidelberg (1984)
3. Baños, A., Barreiro, A.: Reset Control Systems. Springer, London (2012)

4. Beker, O., Hollot, C., Chait, Y., Han, H.: Fundamental properties of reset control systems. Automatica **40**(6), 905–915 (2004)
5. Bhowmick, P., Lanzon, A.: Output strictly negative imaginary systems and its connections to dissipativity theory. In: 2019 IEEE 58th Conference on Decision and Control (CDC), pp. 6754–6759 (2019)
6. Branicky, M.: Multiple Lyapunov functions and other analysis tools for switched and hybrid systems. IEEE Trans. Autom. Control **43**, 475–482 (1998)
7. Brogliato, B., Tanwani, A.: Dynamical systems coupled with monotone set-valued operators: formalisms, applications, well-posedness, and stability. SIAM Rev. **62** (2020)
8. Chu, H., van den Eijnden, S., Heemels, W.: Projection-based controllers with inherent dissipativity properties. In: 2023 62nd IEEE Conference on Decision and Control (CDC), pp 5664–5669 (2023)
9. Clarke, F.H.: Optimization and Nonsmooth Analysis. Wiley (1983)
10. Clegg, J.C.: A nonlinear integrator for servomechanisms. Trans. Am. Inst. Electr. Eng., Part II: Appl. Ind. **77**(1), 41–42 (1958)
11. Cortes, J.: Discontinuous dynamical systems. IEEE Control Syst. Mag. **28**(3), 36–73 (2008)
12. Deenen, D.A., Heertjes, M.F., Heemels, W.P.M.H., Nijmeijer, H.: Hybrid integrator design for enhanced tracking in motion control. In: Proceedings of the American Control Conference (ACC'17), pp. 2863–2868 (2017)
13. Deenen, D.A., Sharif, B., van den Eijnden, S., Nijmeijer, H., Heemels, M., Heertjes, M.: Projection-based integrators for improved motion control: formalization, well-posedness and stability of hybrid integrator-gain systems. Automatica **133**, 109830 (2021)
14. Delimpaltadakis, G., Cortés, J., Heemels, W.P.M.H.: Continuous approximations of projected dynamical systems via control barrier functions. IEEE Trans. Autom. Control **70**(1), 681–688 (2025)
15. Dupuis, P., Nagurney, A.: Dynamical systems and variational inequalities. Ann. Oper. Res. **44**(1), 7–42 (1993)
16. Gelb, A., Vander Velde, W.: Multiple-input Describing Functions and Nonlinear Systems Design. McGraw Hill, New York (1968)
17. Goebel, R., Sanfelice, R., Teel, A.: Hybrid dynamical systems. IEEE Control Syst. **29**, 28–93 (2009)
18. Heemels, W., Brogliato, B.: The complementarity class of hybrid dynamical systems. Eur. J. Control **9**(2), 322–360 (2003)
19. Heemels, W., Weiland, S.: Input-to-state stability and interconnections of discontinuous dynamical systems. Automatica **44**(12), 3079–3086 (2008)
20. Heemels, W., Schumacher, J., Weiland, S.: Projected dynamical systems in a complementarity formalism. Oper. Res. Lett. **27**(2), 83–91 (2000)
21. Heemels, W.P.M.H., Tanwani, A.: Existence and completeness of solutions to extended projected dynamical systems and sector-bounded projection-based controllers. IEEE Control Syst. Lett. **7**, 1590–1595 (2023)
22. Heemels, W.P.M.H., Camlibel, M.K., Heertjes, M.F.: Oblique projected dynamical systems and incremental stability under state constraints. IEEE Control Syst. Lett. **4**(4), 1060–1065 (2020)
23. Henry, C.: An existence theorem for a class of differential equations with multivalued right-hand side. J. Math. Anal. Appl. **41**(1), 179–186 (1973)
24. Horowitz, I., Rosenbaum, P.: Non-linear design for cost of feedback reduction in systems with large parameter uncertainty. Int. J. Control **21**(6), 977–1001 (1975)
25. Khalil, H.K.: Nonlinear Systems, 3rd edn. Prentice-Hall, Upper Saddle River, New Jersey (2002)
26. Krasovskii, N., Subbotin, A.I.: Game-Theoretical Control Problems. Springer (1988)
27. Nagurney, A., Zhang, D.: Projected Dynamical Systems, pp. 9–43. Springer, US, Boston, MA (1996)
28. Nešić, D., Zaccarian, L., Teel, A.: Stability properties of reset systems. Automatica **44**(8), 2019–2026 (2008)
29. Ogata, K.: Discrete-time Control Systems. Prentice Hall (1995)

30. Petersen, I.R.: Negative imaginary systems theory and applications. Annu. Rev. Control **42**, 309–318 (2016)
31. Rantzer, A., Johansson, M.: Piecewise linear quadratic optimal control. IEEE Trans. Autom. Control **45**(4), 629–637 (2000)
32. Sanfelice, R.G., Goebel, R., Teel, A.R.: Generalized solutions to hybrid dynamical systems. ESAIM: Control., Optim. Calc. Var. **14**(4), 699–724 (2008)
33. Sharif, B., Heertjes, M.F., Heemels, W.P.M.H.: Extended projected dynamical systems with applications to hybrid integrator-gain systems. In: IEEE 58th Conference on Decision and Control (CDC'19), pp. 5773–5778 (2019)
34. Sharif, B., Heertjes, M., Nijmeijer, H., Heemels, W.: On the equivalence of extended and oblique projected dynamics with applications to hybrid integrator-gain systems. In: American Control Conference (ACC) 2021, New Orleans, USA (2021)
35. Sharif, B., Alferink, D.W.T., Heertjes, M.F., Nijmeijer, H., Heemels, W.P.M.H.: A discrete-time approach to analysis of sampled-data hybrid integrator-gain systems. In: 2022 IEEE 61st Conference on Decision and Control (CDC), pp. 7612–7617 (2022)
36. Sharif, B., Alferink, D., Heertjes, M., Nijmeijer, H., Heemels, M.: Analysis of sampled-data hybrid integrator-gain systems: a discrete-time approach. Automatica **167**, 111765 (2024)
37. Shevitz, D., Paden, B.: Lyapunov stability theory of nonsmooth systems. IEEE Trans. Autom. Control **39**(9), 1910–1914 (1994)
38. Shi, K., Petersen, I.R.: Digital control of negative imaginary systems: a discrete-time hybrid integrator-gain system approach. In: 2024 European Control Conference (ECC), pp. 2611–2616 (2024)
39. Shi, K., Petersen, I.R.: Hybrid integrator-gain system based integral resonant controllers for negative imaginary systems. In: 2024 IEEE 63rd Conference on Decision and Control (CDC), pp. 2379–2384 (2024)
40. Shi, K., Nikooienejad, N., Petersen, I.R., Moheimani, S.O.R.: Negative imaginary control using hybrid integrator-gain systems: application to mems nanopositioner. IEEE Trans. Control Syst. Technol. **32**(3), 718–730 (2024)
41. Sontag, E.D., Wang, Y.: On characterizations of the input-to-state stability property. Syst. Control Lett. **24**(5), 351–359 (1995)
42. van de Wal, M., Baars, G., Sperling, F., Bosgra, O.: Multivariable h_∞/μ feedback control design for high-precision wafer stage motion. Control Eng. Pract. **10**, 739–755 (2002)
43. van der Schaft, A.: L_2-Gain and Passivity Techniques in Nonlinear Control. Springer (2017)
44. Van Loon, S., Gruntjens, K., Heertjes, M., Wouw, N., Heemels, W.M.: Frequency-domain tools for stability analysis of reset control systems. Automatica **82**, 101–108 (2017)
45. Willems, J.C.: Dissipative dynamical systems. Eur. J. Control **13**(2), 134–151 (2007)

Chapter 3
A Remedy for Performance Limitations in LTI Control?

3.1 Introduction

Fundamental limitations of linear time-invariant (LTI) feedback control design imposed by Bode's gain-phase relationship [19, 21, 23] form an excellent motivation for the use of hybrid controllers in general [8, 15], and hybrid integrator-gain systems (HIGS) in particular [7, 25]. As discussed in the previous chapter, the switching functionality of HIGS gives access to (weak) integrator characteristics with a phase lag of only 38.15° as observed from its describing function. Although the concept of phase is generally defined for linear systems only, these apparent characteristics hint toward the possibility to break free from classical design trade-offs and thus may allow for increased bandwidths and enhanced closed-loop disturbance rejection on the one hand, and desired transient response and robust stability properties on the other hand.

Despite these potential benefits, the fundamental question if (and how) closed-loop LTI systems can truly benefit from feedback control with HIGS in particular in the sense of overcoming LTI performance limitations remains open. The purpose of this chapter is to provide a positive answer to this question. To this end, an LTI example is studied that suffers from inherent time-domain performance limitations. In particular, a third-order single-input single-output (SISO) LTI plant containing a real unstable open-loop pole of which the presence inevitably leads to overshoot in the step response for *any* stabilizing LTI feedback controller is considered. By exploiting HIGS in combination with specifically designed LTI filters, it is shown for the example that overshoot in the step response can be avoided altogether. The results are accompanied by key design considerations, which form important elements in a general design philosophy for control systems including HIGS elements. Systematic design of reset/hybrid controllers so that the performance limitations of LTI controllers can be overcome is acknowledged to be an open research problem, see [27].

S. van den Eijnden et al., *Hybrid Integrator-Gain Systems*, Lecture Notes in Control and Information Sciences 499, https://doi.org/10.1007/978-3-032-18101-5_3

Within the context of this chapter, it must be mentioned that other results are known in the literature that explicitly demonstrate the ability of nonlinear control strategies to meet certain transient time-domain performance specifications that are unattainable with LTI control. For instance, in [2, 8, 27, 28], examples are given of hybrid and reset control strategies that can meet transient performance specifications in terms of overshoot and rise time, which cannot be achieved with any LTI controller. Also, in [11] it is shown by a numerical example that overshoot requirements unattainable by any LTI controller can be met using phase-based variable-gain control, see also [1]. Interestingly, these examples all employ a nonlinear element that produces discontinuous outputs, while the example with HIGS in this chapter uses continuous (non-smooth) signals. Other recent examples in which HIGS is shown to be able to overcome inherent performance limitations of LTI controllers can be found in [7].

This chapter is organized as follows. In Sect. 3.2, fundamental time-domain limitations of LTI systems linking the presence of real unstable and simple open-loop poles to the occurrence of overshoot and performance trade-offs are revisited. The nonlinear control context is discussed in Sect. 3.3 and a numerical example that illustrates the ability of HIGS to overcome this limitation is presented in Sect. 3.4. Important design considerations and insights are provided in Sect. 3.5. The main conclusions of this chapter are summarized in Sect. 3.6.

3.2 Limitations of LTI Control[1]

The example in this chapter involves the class of SISO LTI feedback control systems, depicted in Fig. 3.1. Here, an LTI plant given by the transfer function $P(s)$, $s \in \mathbb{C}$, is placed in feedback interconnection with an LTI controller $C(s)$. The input to the plant consists of the control signal $u(t) \in \mathbb{R}$ and an input disturbance $d(t) \in \mathbb{R}$, both applied at times $t \in \mathbb{R}_{\geq 0}$. The plant output $y(t) \in \mathbb{R}$ is subtracted from the setpoint $r(t) \in \mathbb{R}$ to form the feedback error $e := r - y$.

A third-order SISO LTI plant $P(s)$ that has a real open-loop pole at $s = p$ is considered. For $p > 0$, i.e., in case of an open-loop unstable pole, when placing this plant in feedback interconnection with *any* internally stabilizing LTI controller $C(s)$,

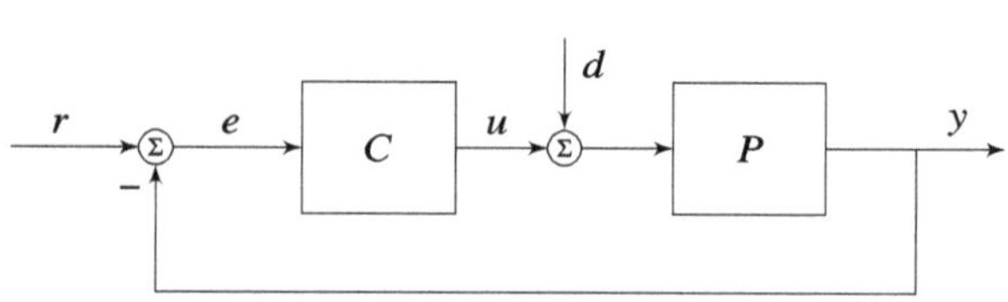

Fig. 3.1 Closed-loop system configuration with SISO LTI plant P and SISO LTI feedback controller C

[1] Some of the text in Sect. 3.2 is reproduced with permission from IEEE Control Systems Letters, Vol 4, van den Eijnden S, Heertjes M, Heemels M, Nijmeijer H, Hybrid integrator-gain systems: A remedy for overshoot limitations in linear control?, 1042–1047. ©2020 IEEE.

a fundamental time-domain limitation exists between the rise time and the amount of overshoot of the closed-loop system's step response when subject to a unit-step input $r(t) = 1$ for $t \in \mathbb{R}_{\geq 0}$, and $r(t) = 0$ otherwise. For $p = 0$, i.e., in case of a simple integrator, an inherent design trade-off exists between overshoot in the step response and zero steady-state error in the presence of constant input disturbances. Before explicitly stating these limitations, necessary definitions for internal stability, rise time, overshoot, and steady-state error are introduced.

Definition 3.1 ([21]) Let the open-loop system in Fig. 3.1 be given by $L(s) = P(s)C(s)$. The closed-loop system is said to be internally stable, if all poles of the transfer function $S(s) = (1 + L(s))^{-1}$ lie in the open left-half complex plane, and there are no unstable pole-zero cancellations in $L(s)$.

Definition 3.2 ([21]) The rise time of a closed-loop system's step response y, when subject to a unit-step input $r(t) = 1$ for $t \in \mathbb{R}_{\geq 0}$, and $r(t) = 0$ otherwise, and for zero initial conditions, is defined as

$$t_r := \sup_{T>0} \left\{ T : y(t) \leq \frac{t}{T} \text{ for all } t \in [0, T] \right\}. \tag{3.1}$$

Definition 3.3 ([21]) The overshoot of a closed-loop system's step response y when subject to a unit-step input $r(t) = 1$ for $t \in \mathbb{R}_{\geq 0}$, and $r(t) = 0$ otherwise, and for zero initial conditions is defined as

$$y_{os} := \sup_{t \in \mathbb{R}_{\geq 0}} (y(t) - r(t)), \tag{3.2}$$

that is, the maximum value by which the step response exceeds the final set-point value.

Definition 3.4 ([28]) The steady-state error of the closed-loop system in Fig. 3.1, when subject to a reference r and an input disturbance d, is defined as the difference between the reference r and the output y as $t \to \infty$, that is

$$e_{ss} = \lim_{t \to \infty} (r(t) - y(t)). \tag{3.3}$$

Proposition 3.1 ([21]) *Suppose that the plant $P(s)$ in Fig. 3.1 has a real pole at $s = p$ with $p > 0$. If the closed-loop system is internally stabilized by any LTI controller $C(s)$, then the step response y resulting from a unit-step input $r(t) = 1$ for $t \in \mathbb{R}_{\geq 0}$, and $r(t) = 0$ otherwise, exhibits overshoot, which satisfies the lower-bound*

$$y_{os} \geq \frac{(pt_r - 1)\exp(pt_r) + 1}{pt_r} \geq \frac{pt_r}{2}. \tag{3.4}$$

Proof The proof can be found in [21, Sect. 1.3]. □

Proposition 3.2 *Suppose that the plant $P(s)$ in Fig. 3.1 contains at least one pole at $s = p$ with $p = 0$. Then, for any stabilizing LTI controller $C(s)$ at least one of the following holds:*

(i) The closed-loop system has non-zero steady-state error $|e_{ss}| > 0$ in the presence of a constant disturbance $d(t) = c$ for all values of $c \in \mathbb{R} \setminus \{0\}$.
(ii) The closed-loop system has overshoot in its step response y resulting from a unit-step input $r(t) = 1$ for $t \in \mathbb{R}_{\geq 0}$, and $r(t) = 0$ otherwise.

Proof By virtue of the internal model principle [23], zero steady-state error in the presence of a constant input disturbance requires the use of integral action in the controller. When using integral control, however, the open-loop system $C(s)P(s)$ has at least two poles at $s = 0$ such that, by virtue of [21, Theorem 1.3.2], the closed-loop system necessarily has overshoot in its step response. Hence, at least one of the conditions stated in the proposition must hold. □

Although both Propositions 3.1 and 3.2 address the presence of overshoot in the step response when using LTI control, there is an essential difference. Namely, whereas *no* internally stabilizing LTI controller exists that leads to non-overshoot in case the plant has an unstable real open-loop pole, for plants containing multiple integrators, one can actually find LTI controllers that lead to no overshoot, but at the cost of not meeting other requirements such as zero steady-state error. Hence, while in the first case one has to accept overshoot, in the second case there is a possibility to balance objectives in a somewhat more desirable, but still limited manner if one restricts oneself to LTI control. In this context, HIGS can offer more favourable results as will be demonstrated in the upcoming parts.

3.3 Nonlinear Control Context

The nonlinear control configuration considered in the numerical examples that demonstrates the ability of HIGS-based control to overcome the limitations stated in Propositions 3.1 and 3.2 is depicted in Fig. 3.2.

In this figure, P is an LTI plant, C_1, C_2 are LTI weighting filters, which are to be specified later, and $\mathcal{H}$ represents HIGS, which has been defined in Sect. 2.2, and for convenience is repeated here as

$$\mathcal{H}: \begin{cases} \dot{x}_h(t) = \omega_h z(t), & \text{if } (z(t), u(t), \dot{z}(t)) \in \mathcal{F}_1, & (3.5a) \\ x_h(t) = k_h z(t), & \text{if } (z(t), u(t), \dot{z}(t)) \in \mathcal{F}_2, & (3.5b) \\ u(t) = x_h(t), & & (3.5c) \end{cases}$$

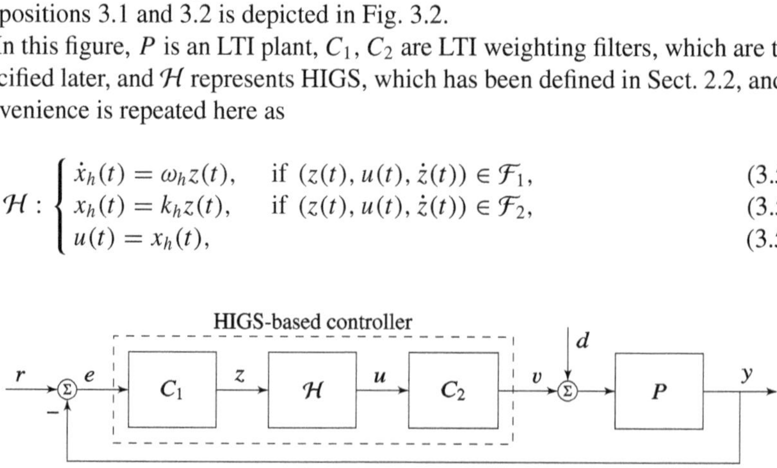

Fig. 3.2 Closed-loop system configuration with SISO LTI plant P and HIGS-based controller as considered in the numerical examples

with state $x_h(t) \in \mathbb{R}$, input $z(t) \in \mathbb{R}$, time-derivative $\dot{z}$ (provided it exists), output $u(t) \in \mathbb{R}$ all at times $t \in \mathbb{R}_{\geq 0}$, and

$$\mathcal{F}_1 := \left\{(z, u, \dot{z}) \in \mathbb{R}^3 \mid (z, u, \dot{z}) \in \mathcal{F} \backslash \mathcal{F}_2\right\}, \tag{3.6a}$$

$$\mathcal{F}_2 := \left\{(z, u, \dot{z}) \in \mathbb{R}^3 \mid (z, u, \dot{z}) \in \mathcal{F} \vee u = k_h z \wedge \omega_h z^2 > k_h \dot{z} z\right\}, \tag{3.6b}$$

where $\mathcal{F} = \left\{(z, u, \dot{z}) \in \mathbb{R} \mid zu \geq \frac{1}{k_h} u^2\right\}$ is the $[0, k_h]$-sector.

As discussed in the previous chapter, the feedback interconnection of HIGS in (3.5) with an LTI system as in the context of Fig. 3.2 admits a piecewise linear (PWL) state space representation of the form

$$\Sigma : \begin{cases} \dot{x}(t) = A_i x(t) + B_i w(t), & \text{if } (x(t), w(t)) \in \mathcal{X}_i, \quad i = \{1, 2\}, \\ y(t) = C x(t) + D w(t), & \end{cases} \tag{3.7}$$

where $x(t) = [x_p^\top(t), x_{c_2}^\top(t), x_h(t), x_{c_1}^\top(t)]^\top \in \mathbb{R}^n$ collects the states x_p of the plant $P(s)$, the states x_{c_1}, x_{c_2} of the LTI filters $C_1(s)$, $C_2(s)$, and the state x_h of HIGS, respectively, $w(t) \in \mathbb{R}^3$ is the vector of exogenous input signals consisting of the reference r and $\dot{r}$ and disturbance d, and $y(t) \in \mathbb{R}$ is in this case the relevant performance output at time $t \in \mathbb{R}_{\geq 0}$. Furthermore, the sets $\mathcal{X}_i$, $i = \{1, 2\}$ are in this case given by

$$\mathcal{X}_i = \left\{(x, w) \in \mathbb{R}^{n+3} \mid Ex + Fw \in \mathcal{F}_i\right\}, \tag{3.8}$$

in which the matrices $E \in \mathbb{R}^{3\times n}$ and $F \in \mathbb{R}^{3\times 3}$ satisfy $Ex + Fw = (z, u, \dot{z})^\top$. An explicit derivation of the relevant system matrices in (3.7) and (3.8) can be found in Sect. B.1. For details on closed-loop system aspects regarding solution concepts and properties, the reader is referred back to Chap. 2.

Before presenting the example in more detail, closed-loop stability of the nonlinear system in Fig. 3.2 is discussed. Although stability is not the main topic in this chapter, it is a prerequisite for performance, and forms an integral part in the limitations stated above. Therefore, it should be addressed here for making a fair comparison between linear and nonlinear control strategies. In this context, the notion of input-to-state stability (ISS) introduced in Definition 2.3 is adopted, as it links to the concept of internal stability given in Definition 3.1.

Theorem 3.1 *Consider the closed-loop system in* (3.7). *Suppose there exist a symmetric $n \times n$ matrix $M \succ 0$ and a number $\mu \in \mathbb{R}$ that satisfy the matrix inequalities*

$$A_1^\top M + M A_1 \prec 0, \quad \text{and} \quad (A_2 + \mu b H)^\top M + M(A_2 + \mu b H) \prec 0 \tag{3.9}$$

with $b = [0, 1, 0]^\top$ and $H = [k_h, -1, 0]E$. Then, the closed-loop system (3.7) *is input-to-state stable in the sense of Definition 2.3.*

The proof of Theorem 3.1 largely proceeds along standard steps that can be found in published works on stability of switched and hybrid systems, see, e.g., [12, 13, 17].

Key in the proof is the use of Finsler's lemma [3] for the second inequality in (3.9) via the gain mode constraint $Hx = k_h z - u = 0$ for $w = 0$. Different (possibly less conservative) conditions for stability are extensively discussed in Chaps. 4 and 5, and the result presented in Theorem 3.1 forms a special case of these more general results. Therefore, an explicit proof is omitted here.

3.4 Plant with a Real Unstable Pole[2]

In this section, an example is presented that illustrates the possibility to overcome the fundamental overshoot limitation in Proposition 3.1. Consider the LTI plant

$$P(s) = \frac{(s+q)\omega^2}{(s-p)(s^2+2\zeta\omega s+\omega^2)}, \tag{3.10}$$

which has a real unstable pole located at $s = p$, with $p = 1$, two stable poles at $s = -\omega(\zeta \pm \sqrt{\zeta^2 - 1})$, where $\omega = 5 \cdot 2\pi$ rad/s and $\zeta = 1.5$, and a zero at $s = -q$ with $q = 3$. Third-order characteristics with a real unstable pole such as in (3.10) may appear for instance in active magnetic bearings [18, 24]. As a control objective, a feedback controller must be designed that *(i) internally stabilizes* the closed-loop system, and *(ii)* achieves a *zero steady-state tracking error* when the system is subject to a unit-step input r *without overshoot.*

3.4.1 LTI Control

When considering an LTI controller $C(s)$, the requirement for internal stability of the closed-loop system removes the possibility for $C(s)$ to introduce an unstable pole-zero cancellation, whereas the objective of having zero steady-state tracking error necessitates integral action in the controller $C(s)$ by virtue of the internal model principle (see, e.g., [23]). A controller $C(s)$ suitable for this purpose is designed on the basis of the Nyquist stability criterion [9, Sect. 6.3], resulting in

$$C(s) = \frac{\omega_i}{s}, \tag{3.11}$$

with integrator frequency $\omega_i = 10$ rad/s. Note that for achieving the given control objectives, other types of LTI controllers can be considered as well (although all would result in overshoot by virtue of Proposition 3.1). To keep the exposition simple

[2] The introductory text in Sect. 3.4 and the text in Sects. 3.4.1 and 3.4.2 is partly reproduced with permission from IEEE Control Systems Letters, Vol 4, van den Eijnden S, Heertjes M, Heemels M, Nijmeijer H, Hybrid integrator-gain systems: A remedy for overshoot limitations in linear control?, 1042–1047. ©2020 IEEE.

and preserve insights in the effects of the controller on the transient response, the choice for an integrator is made here.

According to Proposition 3.1, the closed-loop system with plant (3.10) and LTI controller (3.11) must exhibit overshoot in its step response (with $d(t) = 0$), so that the objective regarding overshoot cannot be met. In this specific example, an aspect that additionally contributes to the presence of overshoot is the necessity for integral control. This can be understood as follows. Assuming zero initial conditions, the error initially satisfies $e(t) > 0$, such that the integrator in (3.11) "sums" the error over time to provide a positive input $u(t) > 0$ for driving the system toward the setpoint. For achieving zero steady-state error, it follows from the final value theorem [19] that

$$\lim_{t\to\infty} u(t) = \lim_{s\to 0} \frac{sC(s)}{1 + C(s)P(s)} \cdot \frac{1}{s} = -\frac{p}{q} < 0. \tag{3.12}$$

Hence, a negative steady-state input is needed for achieving zero steady-state error. This implies a sign change of u, which can only be achieved if e changes sign, thus requiring overshoot. In case $P(s)$ only has stable poles, and thus $p < 0$, the steady-state input should be positive, and overshoot does not necessarily occur.

3.4.2 HIGS-Based Control

Given the previous observations, a HIGS-based integrator is proposed that adopts the functionality of a linear integrator for achieving zero steady-state error, but aims at reducing/eliminating overshoot. Hereto, HIGS in (3.5) is appended with LTI filters C_1 and C_2 in a manner as depicted in Fig. 3.2. These filters are specifically chosen as

$$C_1(s) = \left(\frac{s}{\omega_c} + 1\right), \quad \text{and} \quad C_2(s) = \frac{\omega_i}{s}, \tag{3.13}$$

with $\omega_c = |1 + 4j/\pi|\frac{\omega_h}{k_h}$ rad/s. This particular choice is motivated by using describing function reasoning. The describing function of HIGS, which has been derived in Chap. 2, and which is repeated here as

$$\mathcal{D}(j\omega) = \frac{\omega_h}{j\omega}\left(\frac{\gamma}{\pi} + j\frac{e^{-2j\gamma} - 4e^{-j\gamma} + 3}{2\pi}\right) + k_h\left(\frac{\pi - \gamma}{\pi} + j\frac{e^{-2j\gamma} - 1}{2\pi}\right), \tag{3.14}$$

with $\gamma(\omega) = 2\arctan(k_h\omega/\omega_h)$, implies first-order low-pass magnitude characteristics with cross-over frequency $\omega_c = |1 + 4j/\pi|\frac{\omega_h}{k_h}$ rad/s, accompanied by a phase lag that does not exceed 38.15°. The filter $C_1(s)$ "compensates" the magnitude characteristics of $\mathcal{D}(j\omega)$, and provides additional phase lead up to 90°. As a result, for all $\omega \in \mathbb{R}$ it follows that $\|C_1(j\omega)\mathcal{D}(j\omega)\| \approx 1$, and $\angle(C_1(j\omega)\mathcal{D}(j\omega)) \in (0, 51.85)°$. The describing function characteristics of the HIGS-based integrator, which also includes $C_2(s) = \omega_i/s$, is given by

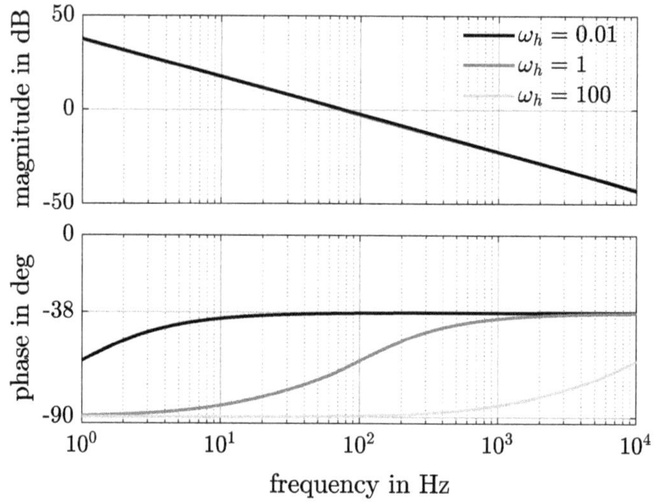

Fig. 3.3 Bode-like representation of a HIGS-based integrator (3.15) for different values of HIGS' integrator frequency ω_h (with $k_h = 1$)

$$C_i(j\omega) = C_1(j\omega)\mathcal{D}(j\omega)C_2(j\omega), \tag{3.15}$$

and demonstrates the characteristic 20 dB/decade magnitude decay similar to the linear integrator in (3.11), but with a phase lag that can be reduced up to 38.15° as shown in Fig. 3.3, i.e., $\angle(C_i(j\omega)) \in (-90, -38.15)°$. A useful feature of this design is that the parameter ω_h (with $k_h = 1$) can be used to vary the phase lag reduction independent from the magnitude, which provides extra freedom in dealing with inherent design limitations. The apparent phase advantage hints toward the possibility for reducing overshoot. Comparable strategies that combine LTI filters with nonlinear elements such as reset integrators [6] or split-path filters [22, 26] for creating new elements with favourable gain and phase properties can be found in the literature, see for instance [14, 16, 20, 29]. Recent strategies related to HIGS that enable improved phase characteristics by directly embedding well-designed LTI filters in the switching conditions are so-called filtered projection elements introduced in [5].

Remark 3.1 For a practical implementation of the non-proper filter $C_1(s)$ in (3.13), the following adaptations can be made

$$C_1(s) = \left(\frac{s}{\omega_c} + 1\right) \cdot \frac{1}{(\tau s + 1)}, \quad \text{and} \quad C_2(s) = \frac{\omega_i}{s} \cdot (\tau s + 1),$$

where $\tau > 0$ is chosen sufficiently small. Both $C_1(s)$ and $C_2(s)$ are proper filters. Note that the describing function characteristics in (3.15) remain unchanged by these transformations.

Remark 3.2 It is interesting to mention that for $\omega_h \to 0$ the describing function characteristics of the HIGS-based integrator as depicted in Fig. 3.3 are equivalent to that of the Clegg integrator [6, 10]. However, there is an essential difference between the HIGS-based integrator and the Clegg integrator. Namely, due to the additional state in $C_2(s)$, the HIGS-based integrator is able to preserve a buffer when $e(t) = 0$ to compensate for constant disturbances. The Clegg integrator is not able to preserve such buffer since due to the resetting mechanism the output of the integrator is reset to zero as soon as $e(t) = 0$, such that a constant disturbance is no longer compensated for by the integrator, see also [28, Remark 8].

3.4.3 Simulation Results

The step response of the plant (3.10) in feedback with the HIGS-based controller, the latter being the series interconnection of the filters $C_1(s)$, and $C_2(s)$ with $\tau = 10^{-4}$ conform Remark 3.1, and HIGS with $k_h = 1$, $\omega_h = 0.8$ rad/s, is shown in Fig. 3.4 (in black), along with the generated control output u. In addition, the step response and control output with the linear controller (3.11) are shown in grey. Overshoot is clearly present for the linear closed-loop system, whereas for the HIGS-based control system there is no overshoot under fairly limited control effort. Note that in both cases the output of the controller attains the steady-state value $u(\infty) = -p/q = -1/3$ that appears necessary for achieving zero steady-state tracking error. It is emphasized that for the linear control system, the overshoot in Fig. 3.4 may be influenced by a different choice of LTI filters, e.g., PID-filters, but overshoot cannot be avoided as a consequence of the fundamental limitation stated in Proposition 3.1. Moreover, given the rise time $t_r = 0.95$ seconds of the HIGS-controlled system, any linear controller resulting in a similar (or larger) rise time yields an overshoot of at least $y_{os} = \frac{pt_r}{2} = 0.475$ according to the lower-bound in (3.4). These results highlight the distinct advantage of the HIGS-based controller over any LTI controller in achieving performance without overshoot.

Before arriving at the final conclusion that for the example discussed, the fundamental limitation in Proposition 3.1 is truly overcome with HIGS, it remains to be shown formally that the corresponding closed-loop system is ISS. Hereto, the matrix conditions in Theorem 3.1 are solved numerically for which the result is provided in Sect. B.1. Given feasibility of the matrix inequalities in (3.9), together with the time-series simulation result in Fig. 3.4, it is concluded that the fundamental limitation presented in Proposition 3.1 is truly overcome with the proposed HIGS-based control strategy for the example discussed.

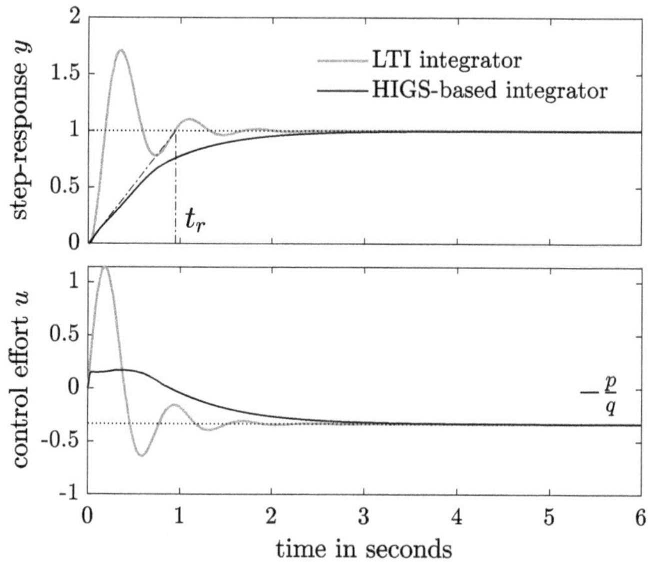

Fig. 3.4 Step response y (top) and control output u (bottom) of the linear (grey) and HIGS-based (black) closed-loop system. Reprinted with permission from IEEE Control Systems Letters, Vol 4, van den Eijnden S, Heertjes M, Heemels M, Nijmeijer H, Hybrid integrator-gain systems: A remedy for overshoot limitations in linear control?, 1042–1047. ©2020 IEEE

3.4.4 Robust Non-overshoot Performance

In order to study robust non-overshoot performance properties of the presented approach in the presence of plant uncertainties, the location of the pole p in (3.10) is varied in the range $p \in [-5, 10]$. The amount of overshoot as a function of the pole location is shown in Fig. 3.5. Because the plant is different, the amount of overshoot, rise-time, and fundamental lower-bound are different. However, for a range of unstable open-loop poles, the qualitative closed-loop system behaviour appears the same, in the sense that overshoot is avoided, thereby indicating a certain amount of robustness to plant uncertainties. Stability of the closed-loop system with HIGS has been formally verified by means of the matrix conditions in Theorem 3.1 for all $p \in [-5, 1.8]$. However, a stable step response is observed in simulations up to $p = 7.8$. Note that for the LTI case, internal stability is guaranteed for all $p \leq 4.5$. These observations hint toward improved robust stability properties with HIGS-based control. One could argue that increased robustness is enabled by the alleged reduction in phase lag by the HIGS-based integrator, as opposed to its linear counterpart (see Fig. 3.3 in this regard).

From Fig. 3.5 zero overshoot is obtained also for $s = p$ with $p = 0$. In case of LTI integral control, the open-loop contains two integrators which inherently leads to overshoot. When using no integrator in the controller, overshoot can be avoided (for instance with $C(s) = 1$) but at the cost of non-zero steady-state error

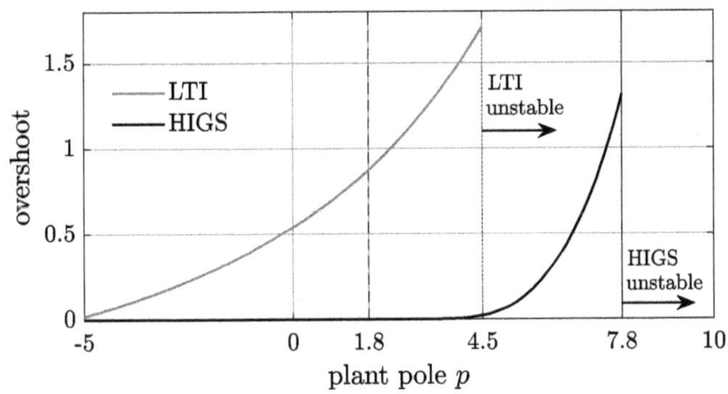

Fig. 3.5 Overshoot as a function of the open-loop pole p for the closed-loop system with an LTI integrator (grey) and a HIGS-based integrator (black)

in the presence of non-zero constant input disturbances. This is stated formally in Proposition 3.2. With the considered HIGS-based integrator, however, both constant disturbance rejection as well as non-overshoot in the step response are achieved. This demonstrates that the fundamental trade-off in Proposition 3.2 is overcome for this example, and indicates the additional design flexibility that is offered with HIGS-based control.

3.5 Design Considerations and Non-overshoot

This section reflects on important design considerations, and discusses the mechanism underlying non-overshoot in the previous example.

3.5.1 Controller Configuration

The nonlinear control strategy applied to the previous example is initially based on HIGS' describing function characteristics. From a quasi-linear perspective, the particular sequence in which the filters in the HIGS-based controller appear is interchangeable. However, from a time-domain perspective this is not the case as switching of HIGS is directly determined by its input z, and, therefore, by the choice for the filter $C_1(s)$ in Fig. 3.2. The effect of filter sequence on time-domain performance has also been pointed out in [4]. The rationale behind the proposed configuration in Fig. 3.2 for reducing overshoot and alleviating performance trade-offs can be explained as follows.

(1) By using the output of HIGS directly as an input to the filter $C_2(s)$ which contains the integrator buffer in the HIGS-based controller, a control mode-switching

mechanism is provided for filling/depleting this buffer at a variable rate, hence, $C_2(s)$ containing the integrator is located directly after HIGS. Despite the integrator mode, HIGS itself cannot maintain a buffer necessary for achieving zero steady-state error due to its sector-boundedness, underlining the necessity for including a pure integrator in the design.

(2) Due to the previous choice, the output of HIGS should have zero crossings for allowing a sign change of the linear integrator's output; recall the earlier discussions regarding LTI control designs. Sign equivalence of HIGS input and output signals then necessitates the input of HIGS to have zero crossings. For achieving non-overshoot performance, by definition, a sign-change of e must be avoided. Hence, e cannot be used directly as an input to HIGS. Rather, $C_1(s)$ as in (3.13) is used for realizing a filtered version of e, i.e., $z = \frac{\dot{e}}{\omega_c} + e$. The weighted combination of e and $\dot{e}$ creates the possibility for z to have zero crossings without e crossing zero, i.e., without inducing overshoot.

3.5.2 Non-overshoot Mechanism

From the previous rationale, in particular the second point, it is clear that the parameter ω_h largely determines whether switching primarily occurs on the basis of the error e or its time-derivative $\dot{e}$, and thus directly influences the interplay between HIGS and the integrator buffer contained in the filter $C_2(s)$. To illustrate the effect of ω_h on such interplay, consider again the example presented in Sect. 3.4 with $p = 1$. Figure 3.6 depicts the time-response of the HIGS state x_h, and the state x_i of the linear integrator in $C_2(s)$ when subject to a step-input for the values $\omega_h = \{0.08, 0.8, 80\}$ rad/s.

The choice $\omega_h = 80$ rad/s (bottom figure) leads to overshoot. This can be explained from the fact that for $\omega_h \to \infty$ the path composed of $C_1(s)$ and HIGS (see Fig. 3.2) effectively reduces to a linear gain. The characteristics of the nonlinear controller tend to that of its linear counterpart, i.e., inherently having overshoot. Hence, choosing ω_h too large results in overshoot. On the other hand, both the choices $\omega_h = 0.08$ rad/s (top figure) and $\omega_h = 0.8$ rad/s (middle figure) result in the absence of overshoot. To understand the underlying mechanism leading to non-overshoot, first note that according to the discussion in Chap. 2, for $\omega_h \to 0$ the characteristics of HIGS tend to that of the Clegg integrator, leading to an arbitrarily fast reduction to zero of the HIGS state x_h whenever z crosses zero. When considering a step response, one initially finds $e(0) > 0$ and $\dot{e}(0) = 0$, such that $z(0) > 0$. In that case, the linear integrator placed behind HIGS starts building up a positive buffer and pushes the system in the direction toward zero error. However, as e is decreasing, at some point the magnitude of $e > 0$ is relatively small compared to the (weighted) magnitude of $\dot{e} < 0$, which results in a sign-change of z. Physically, this means that, relative to the distance toward the set-point, the system is moving fast in the direction of this set-point. To avoid overshoot, the system must rapidly "slow down". This is achieved through HIGS by enforcing Clegg-like switching that either limits the linear integrator buffer build-

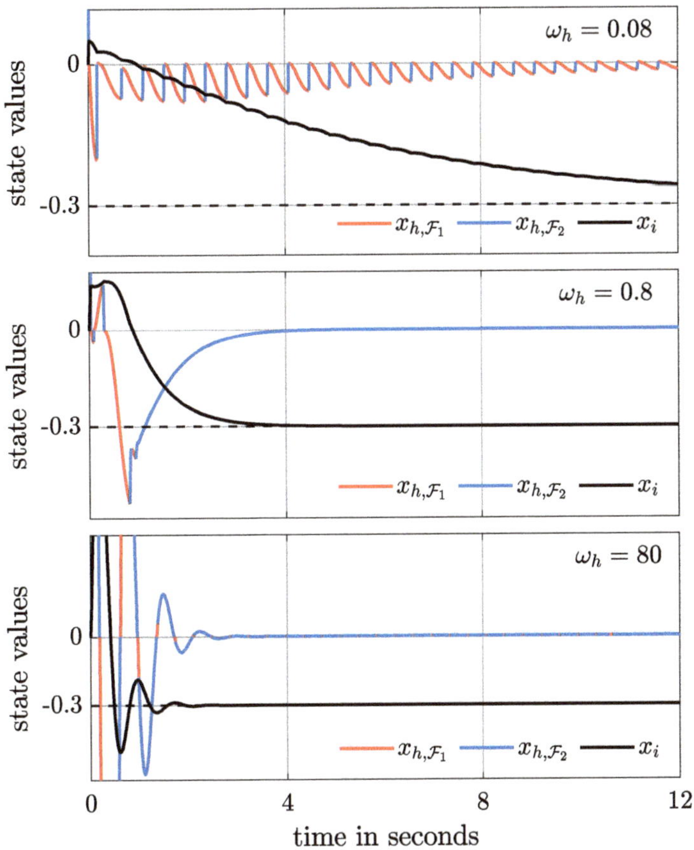

Fig. 3.6 HIGS-state $x_h(t)$ (red: integrator mode, blue: gain mode) and integrator state $x_i(t) = \int_0^t x_h(\tau)d\tau$ as a function of time for $\omega_h = 0.08$ rad/s (top figure), $\omega_h = 0.8$ rad/s (middle figure), and $\omega_h = 80$ rad/s (bottom figure). The final value of the integrator state for attaining zero steady-state error in the example is indicated by the dashed black line and corresponds to $-p/q = -1/3$

up or starts depleting it. Such characteristics are particularly visible for $\omega_h = 0.08$ rad/s in Fig. 3.6 (top). Note, however, that with ω_h small, even a small change in $\dot{e}$ may already induce a sign change in z. Consequently, the Clegg-like mechanism is triggered in a high-frequency manner, which impedes the integrator build-up and can lead to an extremely slow step response. This is also observed in Fig. 3.6 (top) as the integrator state x_i slowly converges to its steady-state value. When choosing $\omega_h = 0.8$ rad/s, non-overshoot is still achieved, but the integrator attains its steady-state value at a much faster rate as can be seen in Fig. 3.6 (middle). In turn, this results in a faster step response. For an additional discussion on the non-overshoot mechanism see [7, 25].

From the above observations, the parameter ω_h appears as a tuning knob for adjusting the overshoot and settling time. As such, a simple tuning guideline for a

HIGS-based integrator design can be derived. One may start with $\omega_h = \infty$, and tune all remaining parameters (ω_i in the example) by means of linear design techniques, e.g., via loop-shaping. Subsequently, one can adjust ω_h to balance the level of overshoot and settling time according to specifications. Note, however, that during such a design process stability of the feedback system needs to be guaranteed. Stability aspects are discussed in more detail in the upcoming chapters.

3.6 Summary

In this chapter, an example is discussed in which HIGS-based control is able to overcome fundamental performance limitations inherent to LTI systems. The example addresses an overshoot performance limitation resulting from the presence of a real unstable or simple open-loop pole in the plant. By interconnecting HIGS in series with well-crafted linear filters, overshoot is avoided, and the limitations are alleviated. Key in the approach is the sequence in which the linear filters appear with respect to HIGS. The obtained insights provide valuable elements for a general HIGS-based control design framework.

Although briefly discussed in this chapter, guaranteeing stability of the closed-loop system remains an issue to be dealt with, particularly in terms of providing easy-to-check and non-conservative conditions. To address this issue, in the next chapter frequency-domain conditions for stability of HIGS-based control systems are presented. Not only are these conditions less conservative than the ones proposed in this chapter, they also provide direction for control design.

References

1. Armstrong, B., Gutierrez, J., Wade, B., Joseph, R.: Stability of phase-based gain modulation with designer-chosen switch functions. Int. J. Robot. Res. **25**, 781–796 (2006)
2. Beker, O., Hollot, C., Chait, Y.: Plant with integrator: An example of reset control overcoming limitations of linear feedback. IEEE Trans. Autom. Control **46**, 1797–1799 (2001)
3. Boyd, S., El Ghaoui, L., Feron, E., Balakrishnan, V.: Linear Matrix Inequalities in System and Control Theory, Studies in Applied Mathematics, vol. 15. SIAM, Philadelphia, PA (1994)
4. Cai, C., Dastjerdi, A.A., Saikumar, N., HosseinNia, S.: The optimal sequence for reset controllers. In: 2020 European Control Conference (ECC), pp. 1826–1833 (2020)
5. Chu, H., van den Eijnden, S., Heertjes, M., Heemels, W.: Filtering in projection-based integrators for improved phase characteristics. In: 2024 IEEE 63rd Conference on Decision and Control (CDC), pp. 2391–2396 (2024)
6. Clegg, J.C.: A nonlinear integrator for servomechanisms. Transactions of the American Institute of Electrical Engineers, Part II: Applications and Industry **77**(1), 41–42 (1958)
7. Dinther, D., Sharif, B., van den Eijnden, S., Nijmeijer, H., Heertjes, M., Heemels, M.: Overcoming performance limitations of linear control with hybrid integrator-gain systems. In: IFAC Conference on Analysis and Design of Hybrid Systems (ADHS) 2021 (2021)

8. Feuer, A., Goodwin, G.C., Salgado, M.: Potential benefits of hybrid control for linear time invariant plants. In: Proceedings of the 1997 American Control Conference (ACC'97), vol. 5, pp. 2790–2794 (1997)
9. Franklin, G., Powell, J., Emami-Naeini, A.: Feedback Control of Dynamic Systems, 5th ed. Prentice Hall (2005)
10. Horowitz, I., Rosenbaum, P.: Non-linear design for cost of feedback reduction in systems with large parameter uncertainty. Int. J. Control **21**(6), 977–1001 (1975)
11. Hunnekens, B., van de Wouw, N., Nešić, D.: Overcoming a fundamental time-domain performance limitation by nonlinear control. Automatica **67**, 277–281 (2016)
12. Johansson, M.: Piecewise Linear Control Systems-A Computational Approach. Lecture Notes in Control and Information Sciences, vol. 284 (2002)
13. Kamenetskiy, V.: Frequency-domain stability conditions for hybrid systems. Autom. Remote. Control. **78**, 2101–2119 (2017)
14. Karybakas, C.A.: Nonlinear integrator with zero phase shift. IEEE Trans. Ind. Electron. Control Instrum. IECI **24**(2), 150–152 (1977)
15. Lau, K., Middleton, R.: Switched integrator control schemes for integrating plants. Eur. J. Control. **9**, 539–559 (2003)
16. Li, Y., Guo, G., Wang, Y.: Reset control for midfrequency narrowband disturbance rejection with an application in hard disk drives. IEEE Trans. Control Syst. Technol. **19**(6), 1339–1348 (2011)
17. Liberzon, D.: Switching in Systems and Control. Birkhauser, Boston (2003)
18. Maslen, E.H., Montie, D.T., Iwasaki, T.: Robustness limitations in self-sensing magnetic bearings. J. Dyn. Syst. Meas. Contr. **128**(2), 197–203 (2005)
19. Middleton, R.: Trade-offs in linear control system design. Automatica **27**(2), 281–292 (1991)
20. Palanikumar, A., Saikumar, N., HosseinNia, S.H.: No more differentiator in PID: development of nonlinear lead for precision mechatronics. In: 2018 European Control Conference (ECC), pp. 991–996 (2018)
21. Seron, M.M., Braslavsky, J.H., Goodwin, G.C.: Fundamental Limitations in Filtering and Control. Springer, Berlin (1997)
22. Sharif, B., van der Maas, A., van de Wouw, N., Heemels, W.P.M.H.: Filtered split-path nonlinear integrator: a hybrid controller for transient performance improvement. IEEE Trans. Control Syst. Technol. 1–13 (2021)
23. Skogestad, S., Postlethwaite, I.: Multivariable Feedback Control: Analysis and Design, 2nd ed. Wiley and Sons (2010)
24. Thibeault, N.M., Smith, R.S.: Magnetic bearing measurement configurations and associated robustness and performance limitations. J. Dyn. Syst. Meas. Contr. **124**(4), 589–598 (2002)
25. van den Eijnden, S.J.A.M., Heertjes, M.F., Heemels, W.P.M.H., Nijmeijer, H.: Hybrid integrator-gain systems: a remedy for overshoot limitations in linear control? IEEE Control Syst. Lett. **4**(4), 1042–1047 (2020)
26. van Loon, S., Hunnekens, B., Heemels, W., van de Wouw, N., Nijmeijer, H.: Split-path nonlinear integral control for transient performance improvement. Automatica **66**, 262–270 (2016)
27. Zhao, G., Nešić, D., Tan, Y., Wang, J.: Open problems in reset control. In: 52nd IEEE Conference on Decision and Control, pp. 3326–3331 (2013)
28. Zhao, G., Nešić, D., Tan, Y., Hua, C.: Overcoming overshoot performance limitations of linear systems with reset control. Automatica **101**, 27–35 (2019)
29. Zheng, Y., Chait, Y., Hollot, C., Steinbuch, M., Norg, M.: Experimental demonstration of reset control design. Control. Eng. Pract. **8**, 113–120 (2000)

Part II
Stability and Performance

Chapter 4
Frequency-Domain Tools for Robust Stability Analysis

4.1 Introduction

The potential of hybrid integrator gain-systems (HIGS) for overcoming fundamental performance limitations of linear time-invariant (LTI) control systems has been shown in the previous chapter. Unfortunately, performance enhancing benefits often do not come without a cost. While linear systems lend themselves well for robust stability and performance analysis using frequency-domain tools such as the Nyquist stability criterion and Bode plots [24], the nonlinear/hybrid nature of a control system featuring HIGS renders many of such tools inapplicable. As the current control practice highly exploits frequency-domain methods, lack of appropriate tooling may compromise broad adoption of nonlinear and hybrid control strategies like HIGS in industry.

To accommodate for this situation, frequency-domain tools for stability analysis and design of switched/hybrid systems in general have been developed in, e.g., [1, 2, 4, 5, 9, 11, 12, 14, 15, 23, 28, 29, 32]. Underlying these tools is the well-known Kalman–Yakubovich–Popov lemma [18, 34], which allows for establishing the equivalence between frequency-domain conditions and the existence of a common quadratic Lyapunov function for switched systems, the latter being posed in the time-domain. In principle, these frequency-domain conditions can be verified using *measured* frequency-response function data of the plant to be controlled, making them useful in practical situations, where accurate parametric models of the plant are often difficult, if not impossible, to obtain.

Although valuable from a practical perspective, classical frequency-domain conditions that are applicable to the class of HIGS-based control systems, such as the ones presented in, e.g., [6, 11, 14, 15, 32] often render a rather conservative estimate on closed-loop stability. To some extent, this conservatism is caused by the fact that the conditions do not sufficiently take into account the particular switching characteristics of HIGS, thereby leading to overly restrictive formulations. For example, the results in [11, 15] are limited to systems with arbitrary switching, which is too

S. van den Eijnden et al., *Hybrid Integrator-Gain Systems*, Lecture Notes in Control and Information Sciences 499, https://doi.org/10.1007/978-3-032-18101-5_4

restrictive for HIGS as it adopts a specific state-dependent switching strategy. In [14], the result is extended by including state-dependent constraints, leading to improved conditions. Such constraints are also considered in [6, 32] for generic sector-bounded nonlinearities. However, in the context of HIGS this means that the underlying internal dynamic behaviour is ignored. It is interesting to remark that in the mentioned works, the quadratic Lyapunov function that results from satisfying the conditions is guaranteed to be positive within the full state space. This may be restrictive too, because trajectories of a closed-loop system with HIGS are confined to a subset of the state space only.

This chapter discusses frequency-domain conditions for stability analysis of the feedback interconnection of an LTI system and HIGS. If satisfied, the conditions guarantee the existence of a quadratic Lyapunov function that is not necessarily positive within the full state space, but rather in a subset of the state space where HIGS is active. The same holds true for negative definiteness of the corresponding time derivatives. In this way, the class of admissible Lyapunov function candidates is extended and the conservatism in the analysis is reduced compared to classical methods. It is shown that the conditions can be verified graphically in a manner that is reminiscent to a Popov plot [7]. The conditions are extended toward robust stability analysis, taking inspiration from the results in [10, 26], where robustified versions of classical circle- and Popov criteria are presented that allow for a graphical verification of robust stability, something which is well appreciated in an industrial setting where plant uncertainty is highly relevant.

This chapter is organized as follows. In Sect. 4.2 the control system setting is discussed, and the problem formulation is given. In Sect. 4.3, frequency-domain stability conditions are presented. The conditions are compared to classical results in a numerical example. An extension toward a robust stability analysis is discussed in Sect. 4.4. Application of the presented results is demonstrated on an experimental motion control set-up in Sect. 4.5. A summary of this chapter is provided in Sect. 4.6.

4.2 System Setting and Problem Formulation

In this section, the generic closed-loop system setting with HIGS as detailed in Chap. 2 is briefly repeated here for convenience, and the problem of deriving frequency-domain conditions for input-to-state stability is discussed.

4.2.1 Closed-Loop System Description

Consider the multi-input multi-output (MIMO) configuration as depicted in Fig. 4.1.

In this figure, the MIMO LTI system G is given by

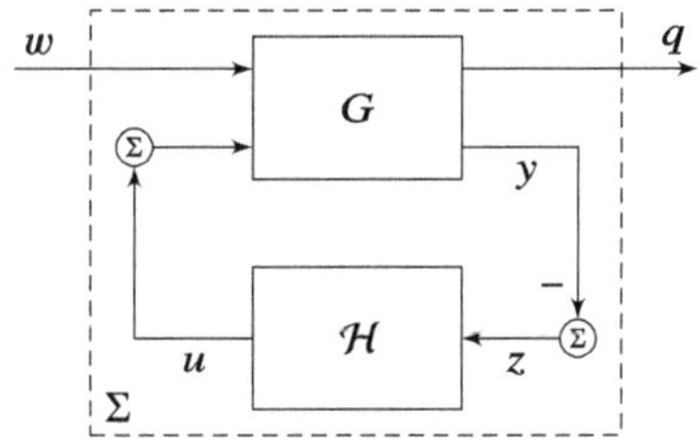

Fig. 4.1 Feedback interconnection of an LTI system G and HIGS $\mathcal{H}$

$$G : \begin{cases} \dot{x}_g(t) &= A_g x_g(t) + B_g u(t) + B_w w(t), \\ y(t) &= C_g x_g(t), \\ q(t) &= C_q x_g(t) + D_q u(t) + D_w w(t), \end{cases} \tag{4.1}$$

with states $x_g(t) \in \mathbb{R}^m$, exogenous input $w(t) \in \mathbb{R}^p$ belonging to the class of piecewise continuous functions ($w \in PC$, conform Chap. 2), controlled output $y(t) \in \mathbb{R}$, control input $u(t) \in \mathbb{R}$, and performance channel $q(t) \in \mathbb{R}^r$, all at time $t \in \mathbb{R}_{\geq 0}$. It is assumed that (A_g, B_g, C_g) is minimal and the corresponding transfer functions from u to y and w to y are denoted by

$$G_{yu}(s) = C_g(sI - A_g)^{-1} B_g, \tag{4.2a}$$
$$G_{yw}(s) = C_g(sI - A_g)^{-1} B_w. \tag{4.2b}$$

Conform Assumption 2.5 the transfer functions in (4.2) have a relative degree of at least two, which implies that $C_g B_g = 0$ and $C_g B_w = 0$.

Recall the hybrid integrator-gain system $\mathcal{H}$ in Fig. 4.1 to be given by

$$\mathcal{H} : \begin{cases} \dot{x}_h(t) = \omega_h z(t), & \text{if } (z(t), u(t), \dot{z}(t)) \in \mathcal{F}_1, & (4.3a) \\ x_h(t) = k_h z(t), & \text{if } (z(t), u(t), \dot{z}(t)) \in \mathcal{F}_2, & (4.3b) \\ u(t) = x_h(t), & & (4.3c) \end{cases}$$

where $x_h(t) \in \mathbb{R}$ denotes the state of the integrator, $z(t) = -y(t) \in \mathbb{R}$ is the input to HIGS, which in the context of Fig. 4.1 (recall Chap. 2) is differentiable, $\dot{z}(t) = -\dot{y}(t)$ denotes the corresponding time-derivative, and $u(t) \in \mathbb{R}$ is the generated output at time $t \in \mathbb{R}_{\geq 0}$. The sets $\mathcal{F}_1$ and $\mathcal{F}_2$ dictating, respectively, the integrator mode and gain mode in (4.3) are given by

$$\mathcal{F}_1 = \left\{(z, u, \dot{z}) \in \mathbb{R}^3 \mid (z, u, \dot{z}) \in \mathcal{F} \backslash \mathcal{F}_2\right\}, \tag{4.4a}$$
$$\mathcal{F}_2 = \left\{(z, u, \dot{z}) \in \mathbb{R}^3 \mid (z, u, \dot{z}) \in \mathcal{F} \vee u = k_h z \wedge \omega_h z^2 > k_h \dot{z} z\right\}, \tag{4.4b}$$

where $\mathcal{F}$ is the $[0, k_h]$-sector, specified as

$$\mathcal{F} = \left\{(z, u, \dot{z}) \in \mathbb{R}^3 \mid zu \geq \frac{1}{k_h} u^2\right\}. \tag{4.5}$$

The closed-loop system in Fig. 4.1 with G as in (4.1), and $\mathcal{H}$ as in (4.3), (4.4) can be written as the PWL system

$$\Sigma : \begin{cases} \dot{x}(t) &= A_i x(t) + Bw(t), \quad \text{if } x(t) \in \mathcal{X}_i,\ i \in \{1, 2\}, \\ q(t) &= Cx(t) + Dw(t), \end{cases} \tag{4.6}$$

with augmented state vector $x(t) = [x_g(t)^\top, x_h(t)]^\top \in \mathbb{R}^n$, where $n = m + 1$. The sets $\mathcal{X}_i$, $i \in \{1, 2\}$, in (4.6) are given by

$$\mathcal{X}_i = \left\{ x \in \mathbb{R}^n \mid Ex \in \mathcal{F}_i \right\}, \tag{4.7}$$

in which the matrix $E \in \mathbb{R}^{3\times n}$ is such that $Ex = (z\ u\ \dot{z})^\top$, and is, therefore, given by

$$E^\top = \begin{bmatrix} -C_g^\top & 0 & -(C_g A_g)^\top \\ 0 & 1 & 0 \end{bmatrix}. \tag{4.8}$$

The mode-dependent system matrices are given by

$$A_1 = \begin{bmatrix} A_g & B_g \\ -\omega_h C_g & 0 \end{bmatrix}, \quad \text{and} \quad A_2 = \begin{bmatrix} A_g & B_g \\ -k_h C_g A_g & 0 \end{bmatrix}, \tag{4.9}$$

and furthermore $B = [B_w^\top\ 0]^\top$, $C = [C_q\ D_q]$, and $D = D_w$. Recall from Chap. 2 that for all inputs $w \in PC$, global existence of solutions to the discontinuous differential equation in (4.6) is guaranteed.

4.2.2 Problem Formulation

Stability properties of the closed-loop system in (4.6) are studied through the notion of input-to-state stability (ISS), recall Definition 2.3. According to Theorem 2.8, the closed-loop system in (4.6) is ISS, if it admits an ISS-Lyapunov function in the sense of Definition 2.5. Linearity of the individual sub-dynamics in (4.6) motivates the choice for considering a candidate ISS-Lyapunov function V to be of a quadratic form, i.e., $V(x) = x^\top Px$ with $P \in \mathbb{S}^{n\times n}$. The next theorem presents sufficient conditions in the form of linear matrix inequalities (LMIs) for guaranteeing the existence of a quadratic ISS-Lyapunov function for the HIGS-controlled system in (4.6), thereby proving ISS.

Theorem 4.1 *Consider the closed-loop system in (4.6). Suppose there exist a matrix $P \in \mathbb{S}^{n\times n}$ and constants $\tau_i \geq 0$, $i \in \{1, 2, 3\}$, that satisfy the LMIs*

$$P - \tau_1 S^\top J S \succ 0, \tag{4.10a}$$

$$A_1^\top P + P A_1 + \tau_2 S^\top J S \prec 0, \tag{4.10b}$$

$$\Theta^\top (A_2^\top P + P A_2 + \tau_3 T^\top J T)\Theta \prec 0, \tag{4.10c}$$

where

$$S = \begin{bmatrix} C_u \\ k_h C_z - C_u \end{bmatrix}, \quad T = \begin{bmatrix} C_u \\ C_z\,(\omega_h I - k_h A_2) \end{bmatrix}, \quad J = \begin{bmatrix} 0 & 1 \\ 1 & 0 \end{bmatrix}, \tag{4.11}$$

with $C_z = [-C_g \ \ 0]$, $C_u = [0 \ \ 1]$, *and* $\Theta^\top = [I \ \ -k_h C_g^\top]$. *Then, the closed-loop system (4.6) is ISS in the sense of Definition 2.3 with an ISS-Lyapunov function given by* $V(x) = x^\top P x$.

Proof The proof is based on showing that, under the hypothesis of the theorem, $V(x) = x^\top P x$ satisfies the conditions in Definition 2.5 thereby qualifying as a suitable ISS-Lyapunov function for the closed-loop system in (4.6).

Feasibility of the LMI in (4.10a) implies $V(x) = x^\top P x > \tau_1 x^\top S^\top J S x$ for any $x \in \mathbb{R}^n \setminus \{0\}$. Since $x^\top S^\top J S x = 2(k_h z - u)u \geq 0$ for all $x \in \mathcal{X}_1 \cup \mathcal{X}_2$, and $\tau_1 \geq 0$, it follows by virtue of the S-procedure that V is positive definite, such that the first condition (see (2.35)) is satisfied.

In a similar manner, feasibility of the LMI in (4.10b) implies

$$\begin{aligned} \nabla V(x)(A_1 x + B w) &= x^\top (A_1^\top P + P A_1) x + 2x^\top P B w \\ &\leq -\epsilon_1 \|x\|^2 - \tau_2 x^\top S^\top J S x + 2x^\top P B w \\ &\leq -\alpha_1 \|x\|^2 + \beta_1 \|w\|^2, \end{aligned} \tag{4.12}$$

for sufficiently small $\epsilon_1 > 0$ and positive values for α_1 and β_1. For obtaining the last inequality in (4.12), use is made of the S-procedure and Young's inequality.

In gain mode one finds $x_h = k_h z = -k_h C_g x_g$ with $x_g \in \mathbb{R}^m$ the states of the LTI part of the system, such that $x = \Theta x_g$. Then, by virtue of Finsler's lemma [3, 16], satisfying inequality (4.10c) implies the existence of some $\mu \in \mathbb{R}$ such that for all $x \in \mathcal{X}_2$

$$\begin{aligned} \nabla V(x)(A_2 x + B w) &= x^\top (A_2^\top P + P A_2) x + 2x^\top P B w \\ &\leq -\epsilon_2 \|x\|^2 - x^\top \left(\tau_3 T^\top J T + \mu H^\top H\right) x + 2x^\top P B w \\ &\leq -\alpha_2 \|x\|^2 + \beta_2 \|w\|^2, \end{aligned} \tag{4.13}$$

with $H = [-k_h C_g \ \ -1]$, and for some sufficiently small $\epsilon_2 > 0$ and positive values for α_2 and β_2. The last inequality in (4.13) follows from use of the S-procedure relaxation terms $x^\top T^\top J T x = 2(\omega_h z - k_h \dot{z})u > 0$ and $Hx = k_h z - u = 0$, which hold for all $x \in \mathcal{X}_2$. Taking $\alpha = \min\{\alpha_1, \alpha_2\}$ and $\beta = \max\{\beta_1, \beta_2\}$, it immediately follows that the second condition in Definition 2.5 is satisfied (see (2.36)), and thus $V(x) = x^\top P x$ is an ISS-Lyapunov function for the closed-loop system in (4.6). Invoking the result in Theorem 2.8 completes the proof. □

The next result demonstrates that the conditions in Theorem 4.1 provide somewhat stronger robust (with respect to arbitrary small state perturbations) stability guarantees for the closed-loop system in (4.6). It is an immediate consequence from the use of a quadratic Lyapunov function, and the fact that the S-procedure relaxation inequalities hold true for all $x \in \overline{\mathcal{X}}_i$, $i = \{1, 2\}$.

Corollary 4.1 *If the conditions in Theorem 4.1 are satisfied, then the Krasovskii regularization of (4.6) (see (2.42) in Sect. 2.3.4) is ISS.*

Although numerically tractable, solving the LMIs in (4.10) requires a parametric model, which, in practice, may be hard to obtain if one desires an accurate model description. Moreover, the conditions provide limited insight in the (re)design of HIGS-based controllers for guaranteed (robust) stability when the set of LMIs in (4.10) turns out to be infeasible. Motivated by these concerns, the main objective in this chapter is to establish frequency-domain conditions that exploit non-parametric models for guaranteeing the existence of a solution to the inequalities in (4.10). This allows for making the transition from time-domain conditions to frequency-domain conditions for guaranteeing ISS.

4.3 Frequency-Domain Conditions for ISS

4.3.1 Classical Conditions

Classical frequency-domain conditions for switched systems originating from the circle- and Popov criterion are tailored to the specific class of HIGS-controlled systems. The equivalence between these conditions and the existence of a *particular* solution to the LMI problem in (4.10) is provided explicitly in the next two theorems, the first which relates to the results in [6, 32], and the second which relates to [11, 14].

Theorem 4.2 *Suppose the matrix A_g in (4.1) is Hurwitz, thereby having that the linear part of the closed-loop system in (4.6) is ISS. The following conditions are equivalent:*

(i) The frequency-domain inequality

$$1 + Re\left\{k_h C_g (j\omega I - A_g)^{-1} B_g\right\} > 0 \tag{4.14}$$

is satisfied for all $\omega \in \mathbb{R} \cup \{\infty\}$.

(ii) The LMIs in (4.10) admit a solution of the form $P = diag(M, m) \succ 0$, with $M \in \mathbb{S}^{(n-1)\times(n-1)}$, $m > 0$, $\tau_1 = 0$, and $\tau_2, \tau_3 \geq 0$.

Proof The proof is given in Sect. A.3.2. □

Discrete-time counterparts of the frequency-domain condition in Theorem 4.2 for sampled-data versions of HIGS-controlled systems can be found in [21, Theorem 5.1] and [22, Theorem 2].

Theorem 4.3 *Suppose the matrix A_1 in (4.9) is Hurwitz, thereby having that the integrator mode subsystem in (4.6) is ISS. The following conditions are equivalent:*

(i) There exist constants $\alpha_1 \geq 0$ and $\alpha_2 \in \mathbb{R}$ such that the frequency-domain inequality

$$1 + Re\left\{c(j\omega I - A_1)^{-1}b\right\} > 0 \tag{4.15}$$

with $c = C_z(\omega_h I - k_h A_2) + \alpha_1 C_z + \alpha_2 H$, $H = k_h C_z - C_u$, and $b = [0\ \ 1]^\top$ is satisfied for all $\omega \in \mathbb{R} \cup \{\infty\}$.

(ii) The LMIs in (4.10) admit a solution of the form $P \succ 0$, $\tau_1 = \tau_2 = 0$, and $\tau_3 \geq 0$.

Proof The proof is given in Sect. A.3.3. □

Similar to the ideas in the classical circle- and Popov-criteria [7], the frequency-domain conditions in (4.14) and (4.15) arise from applying specific loop transformations to the original feedback system. These loop transformations are depicted in Fig. 4.2. Specific properties of HIGS are exploited for rearranging the original feedback interconnection into an equivalent interconnection of two passive subsystems [31, 33]. The loop transformation underlying (4.14) is shown in Fig. 4.2b, and is similar to the one in the classical circle-criterion [7]. In particular, HIGS is transformed into a dynamical system $\bar{\mathcal{H}}$ with input $\bar{z} = z - u/k_h$ and output $\bar{u} = u/k_h$. Sector-boundedness of HIGS then yields $\bar{z}\bar{u} \geq 0$ such that $\bar{\mathcal{H}}$ preserves input-output passivity properties. The LTI part of the system is transformed to

$$\bar{G}_{yu}(s) = 1 + k_h G_{yu}(s) = 1 + k_h C_g(sI - A_g)^{-1}B_g,$$

which, under condition (4.14), is a strictly passive system.

In a similar manner, the loop transformation underlying the frequency-domain condition in (4.15) is depicted in Fig. 4.2c, and bears much resemblance with the loop-transformation used in the classical Popov criterion [7]. In particular, HIGS is transformed into a dynamical system $\bar{\mathcal{H}}$ with input $\bar{z} = k_h\dot{z} + (k + \omega_h)z + \alpha_2 u - \dot{u}$, where $k = -\omega_h - \alpha_1 - k_h\alpha_2$, and output $\bar{u} = \dot{u} - \omega_h z$. Note that $x_h = u$ is differentiable for almost all times $t \in \mathbb{R}_{\geq 0}$. In integrator mode, one finds $\bar{u} = 0$ such that $\bar{z}\bar{u} = 0$, and in gain mode one has $\bar{z}\bar{u} = \alpha_1 z(\omega_h z - k_h\dot{z}) \geq 0$, such that overall $\bar{z}\bar{u} \geq 0$, and the transformed system $\bar{\mathcal{H}}$ preserves passivity properties. The LTI part in Fig. 4.2c is transformed according to

$$\bar{G}_{yu}(s) = 1 + \frac{(k_h s + k)G_{yu}(s) + \alpha_2}{s + \omega_h G_{yu}(s)} = 1 + c(sI - A_1)^{-1}b,$$

which, when satisfying the frequency-domain inequality (4.15), is guaranteed to be a strictly passive system. Both setups allow for applying the passivity theorem (see,

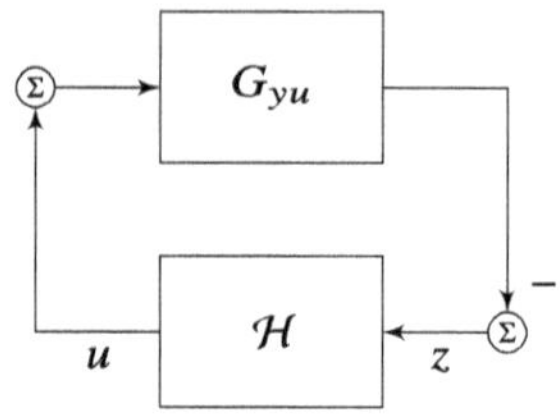

(a) Original system.

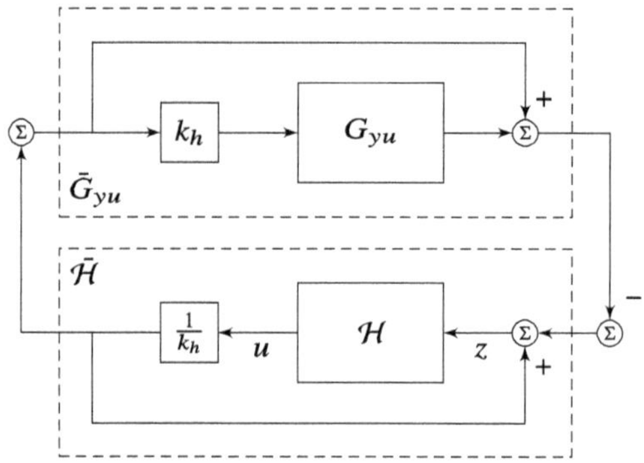

(b) Loop transformation applied in Theorem 4.2.

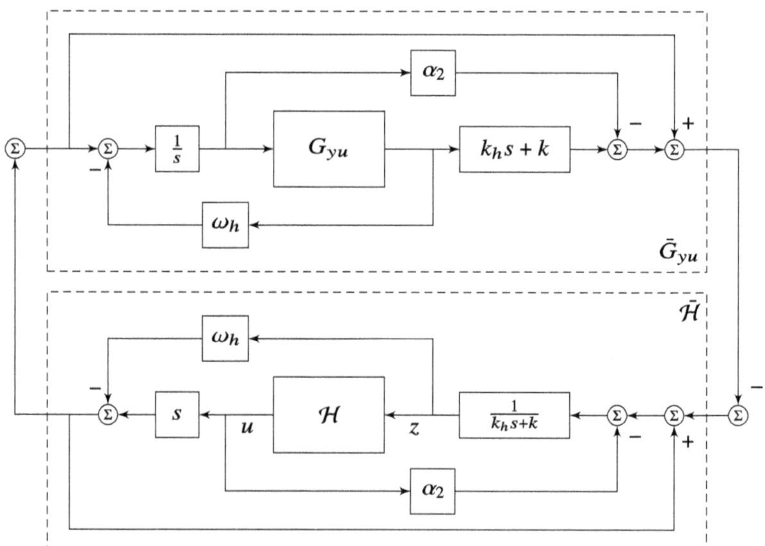

(c) Loop transformation applied in Theorem 4.3 with $k = -\omega_h - \alpha_1 - k_h\alpha_2$.

Fig. 4.2 Loop transformations underlying the frequency-domain inequality (4.14) in Theorem 4.2 and (4.15) in Theorem 4.3, respectively e.g., [13, Theorem 6.3]), which, under some additional observability conditions can imply asymptotic stability of the original closed-loop system.

Regarding Theorems 4.2 and 4.3, additional observations can be made.

(1) As a consequence of the fact that both theorems guarantee the existence of a solution to the LMI problem (4.10) with $\tau_1 = 0$, an ISS-Lyapunov function $V(x) = x^\top Px$ is found that is guaranteed to be positive in the full state space $\mathbb{R}^n$, while the states of the closed-loop system only evolve in part of the state space. This observation reveals possible restrictions on the considered class of Lyapunov function candidates that are used in Theorems 4.2 and 4.3, indicating potential conservatism.

(2) Theorem 4.2 is applicable to HIGS-based control systems for which the LTI part G_{yu} is stable. However, in many applications that may benefit from HIGS-based control, such as motion systems, the linear part of the dynamics contains simple integrators, which renders Theorem 4.2 not straightforwardly applicable as G_{yu} is not stable. As indicated before, the interplay between the LTI dynamics and the integrator dynamics of HIGS is not exploited in the conditions of Theorem 4.2, and HIGS is regarded as a static sector-bounded nonlinearity. This is also visible from the fact that the frequency-domain inequality in (4.14) does not depend on ω_h.

(3) Theorem 4.3 is applicable to HIGS-based control systems having stable linear dynamics in integrator mode. This narrows down the scope of applications for which Theorem 4.3 may be a useful tool for stability analysis. In fact, one may argue that unstable integrator mode dynamics largely contribute to potential performance improvements with HIGS, and, therefore, is a desirable property of a HIGS-based controller design.

(4) Closer inspection of the proofs of Theorems 4.2 and 4.3 reveals that the conditions (4.14) and (4.15) imply the existence of a common quadratic Lyapunov function (CQLF) for the pairs $(A_g, A_g - k_h B_g C_g)$ and $(A_1, A_1 - bc)$, respectively. Existence of a CQLF for $(A_g, A_g - k_h B_g C_g)$ is only possible when the gain mode dynamics are asymptotically stable, i.e., the matrix $A_g - k_h B_g C_g$ is Hurwitz. On the other hand, when considering a CQLF for $(A_1, A_1 - bc)$, one finds that since $A_1 - bc = A_2 - b(\alpha_1 C_z + \alpha_2 H)$, restricting $\alpha_1 = 0$ leads to a CQLF for the integrator mode and gain mode dynamics (recall from Chap. 2, Remark 2.2 that α_2 does not alter the gain mode dynamics). This is equivalent to saying that the switched system (4.6) is asymptotically stable under arbitrary switching. As switching of HIGS is dictated by state-dependent conditions, this possibly leads to conservative results.

In conclusion, although the results in Theorems 4.2 and 4.3 are of interest, they suffer from specific restrictions that are not always beneficial for HIGS-based control design as indicated in the above observations. Therefore, lifting these restrictions in the sense of obtaining less conservative conditions is considered important.

4.3.2 Extended Conditions

To address the previously discussed issues and provide less restrictive conditions, consider the next theorem, which provides a refined frequency-domain condition [30].

Theorem 4.4 *Suppose the matrix $A_g - k_h B_g C_g$ is Hurwitz, thereby having that the gain mode subsystem in (4.6) is ISS. If there exist numbers $\lambda \geq 0$, $k \geq 1$ such that for all $\omega \in \mathbb{R} \cup \{\infty\}$ the frequency-domain inequality*

$$1 + Re\left\{\left(\lambda k C_u + F + \left(\lambda + k\frac{\omega_h}{k_h}\right) H\right)(j\omega I - \mathcal{A}_2)^{-1} b\right\} > 0 \qquad (4.16)$$

with $\mathcal{A}_2 = A_2 + k\frac{\omega_h}{k_h} bH$, $H = k_h C_z - C_u$, and $F = C_z (k_h A_1 - \omega_h I)$ is satisfied, then the LMI problem (4.10) admits a feasible solution of the form $P = (M + \lambda H^\top H) + \tau_1 S^\top J S$ with $M = M^\top \succ 0$, $\tau_1 = \lambda k$, $\tau_2 = \lambda k(k-1)\omega_h/k_h$, and $\tau_3 = 0$.

Proof The proof can be found in Sect. A.3.4. □

To obtain insights in the frequency-domain condition (4.16), similar as before it is instructive to study aspects of the underlying loop transformation depicted in Fig. 4.3, which links to the use of dynamic (Popov) multipliers in classical input-output stability methods [7].

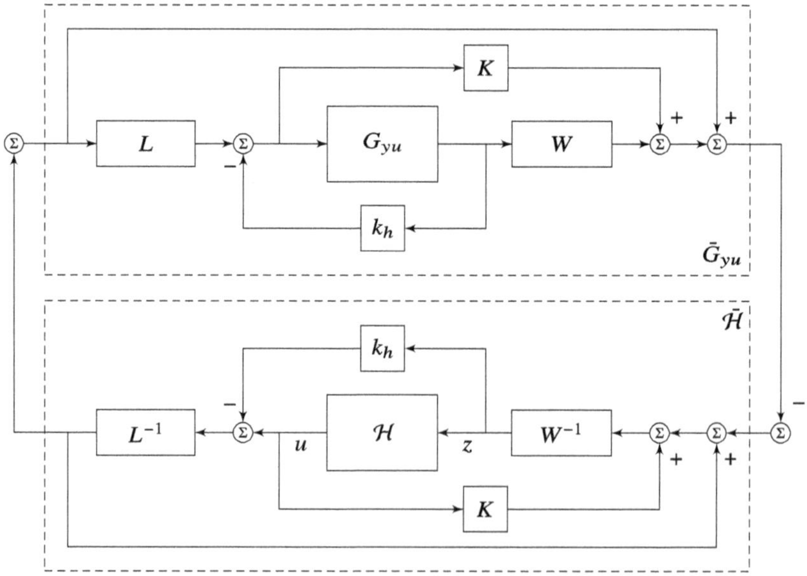

Fig. 4.3 Loop transformation underlying the frequency-domain inequality in (4.16) in Theorem 4.4, where $K = \lambda(k-1) - k\omega_h/k_h$, $L(s) = (s + k\omega_h/k_h)^{-1}$, and $W(s) = \omega_h(1-k) - \lambda k_h - k_h s$

Observe that in Fig. 4.3 the loop-transformed nonlinear system $\bar{\mathcal{H}}$ admits an input $\bar{z} = (W + k_h L^{-1})z - (K + L^{-1})u = (\omega_h z - \dot{u}) - \lambda(k_h z - (1-k)u)$, and output $\bar{u} = L^{-1}(u - k_h z) = \dot{u} - k_h \dot{z} + k\frac{\omega_h}{k_h}(u - k_h z)$. Passivity of $\bar{\mathcal{H}}$ does not directly follow from a similar reasoning as for the transformations studied in the previous section, since non-negativity of the input-output product $\bar{z}\bar{u}$ does not necessarily hold point-wise in time. However, observe that for almost all times $t \in \mathbb{R}_{\geq 0}$ one may write

$$\bar{z}(t)\bar{u}(t) = \tau R(z(t), u(t)) + \frac{d}{dt} U(z(t), u(t)) + Q(z(t), u(t), \dot{z}(t), \dot{u}(t)) \quad (4.17)$$

in which $\tau = \lambda k(k-1)\omega_h / k_h \geq 0$, and

$$R(z, u) = u(k_h z - u), \quad (4.18a)$$

$$U(z, u) = \lambda \left(ku(k_h z - u) + \frac{1}{2}(k_h z - u)^2 \right), \quad (4.18b)$$

and where

$$Q(z, u, \dot{z}, \dot{u}) = (\omega_h z - \dot{u})(\dot{u} - k_h \dot{z} + (K + \lambda)(k_h z - u)). \quad (4.19)$$

From sector-boundedness of HIGS in (4.3), one finds point-wise in time that both $R(z(t), u(t)) \geq 0$, and $U(z(t), u(t)) \geq 0$, and since $\dot{u} = \omega_h z$ in integrator mode and both $u = k_h z$ and $\dot{u} = k_h \dot{z}$ in gain mode, one has $Q(z(t), u(t), \dot{z}(t), \dot{u}(t)) = 0$ for almost all times t. Since $\tau \geq 0$, one then finds (with some abuse of notation for $R(t)$ and $Q(t)$)

$$U(0) + \int_0^T \bar{z}(t)\bar{u}(t)dt = U(T) + \int_0^T (\tau R(t) + Q(t))\, dt \geq 0, \quad (4.20)$$

for all $T \geq 0$, which implies that the transformed system $\bar{\mathcal{H}}$ in Fig. 4.3 preserves input-output passivity properties. Note that the key difference with the transformations employed in Theorems 4.2 and 4.3 is that, besides static input-output properties and knowledge regarding the underlying dynamics, integral properties are exploited here. This is standard in classical notions of passivity [7].

Next, passivity properties of the transformed LTI system $\bar{G}_{yu}(s)$ in Fig. 4.3 are studied and linked to the frequency-domain condition in (4.16). First, note that the transfer function related to (4.16) is given by

$$\hat{G}_{yu}(s) = \mathcal{C}(sI - \mathcal{A}_2)^{-1}\mathcal{B} + \mathcal{D}, \quad (4.21)$$

where $\mathcal{C} = \lambda k C_u + F + (\lambda + k\frac{\omega_h}{k_h})H$, $\mathcal{B} = b$, and $\mathcal{D} = 1$. To rewrite (4.21) into known transfer functions, it is useful to first do a similarity transformation using the matrices

$$T = \begin{bmatrix} I & 0 \\ -k_h C_g & 1 \end{bmatrix}, \quad \text{and} \quad T^{-1} = \begin{bmatrix} I & 0 \\ k_h C_g & 1 \end{bmatrix}. \quad (4.22)$$

Through standard manipulations this leads to the equivalent transfer function

$$\hat{G}_{yu}(s) = \hat{C}(sI - \hat{\mathcal{A}}_2)^{-1}\hat{\mathcal{B}} + \hat{\mathcal{D}}, \tag{4.23}$$

where

$$\left[\begin{array}{c|c} \hat{\mathcal{A}}_2 & \hat{\mathcal{B}} \\ \hline \hat{C} & \hat{\mathcal{D}} \end{array}\right] = \left[\begin{array}{c|c} T\mathcal{A}_2T^{-1} & T\mathcal{B} \\ \hline CT^{-1} & \mathcal{D} \end{array}\right] = \left[\begin{array}{cc|c} A_g - k_h B_g C_g & B_g & 0 \\ 0 & -k\frac{\omega_h}{k_h} & 1 \\ \hline F_g - \lambda k_h k C_g & K & 1 \end{array}\right], \tag{4.24}$$

with $F_g = C_g(\omega_h I - k_h A_g)$ and $K = \lambda(k-1) - k\omega_h/k_h$. Note the upper-triangular structure of the matrix $\hat{\mathcal{A}}_2$, and thus also of $sI - \hat{\mathcal{A}}_2$. The inverse of $sI - \hat{\mathcal{A}}_2$ can be computed as

$$(sI - \hat{\mathcal{A}}_2)^{-1} = \begin{bmatrix} (sI - A_l)^{-1} & (sI - A_l)^{-1}B_g(s + k\frac{\omega_h}{k_h})^{-1} \\ 0 & (s + k\frac{\omega_h}{k_h})^{-1} \end{bmatrix}, \tag{4.25}$$

with $A_l = A_g - k_h B_g C_g$. Using (4.25) in (4.23) yields

$$\begin{aligned} \hat{G}_{yu}(s) &= (F_g - \lambda k_h k C_g)(sI - A_l)^{-1}B_g + K)(s + k\tfrac{\omega_h}{k_h})^{-1} + 1 \\ &= 1 + \left(\frac{K + W(s)G_{yu}(s)}{1 + k_h G_{yu}(s)}\right)L(s) = \bar{G}_{yu}(s), \end{aligned} \tag{4.26}$$

where $W(s) = \omega_h(1-k) - \lambda k_h - k_h s$, $L(s) = (s + k\frac{\omega_h}{k_h})^{-1}$, and equivalence with $\bar{G}_{yu}(s)$ follows from the loop-transformation in Fig. 4.3. As such, the frequency-domain condition in (4.16) implies the LTI part $\bar{G}_{yu}$ in Fig. 4.3 to be (strictly) passive, and thus the interconnection in Fig. 4.3 is (strictly) passive. As mentioned before, (4.26) closely links to multiplier techniques in feedback stability analysis [7], where in this case the LTI filters $W(s)$ and $L(s)$ act as corresponding multipliers.

To see the connection with an ISS-Lyapunov function and feasibility of the LMIs in (4.10), first note that strict passivity of $\bar{G}_{yu}$ is equivalent [13, 31] to the existence of a quadratic storage function $\mathcal{W}(x) = x^\top Mx > 0$ satisfying

$$\begin{aligned} \dot{\mathcal{W}}(x) &= x^\top\left(\mathcal{A}_2^\top M + M\mathcal{A}_2\right)x + x^\top Mb\bar{u} + \bar{u}^\top b^\top Mx + 2x^\top MBw \\ &= -\epsilon x^\top Mx - 2\bar{z}\bar{u} + 2x^\top MBw. \end{aligned} \tag{4.27}$$

This, however, is not an appropriate ISS-Lyapunov function in the sense of Definition 2.5. On the other hand, consider the function

$$V(x) = \mathcal{W}(x) + 2U(z, x_h) > 0, \quad \text{for all} \quad x \in \mathcal{X} = \mathcal{X}_1 \cup \mathcal{X}_2, \tag{4.28}$$

where $U(z, x_h) = \lambda(kx_h(k_h z - x_h) + \frac{1}{2}(k_h z - x_h)^2)$. The inequality in (4.28) follows from the fact that $\mathcal{W}(x) > 0$ for all non-zero $x \in \mathbb{R}^n$ and $U(z, x_h) \geq 0$ for all $x \in \mathcal{X}$. The time-derivative of V satisfies for all $x \in \mathcal{X}$ and almost all times $t \in \mathbb{R}_{\geq 0}$

$$\dot{V}(x) \leq -\epsilon x^\top M x - 2\bar{z}\bar{u} + 2\dot{U}(z, x_h) + 2x^\top M B w < -\alpha\|x\|^2 + \beta\|w\|^2, \tag{4.29}$$

where for obtaining the second inequality it is used that $\bar{z}(t)\bar{u}(t) = \tau R(t) + \frac{d}{dt}U(t) + Q(t)$, with $\tau R(t) + Q(t) \geq 0$. Hence, V is an ISS-Lyapunov function that is guaranteed to be positive definite for all $x \in \mathcal{X}$, and for which the existence is guaranteed by the frequency-domain inequality in (4.16). In fact, one can write $V(x) = x^\top P x$ with $P = M + \lambda H^\top H + \tau_1 S^\top J S$ and $\tau_1 = \lambda k$, which coincides with the specific ISS-Lyapunov function that results from Theorem 4.4. The connection between the frequency-domain inequality and the existence of the quadratic ISS-Lyapunov function (and, therefore, feasibility of the LMIs in (4.10)) is formally shown in the proof of Theorem 4.4 in Sect. A.3.4.

Remark 4.1 Satisfying the inequality in (4.16) with $\lambda = 0$ results in a common quadratic Lyapunov function for the piecewise linear system (4.6), which follows from the fact that this choice yields $\bar{z}\bar{u} = 0$ for all times. Hence, in this case no further information about HIGS is exploited, and the condition is found to be equivalent to (4.15) with $\alpha_1 = 0$ and $\alpha_2 = k\omega_h/k_h$.

4.3.3 Verifying the Frequency-Domain Conditions

While the above discussion helps in establishing links with results from the literature and insights in the existence of a particular ISS-Lyapunov function, at this point in the analysis it is not immediately clear how to verify the frequency-domain condition (4.16) in an effective manner. In order to derive a practical method for verifying the conditions, recall that the relevant transfer function is given by

$$\begin{aligned}
\bar{G}_{yu}(s) &= 1 + \left(\lambda(kC_u + H) + F + k\frac{\omega_h}{k_h}H\right)(sI - \mathcal{A}_2)^{-1}b \\
&= 1 + \left(KS(s) + \frac{1}{k_h}W(s)T(s)\right)L(s) \\
&= 1 + \left(\frac{\omega_h}{k_h}(T(s) - k) - sT(s)\right)L(s) - \lambda(1 - kS(s))L(s),
\end{aligned} \tag{4.30}$$

where $K = \lambda(k-1) - k\omega_h/k_h$, $L(s) = (s + k\omega_h/k_h)^{-1}$, $W(s) = \omega_h(1-k) - \lambda k_h - k_h s$, $S(s) = 1/(1 + k_h G_{yu}(s))$, and $T(s) = k_h G_{yu}(s)S(s)$ with $G_{yu}(s)$ given in (4.2). The latter two transfer functions S and T equal the sensitivity and complementary sensitivity functions in gain mode.

For verifying the conditions of Theorem 4.4, one should first verify if the gain mode dynamics are ISS. This can be done by applying the Nyquist stability criterion [8, Sect. 6.3] to the open-loop characteristics $1 + k_h G_{yu}(j\omega)$. Note that (part of) the characteristics of $G_{yu}(j\omega)$ may be obtained from frequency-response function measurements. Condition (4.16) can be verified by searching for parameters $\lambda \geq 0$, $k \geq 1$ such that the frequency-domain inequality $\mathrm{Re}\left\{\bar{G}_{yu}(j\omega)\right\} > 0$ with $\bar{G}_{yu}(j\omega)$ given in (4.30) holds for all $\omega \in \mathbb{R} \cup \{\infty\}$. This amounts to verifying the inequality

$$1 + X(j\omega, k) - \lambda Y(j\omega, k) > 0, \tag{4.31}$$

where

$$\begin{aligned} X(j\omega, k) &= \mathrm{Re}\left\{\left(\left(\frac{\omega_h}{k_h} - j\omega\right) T(j\omega) - k\frac{\omega_h}{k_h}\right) L(j\omega, k)\right\} \\ &= \frac{\omega_h}{k_h}\mathrm{Re}\left\{(T(j\omega) - k)L(j\omega, k)\right\} + \omega \mathrm{Im}\left\{T(j\omega)L(j\omega, k)\right\}, \end{aligned} \tag{4.32a}$$

and

$$\begin{aligned} Y(j\omega, k) &= \mathrm{Re}\left\{(kT(j\omega) - k + 1)\, L(j\omega, k)\right\} \\ &= k\mathrm{Re}\left\{T(j\omega)L(j\omega, k)\right\} + (1 - k)\mathrm{Re}\left\{L(j\omega, k)\right\}. \end{aligned} \tag{4.32b}$$

For any fixed k, condition (4.31) can be verified graphically by inspecting if the (X, Y)-curve lies to the right of the line $1 + X - \lambda Y = 0$, i.e., a line in $\mathbb{R}^2$ that passes through the point $(-1, 0)$ with a slope of $1/\lambda$. The parameter $k \geq 1$ can be used for shaping the (X, Y)-curve. This graphical test shows resemblance with the classical Popov plot [7].

A few useful insights that follow from the representation in (4.31) are established in the following corollaries.

Corollary 4.2 *If the conditions in Theorem 4.2 are satisfied, i.e., the matrix A_g is Hurwitz and (4.14) holds, then the conditions in Theorem 4.4 are satisfied as well.*

Proof The proof can be found in Sect. A.3.5. □

Corollary 4.2 implies that the conditions in Theorem 4.4 are equal or less restrictive than the conditions in Theorem 4.2.

Corollary 4.3 *Consider the closed-loop system in (4.6) and suppose that the matrix $A_g - k_h B_g C_g$ is Hurwitz. Then for all $\omega_h > 2\mu k_h$, with the H_∞-norm $\mu := \|\omega T(j\omega)\|_\infty$, the conditions of Theorem 4.4 are satisfied, and the closed-loop system in (4.6) is ISS.*

Proof The proof can be found in Sect. A.3.6. □

Remark 4.2 When $T(s)$ depends on ω_h, such as, for example, in the controller designs discussed in Chap. 3, the condition in Corollary 4.3 can be extended to $\omega_h \geq \left\{\omega_h^*, 2\mu^* k_h\right\}$ with $\omega_h^* = \min(\Omega)$, where $\Omega := \left\{\omega_h \in \mathbb{R}_{>0} \mid A_g - k_h B_g C_g\right.$ is Hurwitz$\}$ and $\mu^* = \sup_{\omega_h \in \Omega} \|\omega T(j\omega)\|_\infty$.

The result in Corollary 4.3 is not surprising. As the time-domain behaviour of HIGS tends to a gain for $\omega_h \to \infty$, i.e., arbitrary fast switching, the characteristics of the piecewise linear closed-loop system (4.6) tend to that of the gain mode dynamics. Stability properties of the closed-loop system are then determined by stability properties of the gain mode dynamics. Hence, in this manner the nonlinear contribution of HIGS can be mitigated. This, in turn, shows that it is always possible for a HIGS-based control system to fall back onto a linear design when choosing ω_h sufficiently large, thereby demonstrating a simple mechanism to balance between dominant "linear" and dominant "nonlinear" behaviour of the closed-loop system by tuning the parameter ω_h. These observations also allow for an intuitive tuning approach for HIGS-based control systems with guaranteed stability. Specifically, one can start with choosing ω_h sufficiently large and a stable gain mode design, the latter obtained via, e.g., loop-shaping techniques [25]. By gradually decreasing the value for ω_h, a minimum value ω_h^* can be found for which the frequency-domain condition (4.31) is satisfied. By decreasing ω_h, the effect of the nonlinearity is made more pronounced. Using a (time-domain) performance objective, one can iteratively search for an appropriate value for $\omega_h \geq \omega_h^*$. Small values for ω_h have been observed to benefit transient performance properties, see for instance the example discussed in Chap. 3. Interestingly, for the example considered in Chap. 3, ISS can be verified on the basis of Theorem 4.4.

4.3.4 Example: Mass–Spring–Damper System

To demonstrate the potential improvements resulting from the frequency-domain condition (4.16) in Theorem 4.4, a comparative study is made for the control configuration given in Fig. 4.4.

In this example, the plant is a mass–spring–damper system represented by the transfer function

$$P(s) = \frac{1}{ms^2 + bs + k}, \tag{4.33}$$

with $m = 1$ kg, $b = 0.0564$ Ns/m, and $k = 1$ N/m, and the linear part of the HIGS-based feedback controller is given by a first-order lead filter

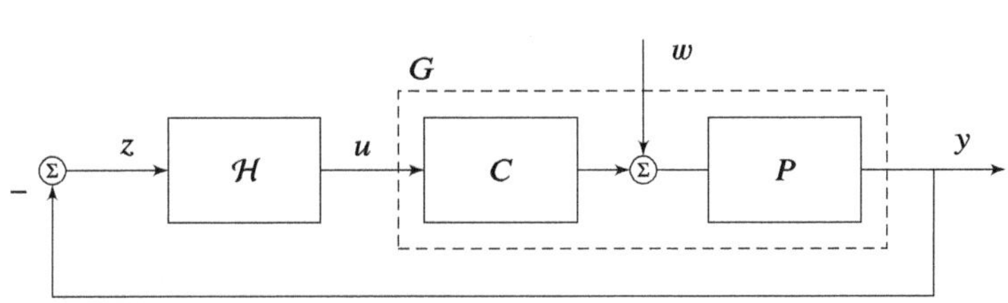

Fig. 4.4 Mass–spring–damper example with HIGS

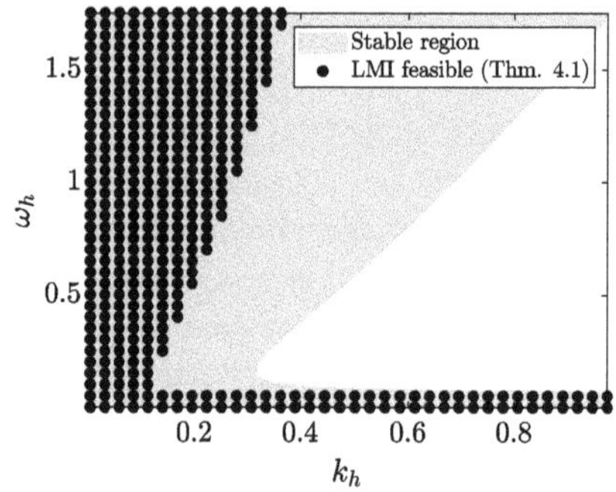

(a) Feasibility of the LMIs (4.10) in Theorem 4.1.

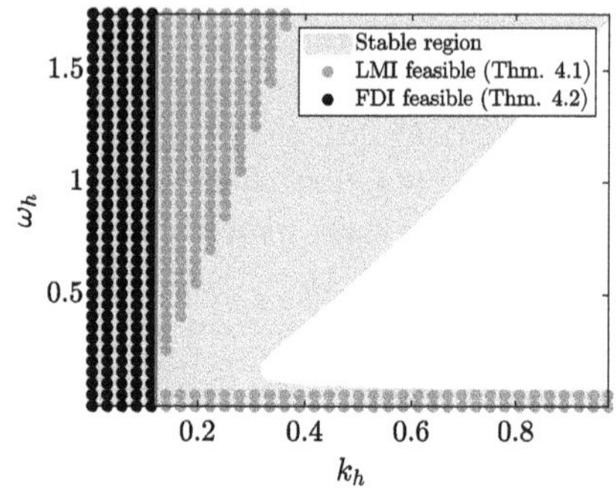

(b) Feasibility of the FDI (4.14) in Theorem 4.2.

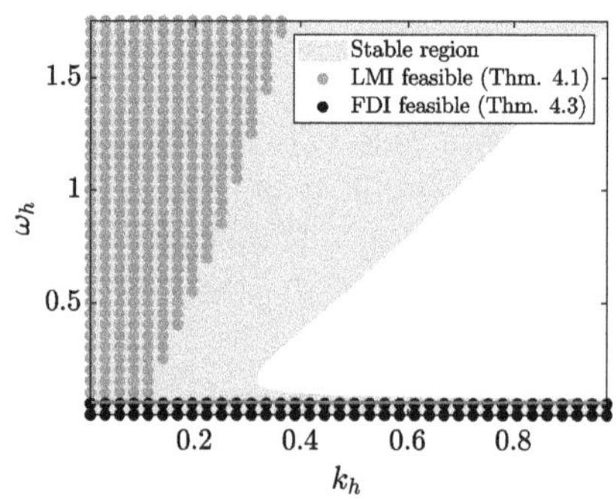

(c) Feasibility of the FDI (4.15) in Theorem 4.3.

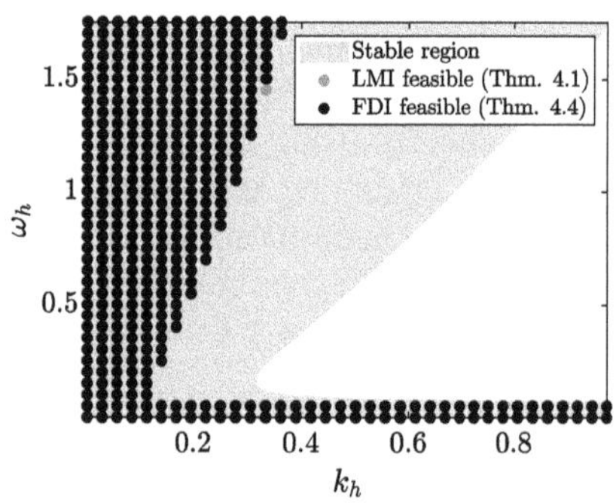

(d) Feasibility of the FDI (4.16) in Theorem 4.4.

Fig. 4.5 A comparison between the time-domain conditions (LMIs) in Theorem 4.1 and the frequency-domain conditions (FDIs) in Theorem 4.2–4.4 for different values of the HIGS parameters ω_h and k_h

$$C(s) = k_p \left(\frac{s + \omega_z}{s + \omega_p} \right), \tag{4.34}$$

with $k_p = 1.4$ Nm, $\omega_z = 5$ rad/s, and $\omega_p = 6.95$ rad/s. The transfer function from u to y is given by $G_{yu}(s) = C(s)P(s)$, and the transfer function from w to y is given by $G_{yw}(s) = P(s)$. In this control configuration, HIGS is deployed as a first-order low-pass filter with less phase-lag than its linear counterpart. For this system, the LMIs in (4.10), as well as the frequency-domain conditions in Theorem 4.2–4.4 are verified for several values of the parameters ω_h and k_h. The corresponding results are shown in Fig. 4.5.

As can be seen from Fig. 4.5b, the conditions in Theorem 4.2 guarantee a solution to the LMI problem in (4.10) for $k_h \leq 0.12$. This value results from the fact that condition (4.14) treats HIGS as a generic sector-bounded nonlinearity, and, therefore, does not take into account its integrator dynamics. This is manifested by the fact that

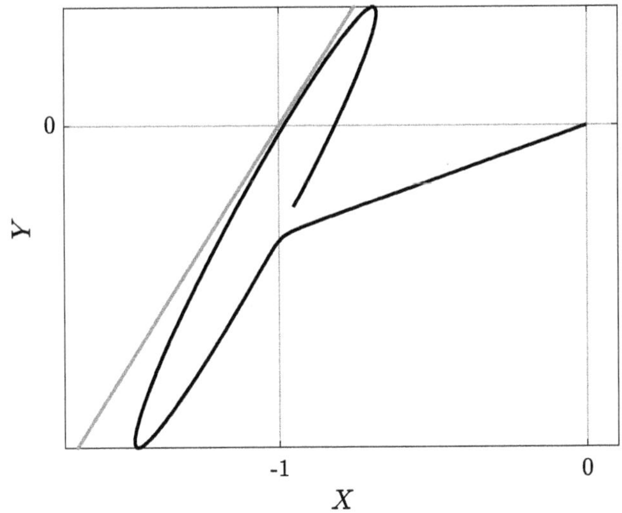

Fig. 4.6 Popov-like plot of the (X, Y)-curve for the system in Fig. 4.4 with $k_h = 0.23$ and $\omega_h = 1.05$. The stability parameters k, λ for satisfying (4.31) are found to be $k = 4$ and $\lambda = 1.35$

(4.14) is independent of ω_h. Since (for values $k_h > 0.12$) a solution to the LMIs in (4.10) clearly depends on ω_h as well, (4.14) can never be satisfied beyond this point.

Similarly, the conditions in Theorem 4.3 are found to guarantee a solution for all $\omega_h \leq 0.056$ rad/s as is shown in Fig. 4.5c. This bound is the result from the fact that for $\omega_h > 0.056$ rad/s, the integrator dynamics are unstable and, therefore, Theorem 4.3 cannot be applied to these cases. Compared to the actual region in Fig. 4.5a where the LMIs in (4.10) admit a solution, the results obtained from Theorems 4.2 and 4.3 appear quite conservative.

On the other hand, as can be seen from Fig. 4.5d, the conditions in Theorem 4.4 lead to a close match with the region where the LMIs in (4.10) have a solution. In fact, the frequency-domain inequality (FDI) in (4.16) and the LMIs in (4.10) almost appear to be identical for the given example (Fig. 4.5a vs. Fig. 4.5d). A graphical check of the conditions in (4.31) through a Popov-like plot as discussed previously is shown in Fig. 4.6 for the pair $k_h = 0.23$ and $\omega_h = 1.05$, with $k = 4$ and $\lambda = 1.35$.

It is stressed that, when compared to the actual stable region obtained by time-domain simulations (grey area), the LMIs in (4.10) resulting from a quadratic Lyapunov function still introduce conservatism in the analysis. Other techniques such as the construction of piecewise quadratic Lyapunov functions [19] should be used for further reducing this conservatism. The construction of piecewise quadratic Lyapunov functions for HIGS-based control systems is discussed in Chap. 5. For the example discussed it will be shown later in Sect. 5.5 that a stability analysis based on such an approach can very closely approximate the grey area in Fig. 4.5.

4.4 Robustifying the Conditions

The transition from time-domain tools toward frequency-domain tools for stability analysis allows for an intuitive extension toward dealing with plant uncertainties. This is particularly useful for industrial applications where one often designs a single controller for a population of systems that includes many system realizations. For the classical circle- and Popov critera, extensions toward robust stability have been made in, e.g., [10, 26]. Inspired by these works, the frequency-domain criteria presented in Theorem 4.4 are extended toward systems with multiplicative uncertainties.

4.4.1 *Stability Criteria for Plants with Multiplicative Uncertainty*

Suppose that the SISO LTI plant $G_{yu}(s)$ in (4.2) is not exactly known, but instead is known to belong to the multiplicative uncertainty set

$$\Pi = \left\{G_{yu}(s) \mid G_{yu}(s) = G_0(s)(1 + W(s)\Delta(s)),\ \Delta(s) \in \mathcal{RH}_\infty \text{ and } \|\Delta\|_\infty \leq 1\right\}, \tag{4.35}$$

where $G_0(s)$ is a nominal model of the plant, $W(s)$ is a proper and stable weighting function that captures magnitude information on uncertainties in the system, and $\Delta(s)$ is the (normalized) uncertainty which can be represented by an arbitrary proper and stable transfer function that satisfies $\|\Delta\|_\infty \leq 1$. For details on uncertainty modelling and appropriate weighting filter selection for SISO LTI systems, see, e.g., [24, Chap. 7].

To verify stability of the HIGS-based control system (4.6) for all possible plants $G_{yu}(s) \in \Pi$, one essentially has to check two conditions: (i) stability of the nominal gain mode dynamics with respect to all allowable perturbations, and (ii) satisfaction of the frequency-domain inequality (4.31) for all $G_{yu}(s) \in \Pi$. In the next proposition, these conditions are formalized and expressed in terms of requirements on the nominal system, i.e., with $\Delta(s) = 0$.

Proposition 4.1 *Suppose that the LTI part of the closed-loop system (4.6) dictating the transfer from u to y satisfies $G_{yu}(s) \in \Pi$ with Π the uncertainty set given in (4.35), and assume that the nominal gain mode dynamics with $\Delta(s) = 0$ are input-to-state stable. If for all $\omega \in \mathbb{R}$ the closed-loop system (4.6) satisfies*

$$\|W(j\omega)L_0(j\omega)\| < \|1 + L_0(j\omega)\|, \tag{4.36}$$

with $L_0(s) = k_h G_0(s)$, and there exist numbers $\lambda \geq 0, k \geq 1$ such that for all $\omega \in \mathbb{R} \cup \{\infty\}$ the frequency-domain inequality

$$1 + X_0(\omega, k) - \lambda Y_0(\omega, k) - \rho(\omega)\sqrt{\left(\frac{\omega_h}{k_h} - \lambda k\right)^2 + \omega^2} > 0, \tag{4.37}$$

is satisfied, where

$$X_0(\omega, k) = \frac{\omega_h}{k_h} Re\left\{(T_0(j\omega) - k)C_l(j\omega)\right\} + \omega Im\left\{T_0(j\omega)C_l(j\omega)\right\}, \tag{4.38a}$$

$$Y_0(\omega, k) = kRe\left\{T_0(j\omega)C_l(j\omega)\right\} + (1-k)Re\left\{C_l(j\omega)\right\}, \tag{4.38b}$$

with

$$T_0(s) = \frac{L_0(s)}{1 + L_0(s)}, \quad \textit{and} \quad C_l(s) = \frac{1}{s + k\frac{\omega_h}{k_h}}, \tag{4.39}$$

and where

$$\rho(\omega) = \frac{\|T_0(j\omega)W(j\omega)C_l(j\omega)\|}{\|1 + L_0(j\omega)\| - \|W(j\omega)L_0(j\omega)\|}, \tag{4.40}$$

then, the overall closed-loop system (4.6) is ISS for all plants $G_{yu}(s) \in \Pi$.

Proof The proof can be found in Sect. A.3.7. □

Condition (4.36) is standard practice in robust control theory for LTI systems, and can be verified by plotting the Nyquist diagram of the nominal open-loop characteristics $L_0(j\omega)$. At each frequency ω superimpose a circle with centre $L_0(j\omega)$ and radius $r(\omega) := \|W(j\omega)L_0(j\omega)\|$. If none of these circles in the complex plane cover the critical point $(-1, 0)$, then (4.36) is satisfied which means that the gain mode dynamics are robustly stable with respect to all allowable plant perturbations contained in Π.

However, due to the nonlinear appearance of λ in (4.37), it is not immediately clear how to check this condition graphically. One possible approach is to plot the nominal (X_0, Y_0) locus and for each frequency ω draw an ellipse centred at (X_0, Y_0) and described by the quadratic equation

$$x^2 - 2\frac{\omega_h}{kk_h}xy + \left(\frac{\omega_h^2}{k^2k_h^2} + \frac{\omega^2}{k^2}\right)y^2 - (\omega\rho(\omega))^2 = 0. \tag{4.41}$$

The resulting hull of the ellipses should lie to the right of the line $1 + x - \lambda y = 0$.

A less involved, but possibly more conservative method, follows from the inequality

$$\sqrt{\left(\frac{\omega_h}{k_h} - \lambda k\right)^2 + \omega^2} \leq \sqrt{\left(\frac{\omega_h}{k_h}\right)^2 + \omega^2} + \lambda k.$$

One can construct a lower-bound for the left-hand side in (4.37) as

$$
\begin{aligned}
1 + \left(X_0(\omega, k) - \rho(\omega)\sqrt{\left(\frac{\omega_h}{k_h}\right)^2 + \omega^2} \right) - \lambda\left(Y_0(\omega, k) + k\rho(\omega)\right) \\
= 1 + X_0'(\omega, k) - \lambda Y_0'(\omega, k).
\end{aligned} \tag{4.42}
$$

Hence, if (4.42) is positive, then the inequality in (4.37) is satisfied. Positiveness of (4.42) can be verified by inspecting if the shifted locus (X_0', Y_0') lies to the right of the straight line defined by $1 + x - \lambda y = 0$ for some $\lambda \geq 0$, $k \geq 1$.

4.4.2 Example: Rotating Beam

To illustrate the applicability of the conditions for robust stability in Proposition 4.1, consider a rotating beam with a mass located at the tip, representing, for example, a robotic arm with end effector [20]. The plant dynamics from input torque τ to the tip deflection θ are not exactly known, but instead are known to belong to the set

$$
\Pi = \{P(s) = P_0(s)\,(1 + W(s)\Delta(s)) \mid \Delta(s) \in \mathcal{RH}_\infty \text{ and } \|\Delta\|_\infty \leq 1\}, \tag{4.43}
$$

where $P_0(s) = 1/s^2$ is a rigid body model of the beam, and

$$
W(s) = \frac{3.5s^2 + 4.9s + 3.5}{s^2 + 27.2s + 289} \tag{4.44}
$$

is a second-order weighting filter that accounts for unmodelled high-frequency dynamics such as, e.g., the flexible modes of the beam or actuator/sensor dynamics [20]. For robustly stabilizing the beam around the horizontal position $\theta^* = 0$, a HIGS-based feedback controller, being the series interconnection of HIGS and an LTI filter, is considered. The closed-loop configuration is depicted in Fig. 4.7.

The linear part of the controller is constructed as the series interconnection of a lead filter, a second-order low-pass filter, and an additional zero, that is,

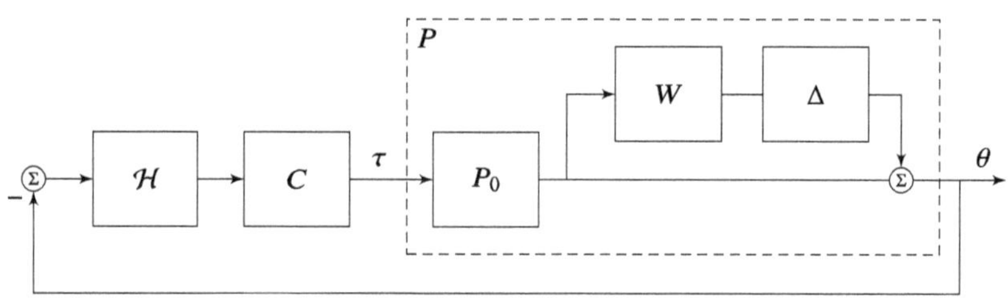

Fig. 4.7 Feedback control configuration for the rotating flexible beam

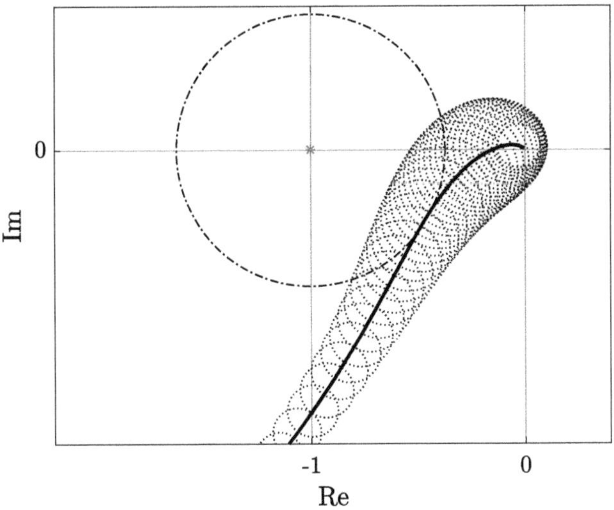

Fig. 4.8 Uncertainty region around the nominal open-loop $L_0(j\omega) = k_h C(j\omega) P_0(j\omega)$ (black) generated by the multiplicative uncertainty. The critical point $(-1, 0)$ is not covered by this region, so the nominal gain mode dynamics are robustly stable against the given uncertainty. The dashed-dotted circle indicates a 4 dB robustness bound (modulus margin) on the nominal system

$$C(s) = k_p \left(\frac{s + \omega_z}{s + \omega_p} \right) \frac{\omega_{lp}^2 (s + \omega_c)}{s^2 + 2\beta\omega_{lp} s + \omega_{lp}^2}. \tag{4.45}$$

The lead filter in (4.45) is used for stabilizing the gain mode dynamics, whereas the second-order low-pass filter is used for providing sufficient roll-off at high frequencies. The zero in (4.45) at ω_c in series with HIGS provides a (local) phase lag reduction in the control loop, intended to improve the transient properties of the closed-loop system, see also the design rationale as discussed earlier in Chap. 3. The controller (4.45) is initially designed on the basis of the nominal rigid-body model $P_0(s)$ in such a manner that the nominal closed-loop in gain mode is stable and satisfies a peaking constraint of 4 dB on the sensitivity function. Note that enforcing such a robustness bound helps in keeping the perturbation term (4.40) small, as this directly relates to the worst-case sensitivity in gain mode. The resulting parameters are given by $k_p = 10$ N/m, $\omega_z = 0.8$ rad/s, $\omega_p = 9$ rad/s, $\omega_{lp} = 11$ rad/s, $\beta = 1.2$ and $\omega_c = \omega_h |1 + 4j\pi|$ rad/s, $k_h = 1$ and $\omega_h = 3.5$ rad/s.

The nominal open-loop characteristics $L_0(j\omega) = k_h C(j\omega) P_0(j\omega)$ together with the uncertainty region generated by the multiplicative uncertainty are shown in Fig. 4.8. From this figure it is concluded that condition (4.36) is satisfied, and the gain mode dynamics are stable for all possible plants $P(s) \in \Pi$.

Verification of the frequency-domain condition in (4.37) is done by means of inspecting the shifted (X_0, Y_0)-locus in (4.42), as well as the uncertainty region around the nominal (X_0, Y_0)-locus generated by ellipsoids of the form described in

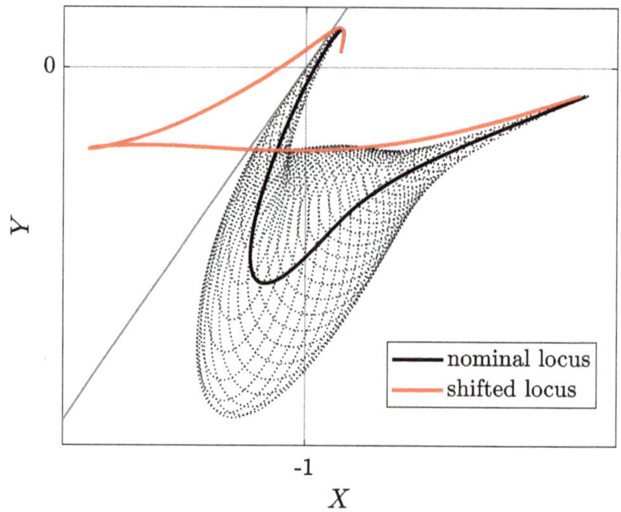

Fig. 4.9 Verifying inequality (4.37) by means of i) the shifted (X_0, Y_0)-locus (red), and ii) the uncertainty region around the nominal (X_0, Y_0)-locus (black) generated by the ellipses described in (4.41). Stability for all $P(s) \in \Pi$ is verified by means of the latter approach with $k = 10$ and $\lambda = 1.05$

(4.41). The corresponding plots (with $k = 10$) are given in Fig. 4.9. First note on the basis of the nominal locus (black curve) in the figure that the nominal HIGS-based control system is ISS. Remark that for this case the conditions in Theorem 4.2 and Theorem 4.3 are not applicable due to the presence of a double integrator in the plant, and the unstable integrator mode dynamics.

Regarding stability of the perturbed system, no conclusions can be drawn on the basis of the shifted (X_0, Y_0)-locus (red curve), as there is no $\lambda \geq 0$ such that this curve lies to the right of the line $1 + x - \lambda y = 0$. On the other hand, the ellipse-based uncertainty region (dotted) around the nominal locus is found to remain to the right of such a line with $\lambda = 1.05$ and thus (4.37) holds with $k = 10$ and $\lambda = 1.05$. This clearly illustrates that, despite its computational advantage, using the shifted locus for verifying inequality (4.37) can be more conservative than the approach using ellipses. Since all conditions in Proposition 4.1 are satisfied, one can conclude that the HIGS-based control system is ISS for all $P(s) \in \Pi$. For a final demonstration, two perturbed plants are considered with $\Delta(s) = -1$ and $\Delta(s) = 1$ for which the relevant plots are given in Fig. 4.10. Conform the previous results, the conditions in Theorem 4.4 are satisfied and the corresponding closed-loop systems are ISS. A numerically obtained solution to the LMI problem in (4.10) that follows from satisfying the frequency-domain conditions in Theorem 4.4 for the aforementioned cases is provided in Sect. B.2.

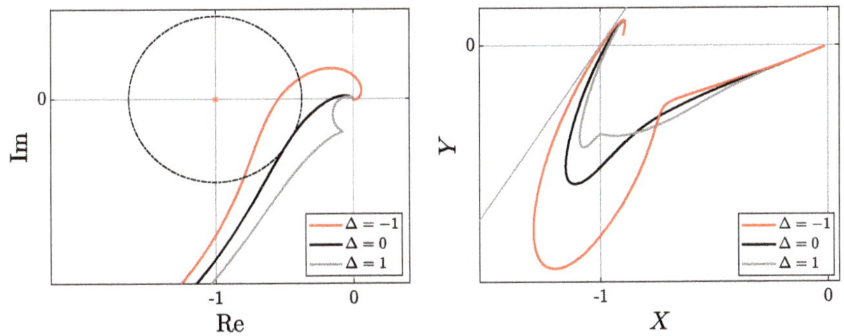

(a) Nyquist plot of the gain mode. (b) Popov-like plot.

Fig. 4.10 Graphical illustration of the conditions in Theorem 4.4 for the nominal plant $P(s) = P_0(s)$, and two perturbed plants $P(s) = P_0(s)(1 - W(s))$ and $P(s) = P_0(s)(1 + W(s))$

4.5 Application to a Motor-Load Motion System

In this section, applicability of the presented tools is demonstrated on an experimental motion control application. The set-up is shown in Fig. 4.11 and consists of two rotating masses connected by a thin, flexible shaft. Non-collocated actuation is considered, i.e., measurements by the encoder at the load side (right-side) are separated from actuation at the motor side (left side), see, e.g., [17] for further details on these type of systems. The measured frequency-response function (FRF) of the plant $P(j\omega)$ from actuator input to encoder position of the load is shown in Fig. 4.12, and is obtained from measurements with a sampling frequency of 2 kHz.

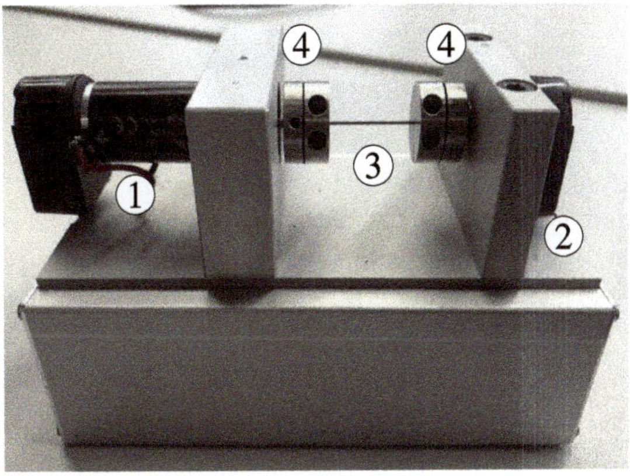

Fig. 4.11 Experimental motor-load motion system set-up with (1): actuator (motor side), (2): encoder (load side), (3): flexible shaft, and (4): rotating masses

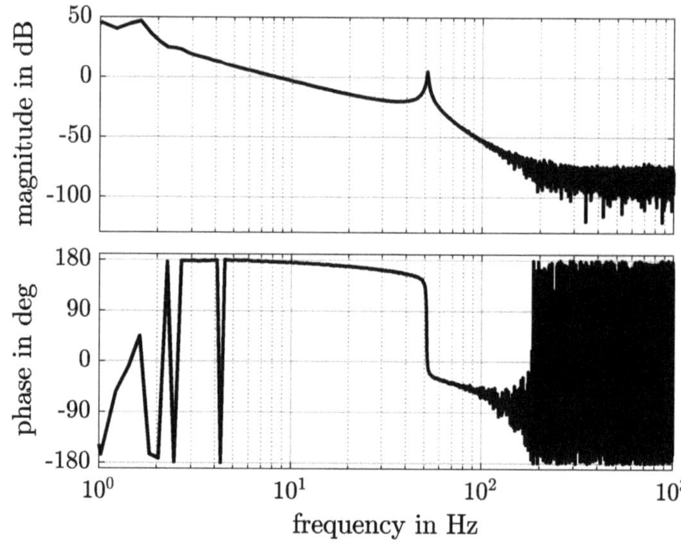

Fig. 4.12 Measured frequency response function (FRF) of the experimental motor-load motion system

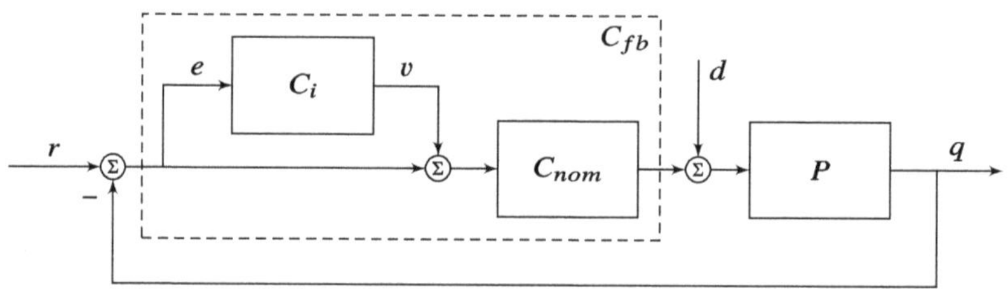

Fig. 4.13 Feedback control scheme for the motor-load motion system

4.5.1 Controller Design

Consider the feedback control scheme as depicted in Fig. 4.13. Given a reference command $r(t) \in \mathbb{R}$, a servo error signal $e(t) \in \mathbb{R}$ is constructed using the relation $e(t) = r(t) - q(t)$, where $q(t) \in \mathbb{R}$ represents the measured output of the motor-load motion system P at time $t \in \mathbb{R}_{\geq 0}$. This system is subject to an input disturbance $d(t) \in \mathbb{R}$ for $t \in \mathbb{R}_{\geq 0}$.

For dealing with external (set-point induced) disturbances, a feedback controller $C_{fb} = (1 + C_i)C_{nom}$ with integral action is designed. The nominal LTI filter $C_{nom}(s)$ is given by

$$C_{nom}(s) = k_p \left(\frac{s + \omega_z}{s + \omega_p} \right) \left(\frac{\omega_{lp}^2}{s^2 + 2\beta_{lp}\omega_{lp}s + \omega_{lp}^2} \right), \tag{4.46}$$

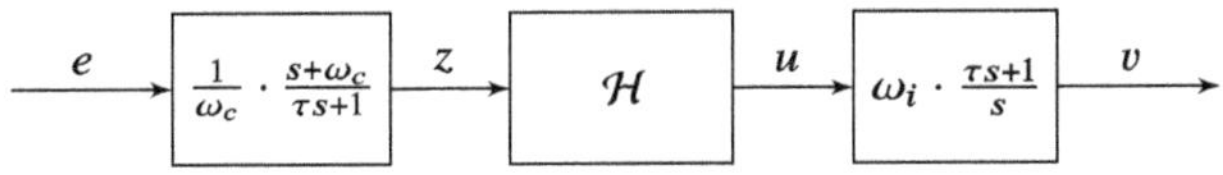

Fig. 4.14 HIGS-based integrator C_i

whereas C_i is a HIGS-based integrator as depicted in Fig. 4.14 with $\omega_c = \omega_h|1 + 4j/\pi|/k_h$ rad/s. This specific HIGS-based integrator design stems from the ideas in Chap. 3 and is intended for balancing transient and steady-state performance properties in a more desirable manner. For a HIGS-based low-pass filter design, see [27].

Tuning of the controller is done on the basis of the approach discussed previously in Sect. 4.3.3 by initially setting $\omega_h \to \infty$ such that C_i effectively reduces to an LTI integrator and the linear scheme $C_{fb}(s) = (1 + \frac{\omega_i}{s})C_{nom}(s)$ is effectively recovered. Using classical loop-shaping techniques, a stable LTI design is obtained with $k_p =$ 3.3 Nm, $\omega_i = 8 \cdot 2\pi$ rad/s, $\omega_z = 2 \cdot 2\pi$ rad/s, $\omega_p = 50 \cdot 2\pi$ rad/s, $\omega_{lp} = 50 \cdot 2\pi$ rad/s, and $\beta_{lp} = 0.5$. After fixing $k_h = 1$ N/m, and $\tau = 2 \cdot 10^{-3}$ s, the HIGS integrator frequency ω_h is reduced to $\omega_h = 5 \cdot 2\pi$ rad/s.

To verify ISS of the closed-loop system design, first observe that due to the simple integrators in C_i and P, one cannot apply Theorem 4.2 and thus one should resort to Theorem 4.3 or Theorem 4.4. The open-loop characteristics of the individual integrator and gain modes, given by

$$L_{int}(j\omega) = \left(1 + \frac{\omega_h}{j\omega}\left(\frac{\omega_i}{\omega_c} + \frac{\omega_i}{j\omega}\right)\right) C_{nom}(j\omega)P(j\omega),$$

$$L_{gain}(j\omega) = \left(1 + k_h\left(\frac{\omega_i}{\omega_c} + \frac{\omega_i}{j\omega}\right)\right) C_{nom}(j\omega)P(j\omega),$$

are inspected through the Nyquist plot depicted in Fig. 4.15.

On the basis of the Nyquist stability criterion, it immediately follows from this figure that the integrator mode is unstable, whereas the gain mode is stable. Hence, one cannot apply Theorem 4.3 either, and, hence, one should resort to the result in Theorem 4.4 for a frequency-domain stability analysis. Note that in the configuration of Fig. 4.13, the relevant transfer function from $y = -z$ to u needed for verifying the frequency-domain inequality in (4.31) is identified as

$$G_{yu}(j\omega) = \left(\frac{\omega_i}{\omega_c} + \frac{\omega_i}{j\omega}\right) \cdot \frac{C_{nom}(j\omega)P(j\omega)}{1 + C_{nom}(j\omega)P(j\omega)}. \tag{4.47}$$

The corresponding Popov-like plot with $k = 15$, and $\lambda = 10$ is shown in Fig. 4.16. Clearly, for these values the frequency-domain inequality in (4.31) is satisfied, such that, together with stability of the gain mode subsystem, all conditions of Theorem 4.4 are satisfied. Hence, the closed-loop system is ISS. Interestingly, the smallest value for which ISS could be shown on the basis of Theorem 4.4 was found to

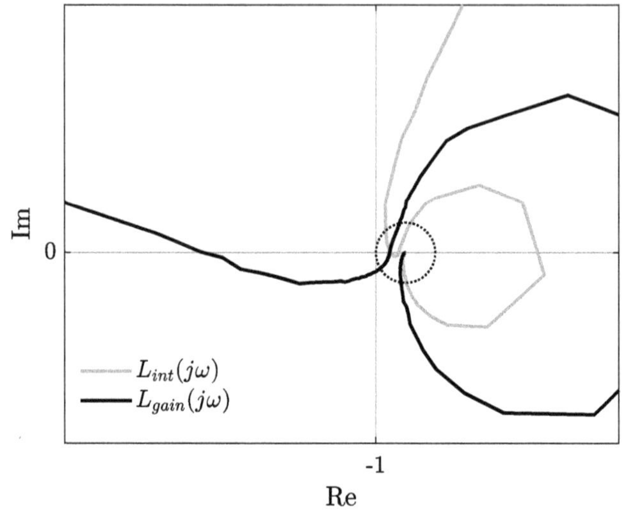

Fig. 4.15 Nyquist plot of the open-loop characteristics of the integrator- and gain mode using the measured FRF of the motor-load motion system

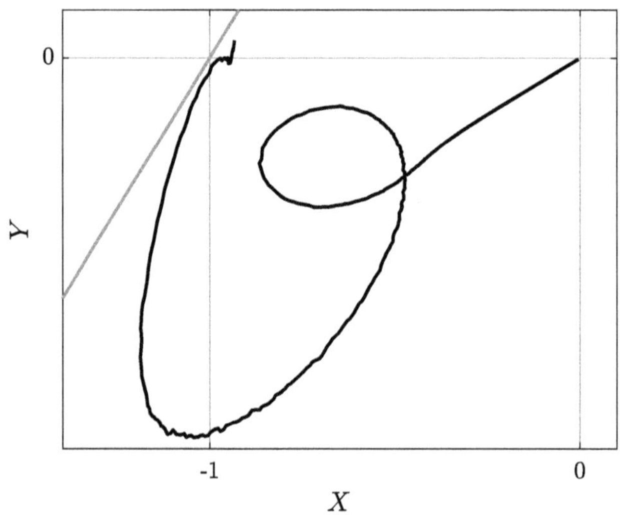

Fig. 4.16 Popov-like plot constructed on the basis of the measured FRF of the motor-load motion system and parameter values $k = 15$, $\lambda = 10$

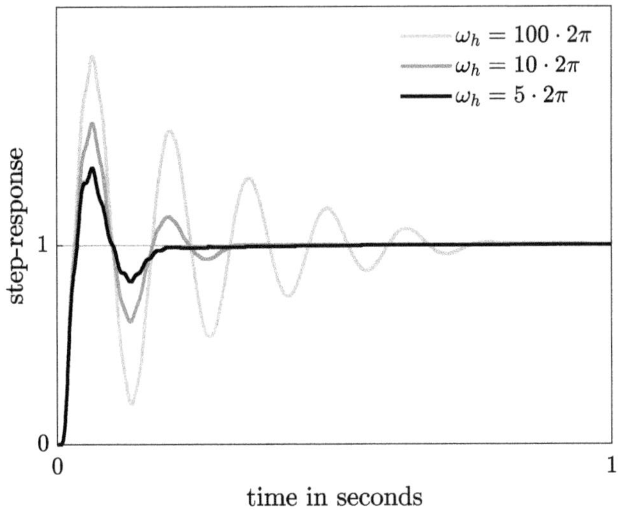

Fig. 4.17 Measured time-responses of the closed-loop system with $\omega_h \in \{5, 10, 100\} \cdot 2\pi$ rad/s when subject to a step input $r(t) = 1$, $t \in \mathbb{R}_{\geq 0}$

be $\omega_h = 1.5 \cdot 2\pi$ rad/s. For smaller values of ω_h, satisfaction of the conditions in Theorem 4.4 is limited by the fact that the gain mode dynamics become unstable. Nevertheless, for smaller values stable closed-loop behaviour was observed during experiments. For improving upon this aspect, it may be useful to refine the conditions in Theorem 4.4 to allow for unstable gain mode dynamics as well.

Typical step-responses of the closed-loop system for $\omega_h \in \{5, 10, 100\} \cdot 2\pi$ rad/s are shown in Fig. 4.17, and clearly demonstrate the positive effect of reducing ω_h on transient properties such as overshoot and settling times.

4.6 Summary

In this chapter, frequency-domain conditions for input-to-state stability (ISS) of HIGS-based control systems are discussed. The conditions result from exploiting integral properties of HIGS, and guarantee the existence of a quadratic Lyapunov function that is strictly positive only within the relevant subregion of the state space where the dynamics of HIGS are active. As such, these conditions may provide an advantage over classical results such as the circle-criterion in the sense of reducing conservatism in the analysis. The frequency-domain conditions can be verified graphically in a manner that is comparable to a Popov plot. An extension is made for verifying ISS of systems that are subject to multiplicative uncertainty, and several design insights for (robust) HIGS-based controller design are addressed. Application of the tools is demonstrated by means of an experimental case-study on a benchmark motion control system.

Although the presented frequency-domain conditions for (robust) stability are particularly valuable for practical design and analysis purposes, at the same time these can still be conservative. This conservatism is a result from the fact that underlying the conditions is the use of a quadratic Lyapunov function for an essentially piecewise linear system. In the next chapter, stability and performance analysis based on *piecewise* quadratic Lyapunov functions is pursued. Such functions are known to introduce a significant amount of flexibility in the analysis, which can lead to reduced conservatism, though potentially at the expense of losing the more practically verifiable frequency-domain interpretation.

References

1. Arcak, M., Larsen, M., Kokotović, P.: Circle and popov criteria as tools for nonlinear feedback design. Automatica **39**(4), 643–650 (2003)
2. Beker, O., Hollot, C., Chait, Y., Han, H.: Fundamental properties of reset control systems. Automatica **40**(6), 905–915 (2004)
3. Boyd, S., El Ghaoui, L., Feron, E., Balakrishnan, V.: Linear Matrix Inequalities in System and Control Theory, Studies in Applied Mathematics, vol. 15. SIAM, Philadelphia, PA (1994)
4. Dastjerdi, A.A., Astolfi, A., HosseinNia, S.H.: A frequency-domain stability method for reset systems. In: 2020 59th IEEE Conference on Decision and Control (CDC'20), pp. 5785–5791 (2020)
5. Deenen, D.A., Heertjes, M.F., Heemels, W.P.M.H., Nijmeijer, H.: Hybrid integrator design for enhanced tracking in motion control. In: Proceedings of the American Control Conference (ACC'17), pp. 2863–2868 (2017)
6. Deenen, D.A., Sharif, B., van den Eijnden, S., Nijmeijer, H., Heemels, M., Heertjes, M.: Projection-based integrators for improved motion control: formalization, well-posedness and stability of hybrid integrator-gain systems. Automatica **133**, 109830 (2021)
7. Desoer, C.A., Vidyasagar, M.: Feedback Systems. Society for Industrial and Applied Mathematics (2009)
8. Franklin, G., Powell, J., Emami-Naeini, A.: Feedback Control of Dynamic Systems, 5th ed. Prentice Hall (2005)
9. Griggs, W.M., King, C.K., Shorten, R.N., Mason, O., Wulff, K.: Quadratic Lyapunov functions for systems with state-dependent switching. Linear Algebra Appl. **433**(1), 52–63 (2010)
10. Impram, S.T., Munro, N.: A note on absolute stability of uncertain systems. Automatica **37**(4), 605–610 (2001)
11. Kamenetskiy, V.: Frequency-domain stability conditions for hybrid systems. Autom. Remote. Control. **78**, 2101–2119 (2017)
12. Kamenetskiy, V.: Switched systems, lur'e systems, absolute stability, aizerman problem. Autom. Remote. Control. **80**, 1375–1389 (2019)
13. Khalil, H.K.: Nonlinear Systems, 3rd edn. Prentice-Hall, Upper Saddle River, New Jersey (2002)
14. King, C., Griggs, W., Shorten, R.: A kalman-yakubovich-popov-type lemma for systems with certain state-dependent constraints. Automatica **47**, 2107–2111 (2011)
15. Kunze, M., Karimi, A., Longchamp, R.: Frequency domain controller design by linear programming guaranteeing quadratic stability. In: 2008 47th IEEE Conference on Decision and Control (CDC'08), pp. 345–350 (2008)

16. Meijer, T.J., Scheres, K.J.A., van den Eijnden, S., Holicki, T., Scherer, C.W., Heemels, W.P.M.H.: A unified non-strict finsler lemma. IEEE Control Syst. Lett. **8**, 1955–1960 (2024)
17. Preumont, A.: Collocated Versus Non-collocated Control, pp. 60–74. Springer, Netherlands, Dordrecht (1997)
18. Rantzer, A.: On the kalman-yakubovich-popov lemma. Syst. Control Lett. **28**(1), 7–10 (1996)
19. Rantzer, A., Johansson, M.: Piecewise linear quadratic optimal control. IEEE Trans. Autom. Control **45**(4), 629–637 (2000)
20. Sciavicco, L., Siciliano, B.: Modelling and Control of Robot Manipulators. Springer, London (2000)
21. Sharif, B., Alferink, D.W.T., Heertjes, M.F., Nijmeijer, H., Heemels, W.P.M.H.: A discrete-time approach to analysis of sampled-data hybrid integrator-gain systems. In: 2022 IEEE 61st Conference on Decision and Control (CDC), pp. 7612–7617 (2022)
22. Sharif, B., Alferink, D., Heertjes, M., Nijmeijer, H., Heemels, M.: Analysis of sampled-data hybrid integrator-gain systems: A discrete-time approach. Automatica **167**, 111765 (2024)
23. Shorten, R., Corless, M., Wulff, K., Klinge, S., Middleton, R.: Quadratic stability and singular siso switching systems. IEEE Trans. Autom. Control **54**(11), 2714–2718 (2009)
24. Skogestad, S., Postlethwaite, I.: Multivariable Feedback Control: Analysis and Design, 2nd ed. Wiley and Sons (2010)
25. Steinbuch, M., Norg, M.: Advanced motion control: an industrial perspective. Eur. J. Control. **4**(4), 278–293 (1998)
26. Tsypkin, Y.Z., Polyak, B.T.: Robust absolute stability of continuous systems. In: Mansour, M., Balemi, S., Truöl, W. (eds.) Robustness of Dynamic Systems with Parameter Uncertainties, pp. 113–121. Birkhäuser Basel, Basel (1992)
27. van den Eijnden, S., Knops, Y., Heertjes, M.: A hybrid integrator-gain based low-pass filter for nonlinear motion control. In: 2018 IEEE Conference on Control Technology and Applications (CCTA), pp. 1108–1113 (2018)
28. van den Eijnden, S., Sharif, B., Heertjes, M., Heemels, W.: Frequency-domain stability conditions for split-path nonlinear systems. IFAC-PapersOnLine **56**(1), 301–306 (2023)
29. van den Eijnden, S.J.A.M., Heertjes, M.F., Heemels, W.P.M.H., Nijmeijer, H.: Frequency-domain tools for stability analysis of hybrid integrator-gain systems. In: 2021 European Control Conference (ECC), pp. 1895–1900 (2021)
30. van den Eijnden, S.J.A.M., Heertjes, M.F., Nijmeijer, H., Heemels, W.P.M.H.: Stability analysis of hybrid integrator-gain systems: a frequency-domain approach. Automatica **164**, 111641 (2024)
31. van der Schaft, A.: L_2-Gain and Passivity Techniques in Nonlinear Control. Springer International Publishing (2017)
32. Van Loon, S., Gruntjens, K., Heertjes, M., Wouw, N., Heemels, W.M.: Frequency-domain tools for stability analysis of reset control systems. Automatica **82**, 101–108 (2017)
33. Willems, J.C.: Dissipative dynamical systems. Eur. J. Control. **13**(2), 134–151 (2007)
34. Yakubovich, V.: Solution of certain matrix inequalities in the stability theory of nonlinear control systems (in Russian). Dokl. Akad. Nauk SSSR **143**, 1304–1307 (1962)

Chapter 5
LMI-Based Tools for Stability and Performance Analysis

5.1 Introduction

In the previous chapter, frequency-domain conditions for (robust) stability and performance of hybrid integrator–gain systems (HIGS) have been derived. The power of these conditions is represented by the fact that these conditions do not require the use of parametric models, but rather can be verified (often graphically) on the basis of easy-to-measure frequency-response function (FRF) data. Moreover, these conditions allow for intuitive controller tuning procedures. Such aspects make the frequency-domain conditions valuable tools in an industrial environment. The advantages, however, can possibly come at the expense of increased conservatism in the analysis, which is mainly attributed to the underlying use of a *quadratic* Lyapunov function for an essentially piecewise linear (PWL) system. It is known, see, e.g., [15, 16, 26] that some stable PWL systems do not admit a quadratic Lyapunov function, thereby narrowing down applicability of these tools. To deal with this issue and improve upon the previous results, in this chapter use of more flexible *piecewise quadratic* (PWQ) Lyapunov functions is considered [2, 5, 13–16, 25, 26, 30, 31].

The use of PWQ Lyapunov functions for stability and performance analysis of switched/hybrid systems has been successfully applied to numerous fields of control applications, including reset control systems, see, e.g., [1, 19, 21, 33]. Key in the works on constructing PWQ Lyapunov functions for reset systems is a partitioning of subsets of the state space that dictate flow of the reset control system into smaller regions, each to which a different quadratic function is assigned. Since each function has to satisfy certain demands in the associated subregions only, the set of admissible Lyapunov functions may be significantly increased, thereby reducing conservatism.

Inspired by the approaches mentioned above, in this chapter an extension toward HIGS-based control systems is made. In particular, the three-dimensional subspace determining the active dynamics of HIGS is partitioned into convex polyhedral cones. There are many possible choices for the sizes and shapes of the polyhedral cones. For example, in [9] a polyhedral partitioning based on a spherical division of the

S. van den Eijnden et al., *Hybrid Integrator-Gain Systems*, Lecture Notes in Control and Information Sciences 499, https://doi.org/10.1007/978-3-032-18101-5_5

three-dimensional region of HIGS is proposed. In this chapter, however, an alternative approach is detailed in which the relevant subspace is divided into three-dimensional *simplicial* cones, i.e., polyhedral convex cones that are generated by the intersection of three half-spaces [8, 14]. This choice is mainly motivated by numerical considerations. Namely, when considering polyhedral cones that result from a larger number of intersecting half-spaces such as in [9], ensuring continuity of the PWQ function in a numerical search procedure should typically be done by posing explicit equality constraints. From a computational point-of-view, such constraints are particularly hard to satisfy as numerical solvers work with finite precision and can, therefore, only approximate a solution. A result that most likely violates the equality constraints is returned, potentially leading to false conclusions on stability [23]. When using simplicial cones, one can exploit a compact matrix parametrization of piecewise quadratic functions on polyhedral partitions in a manner as introduced in [15]. This parametrization allows for the continuity requirement to be directly incorporated in the description of each local quadratic function, thereby guaranteeing continuity a priori, and avoid the need for equality constraints in the eventual set of constraints.

On the basis of the above partitioning strategy, conditions for stability and performance of HIGS-based control systems are formulated in this chapter. Two performance measures are considered, namely the $\mathcal{L}_2$-gain, which characterizes the worst-case energy gain of the system with respect to a specific class of inputs, and the $\mathcal{H}_2$-norm, which reflects the transient behaviour through the energy in the system's impulse response. The relevance of these particular measures for systems with HIGS comes from the possibility to improve disturbance suppression properties and transient properties as illustrated in Chap. 3. By exploiting a tailored partitioning with PWQ functions, the conditions for stability and performance of HIGS-based control systems are formulated in terms of linear matrix inequalities (LMIs) in a manner that is robust from a numerical perspective. In addition, algebraic conditions are presented that provide relevant insights regarding (in)feasibility of the LMIs. These results rigorously demonstrate the benefits obtained from exploiting partition refinements, and help in guiding such refinements as to increase the possibility of finding a solution to the presented LMI conditions.

Next to the use of PWQ Lyapunov functions, the use of continuous piecewise affine (PWA) Lyapunov functions for stability and performance analysis of HIGS-based control systems has been considered recently in [24]. The search for PWA functions can be cast as a numerically efficient linear programming problem, offering an interesting alternative to the LMI conditions discussed in this chapter.

This chapter is organized as follows. In Sect. 5.2 the relevant closed-loop system setting with HIGS is recalled. In Sect. 5.3, the partitioning strategy for the construction of PWQ functions is discussed. Time-domain tools for stability and performance analysis are given in Sect. 5.4, and applicability of the tools is demonstrated through numerical examples in Sect. 5.5. Section 5.6 provides a summary.

5.2 System Setting and Problem Formulation

For the sake of readability, the generic closed-loop system setting with HIGS that is considered throughout this book and is discussed in detail in Chap. 2 is repeated in this section. Furthermore, the problem of finding conditions for stability and performance with a minimum amount of conservatism is formulated in detail.

5.2.1 Closed-Loop System Description

Consider the feedback interconnection depicted in Fig. 5.1. The multi-input multi-output (MIMO) LTI system G is represented in state space form as

$$G : \begin{cases} \dot{x}_g(t) &= A_g x_g(t) + B_g u(t) + B_w w(t), \\ y(t) &= C_g x_g(t), \\ q(t) &= C_q x_g(t) + D_q u(t) + D_w w(t), \end{cases} \tag{5.1}$$

with states $x_g(t) \in \mathbb{R}^m$, outputs $q(t) \in \mathbb{R}^r$, and inputs $w(t) \in \mathbb{R}^p$ at time $t \in \mathbb{R}_{\geq 0}$. Conform Chap. 2, w is assumed to belong to the class of piecewise continuous functions ($w \in PC$). The input to and output of HIGS $\mathcal{H}$ are denoted by $z(t) = -y(t) \in \mathbb{R}$, and $u(t) \in \mathbb{R}$, respectively. Recall that the matrices (A_g, B_g, C_g) are minimal and by virtue of Assumption 2.5 on the relative degree one has $C_g B_g = 0$ and $C_g B_w = 0$.

In Fig. 5.1, the HIGS element $\mathcal{H}$ is given by

$$\mathcal{H} : \begin{cases} \dot{x}_h(t) = \omega_h z(t), & \text{if } (z(t), u(t), \dot{z}(t)) \in \mathcal{F}_1, \quad (5.2a) \\ x_h(t) = k_h z(t), & \text{if } (z(t), u(t), \dot{z}(t)) \in \mathcal{F}_2, \quad (5.2b) \\ u(t) = x_h(t), & \quad (5.2c) \end{cases}$$

where $x_h(t) \in \mathbb{R}$ denotes the state of the integrator, $z(t) \in \mathbb{R}$ is the input, which in the closed-loop setting of Fig. 5.1 is at least one time differentiable, $\dot{z}(t) \in \mathbb{R}$ is the

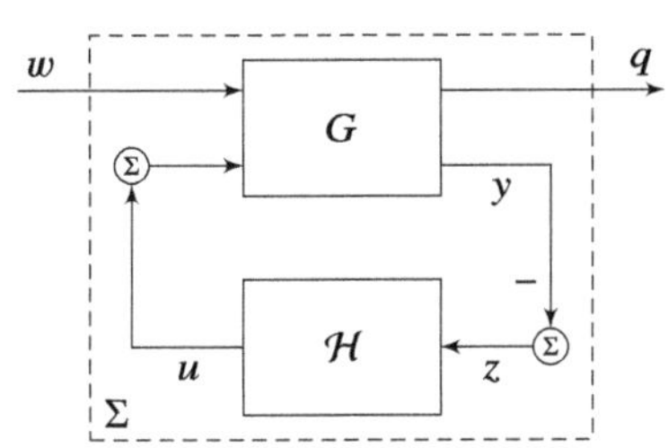

Fig. 5.1 Feedback interconnection of a MIMO LTI system G and a SISO HIGS element $\mathcal{H}$ with inputs w and performance channel q

corresponding time-derivative, and $u(t) \in \mathbb{R}$ is the generated output. The sets $\mathcal{F}_1$ and $\mathcal{F}_2$ are given by

$$\mathcal{F}_1 = \left\{(z, u, \dot{z}) \in \mathbb{R}^3 \mid (z, u, \dot{z}) \in \mathcal{F} \setminus \mathcal{F}_2\right\}, \tag{5.3a}$$

$$\mathcal{F}_2 = \left\{(z, u, \dot{z}) \in \mathbb{R}^3 \mid (z, u, \dot{z}) \in \mathcal{F} \vee u = k_h z \wedge \omega_h z^2 > k_h \dot{z} z\right\}, \tag{5.3b}$$

where the $[0, k_h]$-sector $\mathcal{F}$ is given by

$$\mathcal{F} = \left\{(z, u, \dot{z}) \in \mathbb{R}^3 \mid zu \geq \frac{1}{k_h} u^2\right\}. \tag{5.4}$$

For a visualization of the sets $\mathcal{F}_1, \mathcal{F}_2$, and $\mathcal{F}$, and further discussion on the rationale, the reader is referred back to Chap. 2.

The interconnection of the LTI system G as in (5.1) and HIGS $\mathcal{H}$ in (5.2) is written as the PWL system

$$\Sigma : \begin{cases} \dot{x}(t) &= A_i x(t) + B w(t), \quad \text{if } x(t) \in \mathcal{X}_i,\ i \in \{1, 2\}, \\ q(t) &= C x(t) + D w(t), \end{cases} \tag{5.5}$$

with augmented state vector $x(t) = [x_g(t)^\top, x_h(t)]^\top \in \mathbb{R}^n$, where $n = m + 1$. The sets $\mathcal{X}_i$, $i \in \{1, 2\}$ in (5.5) are given by

$$\mathcal{X}_i = \left\{x \in \mathbb{R}^n \mid Ex \in \mathcal{F}_i\right\}, \tag{5.6}$$

where the matrix E is such that $Ex = (z\ u\ \dot{z})^\top$, and for that reason is given by

$$E^\top = \begin{bmatrix} -C_g^\top & 0 & -(C_g A_g)^\top \\ 0 & 1 & 0 \end{bmatrix}. \tag{5.7}$$

Note that $\mathcal{X}_1 \cup \mathcal{X}_2 = \mathcal{X} := \{x \in \mathbb{R}^n \mid Ex \in \mathcal{F}_1 \cup \mathcal{F}_2\}$. The mode-dependent system matrices are given by

$$A_1 = \begin{bmatrix} A_g & B_g \\ -\omega_h C_g & 0 \end{bmatrix}, \quad \text{and} \quad A_2 = \begin{bmatrix} A_g & B_g \\ -k_h C_g A_g & 0 \end{bmatrix}, \tag{5.8}$$

and furthermore $B = [B_w^\top\ 0]^\top$, $C = [C_q\ D_q]$, and $D = D_w$. Recall again from Chap. 2 that for $w \in PC$, global existence of solutions to the discontinuous differential equation in (5.5) is guaranteed.

5.2.2 Problem Formulation

This chapter is mainly concerned with formulating potentially less conservative conditions for stability and performance of a HIGS-controlled system as in (5.5). Stability is meant here in the sense of input-to-state stability (ISS), for which the definition has been given earlier in Definition 2.3. Performance is quantified in terms of two measures, being the $\mathcal{L}_2$-gain and the $\mathcal{H}_2$-norm.

Definition 5.1 The closed-loop system in (5.5) is said to have an $\mathcal{L}_2$-gain from w to q smaller than or equal to γ, if there exists a $\mathcal{K}_\infty$-function ν such that for any square integrable $w \in PC$, and any initial condition $x(0) = x_0 \in \mathcal{X}$, all corresponding solutions to (5.5) satisfy

$$\|q\|_2 \leq \nu(\|x_0\|) + \gamma\|w\|_2. \tag{5.9}$$

Definition 5.2 The closed-loop system in (5.5) with $w = 0$ is said to have an $\mathcal{H}_2$-norm smaller than or equal to γ if for a specific initial value $x(0) = x_0 \in \mathcal{X}$ the corresponding solutions to (5.5) satisfy

$$\|q\|_2 \leq \gamma. \tag{5.10}$$

One of the possible interpretations of the $\mathcal{H}_2$-norm is the total energy of the output q due to an impulsive input w. For LTI systems, the response to such impulsive input is equivalent to the response when the system is subject to an initial condition $x_0 = B$, where B denotes the input column (see [1] in the context of reset control systems).

As in the previous chapter, ISS can be verified by finding a suitable ISS-Lyapunov function, see Definition 2.5. For assessing performance in terms of the $\mathcal{L}_2$-gain and the $\mathcal{H}_2$-norm, the concept of dissipativity as introduced in Definition 2.6 is useful. Recall that the closed-loop system in (5.5) admits a finite $\mathcal{L}_2$-gain from w to q smaller than or equal to γ, if it is dissipative with respect to the supply function $s(w, q, x) = \gamma^2\|w\|^2 - \|q\|^2$. By a comparable reasoning, it has an $\mathcal{H}_2$-norm smaller than or equal to γ for some initial condition x_0, if it is dissipative with respect to the supply function $s(w, q, x) = -\|q\|^2$. Hence, determining upper-bounds on the $\mathcal{L}_2$-gain and $\mathcal{H}_2$-norm amounts to finding appropriate storage functions that satisfy the conditions in Definition 2.6 with the corresponding supply functions. This chapter provides computationally tractable conditions for the existence of ISS-Lyapunov/storage functions to characterize stability and performance of the closed-loop system in (5.5) under an as small as possible level of conservatism. For that purpose, an approach is pursued that exploits piecewise quadratic (PWQ) functions.

5.3 Partitioning Strategy and PWQ Functions

Key in the construction of PWQ functions is the division of (a subset of) the state space of the considered system into smaller subregions. This section discusses the specific partitioning strategy for HIGS, and provides a procedure for efficiently constructing the matrices associated with the computation of PWQ functions. Before going into detail, some definitions regarding cones and conical partitions are introduced.

Definition 5.3 (*Positive hull,* [8]) The positive hull of a set $\mathcal{R} \subset \mathbb{R}^n$ is the set of all positive combinations of elements from $\mathcal{R}$, i.e.,

$$\text{pos}\,(\mathcal{R}) = \left\{ \sum_{m=1}^{M} \lambda_m r_m \mid r_m \in \mathcal{R}, \lambda_m \geq 0, m = 1, \ldots, M, \text{ for some } M \in \mathbb{N}_{>0} \right\}.$$

Definition 5.4 (*Polyhedral cone,* [8]) A polyhedral cone is the positive hull of a minimal (non-redundant) set $\mathcal{R} := \{r_1, \ldots, r_M\}$ with a finite number of elements. In this case $\mathcal{R}$ is called the generating set of the polyhedral cone, and $r_1, \ldots, r_M$ are called the generators.

Definition 5.5 (*Simplicial cone,* [8]) A simplicial cone in $\mathbb{R}^n$ is the positive hull of a set $\mathcal{R}$ with n linearly independent generators.

Polyhedral and simplicial cones can be written as

$$\mathcal{S} = \left\{ x \in \mathbb{R}^n \mid Sx \geq 0 \right\},$$

for some suitable matrix $S \in \mathbb{R}^{q \times n}$ of full rank [8]. For simplicial cones, it follows that the matrix S is square and invertible. The notation $-\mathcal{S}$ indicates the set $\{x \in \mathbb{R}^n \mid -x \in \mathcal{S}\}$, also denoted by its negative complement. In the remainder, a set $\mathcal{F}$ is called *symmetric* if it is formed as the union between a set and its negative complement, that is, $\mathcal{F} = \mathcal{S} \cup -\mathcal{S}$. For an in-depth discussion on the topic of polyhedral and simplicial cones, see, e.g., [8].

Definition 5.6 (*Simplicial partition,* [14]) Let a set $\mathcal{F} \subseteq \mathbb{R}^n$ and an integer $N > 0$ be given. A simplicial partition of $\mathcal{F}$ is a family $\mathcal{P}_{\mathcal{F}} = \{\mathcal{S}_1, \ldots, \mathcal{S}_N\}$ consisting of a finite number of simplicial cones $\mathcal{S}_n$, $n \in \mathcal{N} := \{1, \ldots, N\}$, that satisfy $\mathcal{F} = \bigcup_{n \in \mathcal{N}} \mathcal{S}_n$, and $\text{int}(\mathcal{S}_n) \cap \text{int}(\mathcal{S}_m) = \emptyset$ for $n \neq m$.

5.3.1 *Simplicial Partitioning of $\mathcal{F}$*

As discussed in the introduction, for the approach in this chapter, the three-dimensional region $\mathcal{F}$ as defined in (5.4) and shown in Fig. 5.2 is partitioned into simplicial cones, see Definition 5.6. A simplicial partition of the region $\mathcal{F}$ is

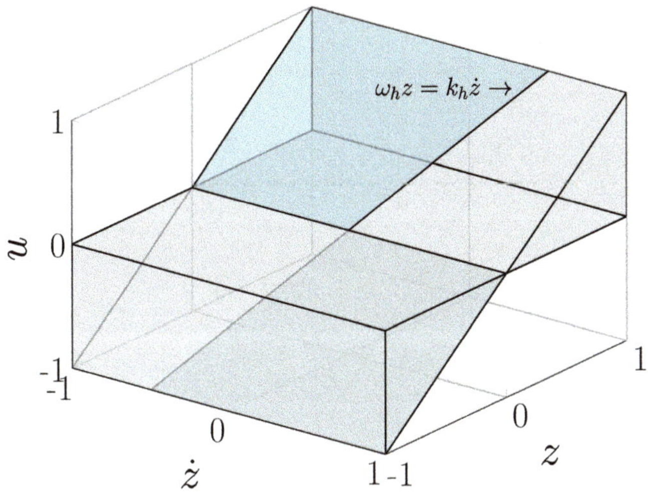

Fig. 5.2 Three-dimensional region $\mathcal{F}$ as defined in (5.4). In this figure, the integrator mode region $\mathcal{F}_1$ in (5.3a) is depicted in grey, and the gain mode region $\mathcal{F}_2$ in (5.3b) is depicted in blue

constructed in the following manner. From (5.4) and Fig. 5.2, it can be seen that $\mathcal{F}$ is formed as the union of two polyhedral cones $\hat{\mathcal{S}}$ and $-\hat{\mathcal{S}}$, that is, $\mathcal{F} = \hat{\mathcal{S}} \cup -\hat{\mathcal{S}}$, with

$$\hat{\mathcal{S}} := \left\{ \xi \in \mathbb{R}^3 \mid \hat{S}\xi \geq 0 \right\}, \tag{5.11}$$

where $\xi = (z, u, \dot{z})^\top$, and

$$\hat{S} = \begin{bmatrix} 0 & 1 & 0 \\ k_h & -1 & 0 \end{bmatrix}. \tag{5.12}$$

The intersection of $\hat{\mathcal{S}}$ and $-\hat{\mathcal{S}}$ is defined by $\hat{S}\xi = 0$, and coincides with the line $\xi = \left\{ (0, 0, \dot{z})^\top \mid \dot{z} \in \mathbb{R} \right\}$, see Fig. 5.2. According to [8, Lemma 1.40], polyhedral cones with non-empty interior can always be partitioned into a finite number of simplicial cones. This result allows the polyhedral cone $\hat{\mathcal{S}}$ in (5.11) to be partitioned into N simplicial cones $\hat{\mathcal{S}}_n$ given by

$$\hat{\mathcal{S}}_n = \left\{ \xi \in \mathbb{R}^3 \mid \hat{S}_n\xi \geq 0 \right\}, \quad n \in \mathcal{N} := \{1, \ldots, N\}, \tag{5.13}$$

with $\hat{S}_n \in \mathbb{R}^{3\times 3}$ an invertible matrix. An illustration of a simplicial cone $\hat{\mathcal{S}}_n$ is provided in Fig. 5.3. The cone is spanned by three unit vectors r_n^1, r_n^2, r_n^3, denoted as the generators such that $\hat{\mathcal{S}}_n = \text{pos}\left\{ r_n^1, r_n^2, r_n^3 \right\}$, and the generating set (see Definition 5.4) is given by $\mathcal{R}_n = \left\{ r_n^1, r_n^2, r_n^3 \right\}$.

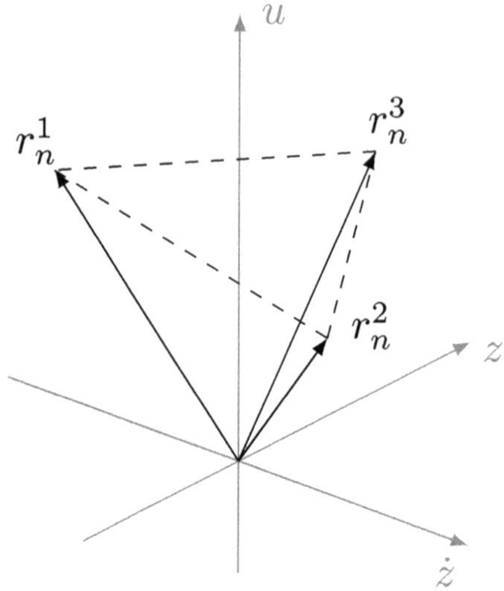

Fig. 5.3 Illustration of a simplicial cone $\hat{S}_n$, which is spanned by three vectors r_n^1, r_n^2, r_n^3, referred to as generators such that $\hat{S}_n = \text{pos}\left\{r_n^1, r_n^2, r_n^3\right\}$. The generating set of the cone is given by $\mathcal{R}_n = \left\{r_n^1, r_n^2, r_n^3\right\}$

Denote the simplicial partition of $\hat{S}$ into N simplicial cones by

$$\mathcal{P}_{\hat{S}} := \left\{\hat{S}_1, \ldots, \hat{S}_N\right\}. \tag{5.14}$$

A generating set $\mathcal{R}$ of the partition $\mathcal{P}_{\hat{S}}$ is constructed as the union of the generating sets $\mathcal{R}_n$ of each simplicial cone $\hat{S}_n \in \mathcal{P}_{\hat{S}}$, i.e., $\mathcal{R} = \bigcup_{n \in \mathcal{N}} \mathcal{R}_n$. In what follows, it is assumed that the simplicial partition $\mathcal{P}_{\hat{S}}$ is constructed in such a manner that the unit vectors $[0, 0, 1]^\top$ and $[0, 0, -1]^\top$ are generators, and thus $\{[0, 0, 1]^\top, [0, 0, -1]^\top\} \in \mathcal{R}$. Moreover, it is assumed that the vector defining the line $u = k_h z, \omega_h = k_h \dot{z}$ describing a common boundary of the set $\mathcal{F}_1$ and $\mathcal{F}_2$ is a generator of the partition as well. Clearly, a partition of $-\hat{S}$ directly results from a partitioning of $\hat{S}$, and is denoted by $-\mathcal{P}_{\hat{S}} := \{-\hat{S}_1, \ldots, -\hat{S}_N\}$. Since $\mathcal{F}$ is symmetric, a simplicial partition of $\mathcal{F}$ is then given by $\mathcal{P}_{\mathcal{F}} := \mathcal{P}_{\hat{S}} \cup -\mathcal{P}_{\hat{S}}$. An impression of a possible partition of $\mathcal{F}$ into simplicial cones is illustrated in Fig. 5.4. Several efficient numerical procedures can be employed for constructing an appropriate simplicial division of an n-dimensional space, see, e.g., the procedures discussed in [4, 12, 14, 32].

Since $\overline{\mathcal{F}}_1 = \mathcal{F}$, a partition of the integrator mode region $\mathcal{F}_1$ in (5.3a), denoted by $\mathcal{P}_{\mathcal{F}_1}$ is equivalent to a partition of the full region $\mathcal{F}$, i.e., $\mathcal{P}_{\mathcal{F}_1} = \mathcal{P}_{\mathcal{F}}$. As $\mathcal{F}_2$ forms a "boundary set" of $\mathcal{F}$, it follows that a partitioning of $\mathcal{F}$ induces a simplicial partitioning of $\mathcal{F}_2$, denoted by $\mathcal{P}_{\mathcal{F}_2}$. The cones belonging to the partitioning $\mathcal{P}_{\mathcal{F}_2}$ are described by $\hat{\mathcal{T}}_m = \text{pos}\left\{r_m^i, r_m^j\right\}$, $m \in \mathcal{M}$, $(i, j) \in \{1, 2, 3\}$, where the set $\mathcal{M} \subset \mathcal{N}$ contains the indices of the cones $\hat{S}_m$, $m \in \mathcal{N}$ for which $\mathcal{F}_2 \cap \hat{S}_m \neq \emptyset$, and r_m^i, r_m^j are those

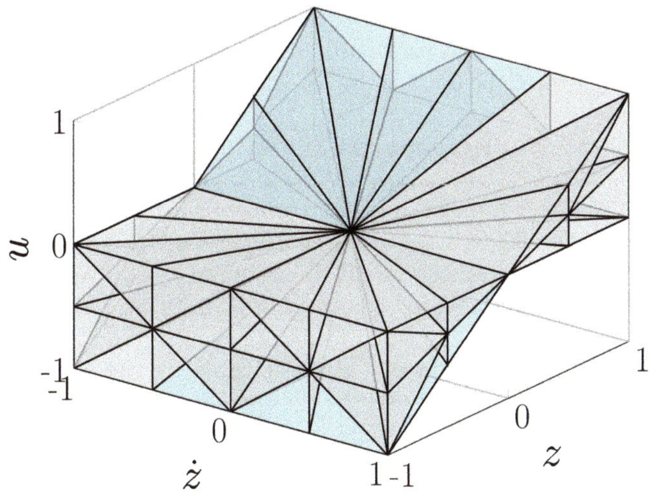

Fig. 5.4 Impression of a simplicial partition of $\mathcal{F}$ into N simplicial cones $\hat{\mathcal{S}}_n$. The blue shaded regions correspond to "boundary planes" $\hat{\mathcal{T}}_m, m \in \mathcal{M}$ that partition the gain mode region $\mathcal{F}_2$

generators of $\hat{\mathcal{S}}_m$ that satisfy $r_m^i, r_m^j \in \overline{\mathcal{F}}_2$. As such, the generating set of $\hat{\mathcal{T}}_m$ is given by $\mathcal{R}'_m = \left\{r_m^i, r_m^j\right\}$. Clearly, $\hat{\mathcal{T}}_m$ are polyhedral cones in $\mathbb{R}^3$, and can alternatively be described by

$$\hat{\mathcal{T}}_m = \left\{\xi \in \mathbb{R}^3 \mid \hat{T}_m \xi \geq 0 \wedge h\xi = 0\right\}, \quad m \in \mathcal{M} \subset \mathcal{N}, \tag{5.15}$$

where $\hat{T}_m \in \mathbb{R}^{2\times 3}$ is a full row rank matrix, and $h = [k_h, -1, 0]$. The partition $\mathcal{P}_{\mathcal{F}_2}$ of the gain mode region is then given by

$$\mathcal{P}_{\mathcal{F}_2} = \left\{\hat{\mathcal{T}}_{m_1}, \ldots, \hat{\mathcal{T}}_{m_M}\right\} \cup \left\{-\hat{\mathcal{T}}_{m_1}, \ldots, -\hat{\mathcal{T}}_{m_M}\right\} = \mathcal{P}_{\hat{\mathcal{T}}} \cup -\mathcal{P}_{\hat{\mathcal{T}}}, \tag{5.16}$$

where $\{m_1, \ldots, m_M\} = \mathcal{M}$ and $M \leq N$ denotes the number of elements in $\mathcal{M}$. The generating set $\mathcal{R}'$ of the partition $\mathcal{P}_{\mathcal{F}_2}$ is formed as $\mathcal{R}' := \bigcup_{m\in\mathcal{M}} \mathcal{R}'_m$, and contains all generators $r_n, n \in \mathcal{N}$ of the partition $\mathcal{P}_{\mathcal{F}}$ that satisfy $r_n \in \overline{\mathcal{F}}_2$.

5.3.2 PWQ Function Parametrization and Constraint Matrices

The main advantages of using a simplicial partitioning as discussed above, are (i) the possibility to directly incorporate continuity properties in the definition of the PWQ function, and (ii) doing so without introducing additional conservatism. Particularly, by exploiting a compact matrix parametrization of piecewise quadratic functions on polyhedral partitions, adopted from [16, 26] (see also [2, 14]), and defined as

$$V_i(x) = x^\top P_i x = x^\top F_i^\top \Phi F_i x, \tag{5.17}$$

where $F_i \in \mathbb{R}^{r\times n}, i \in \mathcal{N}$ form the so-called *continuity matrices* satisfying $F_i x = F_j x$ for all x on a boundary shared by two polyhedral regions, and $\Phi \in \mathbb{S}^{r\times r}$ is a symmetric matrix that contains the decision variables, continuity of the PWQ functions over cell boundaries is guaranteed (see Lemma 4.2 in [15]). In this formulation, explicit equality constraints become obsolete, which has computational advantages.

Based on the partitioning strategy as illustrated in Fig. 5.4, the following two propositions present useful results for the construction of the matrices $\hat{S}_i$ in (5.13) and $\hat{T}_i$ in (5.15) describing the simplicial cones, and the continuity matrices used in (5.17). These results are based on the ideas outlined in the seminal works [15, 16]. In formulating the propositions, it will be useful to collect the generators of a simplicial cone $\hat{\mathcal{S}}_i \in \mathcal{P}_{\hat{\mathcal{S}}}$ in a matrix $R_i := [r_i^1, r_i^2, r_i^3] \in \mathbb{R}^{3\times 3}$, $i \in \mathcal{N}$, and the generators of a cone $\hat{\mathcal{T}}_j \in \mathcal{P}_{\hat{\mathcal{T}}}$ in a matrix $R'_j := [r_j^1, r_j^2] \in \mathbb{R}^{3\times 2}$, $j \in \mathcal{M}$.

Proposition 5.1 *Consider a simplicial partition $\mathcal{P}_{\hat{\mathcal{S}}}$. The matrices*

$$\hat{S}_i = R_i^{-1}, \qquad i \in \mathcal{N} \tag{5.18}$$

$$\hat{T}_j = \left((R'_j)^\top R'_j\right)^{-1} (R'_j)^\top, \qquad j \in \mathcal{M} \tag{5.19}$$

and $h = [k_h, -1, 0]$ satisfy the inequalities in (5.13) and (5.15), respectively.

Proof The proof can be found in Sect. A.4.1. □

Proposition 5.2 *Consider a simplicial partition $\mathcal{P}_{\hat{\mathcal{S}}}$. Define the set $\bar{\mathcal{R}} := \mathcal{R} \setminus \{[0, 0, -1]^\top\}$, and collect its elements in the matrix $\bar{R} := [r_1, \dots r_p] \in \mathbb{R}^{3\times p}$. Let the matrices $E_i \in \mathbb{R}^{p\times 3}$, $i \in \mathcal{N}$ have its q-th row equal to zero for all $q \in \{1, \dots, p\}$ such that $r_q \notin \hat{\mathcal{S}}_i$, and the remaining rows equal to the rows of a three-dimensional identity matrix such that $\bar{R}E_i = R_i$, in which no distinction is made between the generators $[0, 0, 1]^\top$ and $[0, 0, -1]^\top$. Then, the matrices*

$$\hat{F}_i = E_i \left(\bar{R}E_i\right)^{-1}, \quad i \in \mathcal{N}, \tag{5.20}$$

satisfy the property $\hat{F}_i\xi = \hat{F}_j\xi$ for all $\xi \in \hat{\mathcal{S}}_i \cap \hat{\mathcal{S}}_j$ and $\xi \in -\hat{\mathcal{S}}_i \cap -\hat{\mathcal{S}}_j$, $i, j \in \mathcal{N} \times \mathcal{N}$, $i \neq j$, and all $\xi \in \hat{\mathcal{S}}_i \cap -\hat{\mathcal{S}}_j$, $i, j \in \mathcal{N} \times \mathcal{N}$.

Proof A proof is given in Sect. A.4.2. □

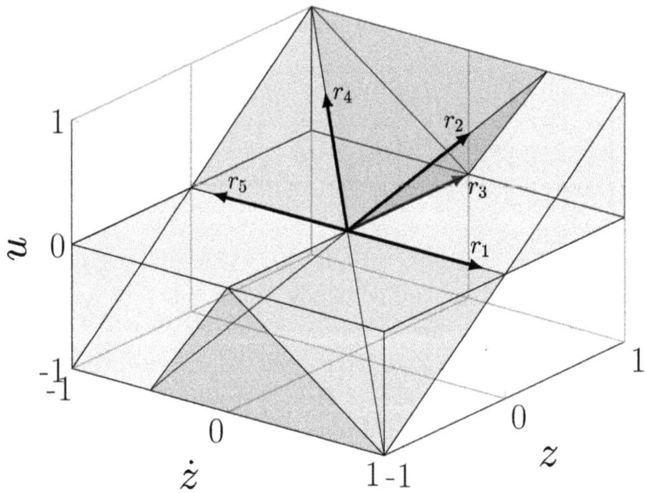

Fig. 5.5 Example of a simplicial partitioning of the set $\mathcal{F}$. The regions $\hat{\mathcal{S}}$ and $-\hat{\mathcal{S}}$ are partitioned into three simplicial cones $\hat{\mathcal{S}}_n, n = \{1, 2, 3\}$. The unit vectors $r_i, i = \{1, \ldots, 5\}$ are the generators of the partitioning, and the simplicial cones are given by $\hat{\mathcal{S}}_1 = \text{pos}\{r_1, r_2, r_3\}$ (grey), $\hat{\mathcal{S}}_2 = \text{pos}\{r_2, r_3, r_4\}$ (blue), and $\hat{\mathcal{S}}_3 = \text{pos}\{r_3, r_4, r_5\}$ (green)

Example 5.1 In order to emphasize the subtlety in Proposition 5.2 of making no distinction between the generators $[0, 0, 1]^\top$ and $[0, 0, -1]^\top$ for constructing the continuity matrices, consider Fig. 5.5 which shows an example of a simplicial partitioning of the three-dimensional region $\hat{\mathcal{S}}$ (respectively $-\hat{\mathcal{S}}$) into three simplicial cones.

The generating set $\mathcal{R}$ of the partitioning is given by $\mathcal{R} = \{r_1, r_2, r_3, r_4, r_5\}$, with generators

$$r_1 = \begin{bmatrix} 0 \\ 0 \\ 1 \end{bmatrix}, r_2 = \frac{1}{\sqrt{3}} \begin{bmatrix} 1 \\ 1 \\ 1 \end{bmatrix}, r_3 = \begin{bmatrix} 1 \\ 0 \\ 0 \end{bmatrix}, r_4 = \frac{1}{\sqrt{3}} \begin{bmatrix} 1 \\ 1 \\ -1 \end{bmatrix}, \quad \text{and} \quad r_5 = \begin{bmatrix} 0 \\ 0 \\ -1 \end{bmatrix}.$$

The reduced set $\bar{\mathcal{R}}$ is given by $\bar{\mathcal{R}} = \mathcal{R} \setminus \{[0, 0, -1]^\top\} = \{r_1, r_2, r_3, r_4\}$. The simplicial cones $\hat{\mathcal{S}}_n, n = \{1, 2, 3\}$ are formed by $\hat{\mathcal{S}}_1 = \text{pos}\{r_1, r_2, r_3\}$, $\hat{\mathcal{S}}_2 = \text{pos}\{r_2, r_3, r_4\}$, and $\hat{\mathcal{S}}_3 = \text{pos}\{r_3, r_4, r_5\}$, see Fig. 5.5. The selection matrices E_n are then constructed as

$$E_1 = \begin{bmatrix} 1\,0\,0 \\ 0\,1\,0 \\ 0\,0\,1 \\ 0\,0\,0 \end{bmatrix}, \quad E_2 = \begin{bmatrix} 0\,0\,0 \\ 1\,0\,0 \\ 0\,1\,0 \\ 0\,0\,1 \end{bmatrix}, \quad \text{and} \quad E_3 = \begin{bmatrix} 0\,0\,1 \\ 0\,0\,0 \\ 1\,0\,0 \\ 0\,1\,0 \end{bmatrix},$$

and the continuity matrices as in (5.20) follow as

$$\hat{F}_1 = \begin{bmatrix} 0 & -1 & 1 \\ 0 & \sqrt{3} & 0 \\ 1 & -1 & 0 \\ 0 & 0 & 0 \end{bmatrix}, \quad \hat{F}_2 = \begin{bmatrix} 0 & 0 & 0 \\ 0 & \frac{\sqrt{3}}{2} & \frac{\sqrt{3}}{2} \\ 1 & -1 & 0 \\ 0 & \frac{\sqrt{3}}{2} & -\frac{\sqrt{3}}{2} \end{bmatrix}, \quad \text{and} \quad \hat{F}_3 = \begin{bmatrix} 0 & 1 & 1 \\ 0 & 0 & 0 \\ 1 & -1 & 0 \\ 0 & \sqrt{3} & 0 \end{bmatrix}.$$

Consider the point $\xi = [z, u, \dot{z}]^\top = [2, 1, 1]^\top$, which is located on the boundary between $\hat{\mathcal{S}}_1$ and $\hat{\mathcal{S}}_2$. For this point one finds $\hat{F}_1\xi = \hat{F}_2\xi = [0, \sqrt{3}, 1, 0]^\top$. For the point $\xi = [2, 1, -1]^\top$ located on the boundary between $\hat{\mathcal{S}}_2$ and $\hat{\mathcal{S}}_3$ one finds $\hat{F}_2\xi = \hat{F}_3\xi = [0, 0, 1, \sqrt{3}]^\top$. For $\xi = [0, 0, -1]^\top$, which is on the boundary between $\hat{\mathcal{S}}_3$ and $-\hat{\mathcal{S}}_1$, one finds $\hat{F}_1\xi = \hat{F}_3\xi = [-1, 0, 0]^\top$. If the vector r_5 would have been taken into account for constructing the continuity matrices $\hat{F}_i$, $i = \{1, 2, 3\}$, then for $\xi = [0, 0, -1]^\top$ one finds $\hat{F}_1\xi = [1, 0, 0, 0, 0]^\top$ and $\hat{F}_3\xi = [0, 0, 0, 0, -1]^\top$. Hence, $\hat{F}_1\xi \neq \hat{F}_3\xi$ for some $\xi \in \{\hat{\mathcal{S}}_1 \cap -\hat{\mathcal{S}}_3\} \cup \{-\hat{\mathcal{S}}_1 \cap \hat{\mathcal{S}}_3\}$. The reason for this is that one of the common boundaries is described by both r_1 and $r_5 = -r_1$. Eliminating r_5 (or r_1) ensures that this boundary is represented uniquely, i.e., by only one vector in $\bar{\mathcal{R}}$.

Using $\xi = Ex$, with E given in (5.7), the sets $\hat{\mathcal{S}}_i$, $i \in \mathcal{N}$ in (5.13) and $\hat{\mathcal{T}}_j$, $j \in \mathcal{M}$ in (5.15) are transformed to

$$\mathcal{S}_i = \left\{x \in \mathbb{R}^n \mid S_i x \geq 0\right\}, \tag{5.21a}$$

$$\mathcal{T}_j = \left\{x \in \mathbb{R}^n \mid T_j x \geq 0 \wedge Hx = 0\right\}, \tag{5.21b}$$

where

$$S_i = \hat{S}_i E, \quad T_j = \hat{T}_j E, \quad \text{and} \quad H = hE, \tag{5.22}$$

for describing the polyhedral partitions in the state space of the closed-loop system (5.5). The continuity matrices are extended to

$$F_i = \left[(\hat{F}_i E)^\top, I_n\right]^\top, \tag{5.23}$$

where $\hat{F}_i$ is defined in (5.20), and I_n is an $n \times n$ identity matrix that is added for the purpose of ensuring full rank of F_i. Alternatively, to ensure full rank one can append $\hat{F}_i E$ with a matrix Y satisfying $\mathrm{im} Y^\top \supseteq \mathrm{im} E_\perp$ where $E_\perp$ denotes a matrix whose columns form a basis of $\ker E$. Note that $\bigcup_{i \in \mathcal{N}} \{\mathcal{S}_i \cup -\mathcal{S}_i\} = \overline{\mathcal{X}}_1 = \mathcal{X}$, and $\bigcup_{j \in \mathcal{M}} \{\mathcal{T}_j \cup -\mathcal{T}_j\} = \overline{\mathcal{X}}_2$, where $\mathcal{X} = \mathcal{X}_1 \cup \mathcal{X}_2$ follows from (5.6).

It may appear that the parametrization (5.17) with the continuity matrices as in (5.23) can introduce some conservatism in the analysis due to fixing the structure of the matrices P_i a priori. This is indeed true when the region $\mathcal{Z} := \left\{(z, u, \dot{z}) \in \mathbb{R}^3 \mid z \geq 0, u = 0\right\}$, (alternatively $-\mathcal{Z}$), i.e., the $(z, \dot{z})$-plane with $z \geq 0$ (or $z \leq 0$) is divided into only two regions. In that case, one finds for $u = 0$ and $v = (z, \dot{z}) \in \mathbb{R}^2$ that $F_i H v = F_j H v = F v$, where $\mathrm{im} H = \ker U$ with $U = [0, 1, 0]$.

Hence, in that case the PWQ function V as in (5.17) degenerates to a quadratic function in that plane, i.e., $V = v^\top H^\top P_i H v = v^\top H^\top P_j H v = v^\top F^\top \Phi F v = v^\top P v$. This follows from the fact that F_i and F_j are based on the same generating vectors in $\mathcal{Z}$. It is relatively simple to construct examples for which a piecewise quadratic function can still be constructed with this partitioning, and thus using a quadratic one introduces conservatism. A degenerate case (i.e., a partitioning of $\mathcal{Z}$ into only two regions) can be found in Example 5.1. When dividing $\mathcal{Z}$ into more than two regions, however, ensuring the continuity property through the parametrization in (5.17) is done in a non-conservative manner, as follows from the next proposition.

Proposition 5.3 *Consider a symmetric simplicial partitioning of $\mathcal{F}$ in which the region $\mathcal{Z} = \{(z, u, \dot{z}) \in \mathbb{R}^3 \mid z \geq 0, u = 0\}$ is divided into more than two regions. Then, the following statements are equivalent:*

(i) A PWQ function of the form

$$V(x) = x^\top P_i x, \quad \text{if } x \in \mathcal{S}_i \cup -\mathcal{S}_i, \quad i \in \mathcal{N}, \tag{5.24}$$

with $P_i = P_i^\top$ defined over a polyhedral partition $\mathcal{P}_\mathcal{X} = \{\mathcal{S}_1, \ldots, \mathcal{S}_N\} \cup \{-\mathcal{S}_1, \ldots, -\mathcal{S}_N\}$, with $\mathcal{S}_i$ given in (5.21a), is continuous.

(ii) There exists a symmetric matrix $\Phi \in \mathbb{R}^{r\times r}$ such that

$$P_i = F_i^\top \Phi F_i, \quad i \in \mathcal{N}, \tag{5.25}$$

with F_i given in (5.23).

(iii) There exist symmetric matrices $P_i = P_i^\top$, $i \in \mathcal{N}$ that satisfy

$$Z_{ij}^\top (P_i - P_j) Z_{ij} = 0, \tag{5.26}$$

where $\mathrm{im} Z_{ij} = \ker H_{ij}$, and H_{ij} describes the boundary between $\mathcal{S}_i$ and $\mathcal{S}_j$.

Proof See Sect. A.4.3. □

It is also worth mentioning in this regard that compared to a generic polyhedral partitioning strategy, a simplicial partitioning strategy itself does not introduce more conservatism in the analysis. Namely, since each polyhedral cone with non-empty interior can be partitioned into a finite number of simplicial cones [8, Lemma 1.40], then when defining a continuous function over a polyhedral partitioning such as discussed in [9], this same function can be defined over a simplicial partitioning as well. The converse, however, not necessarily holds true.

5.4 LMI Conditions for Stability and Performance

Equipped with the above notions and definitions, this section presents LMI conditions for stability and performance of the HIGS-based feedback system in (5.5). Additional insights on (in)feasibility of the LMIs revealing the need for partition refinements are discussed at the end of this section.

5.4.1 LMI Conditions

The next result presents LMI conditions which, if feasible, guarantee the existence of PWQ functions certifying ISS and performance in terms of the $\mathcal{L}_2$-gain and $\mathcal{H}_2$-norm of the closed-loop system in (5.5).

Theorem 5.1 *Consider the closed-loop system in (5.5) with square integrable $w \in PC$, and let a polyhedral symmetric partition $\mathcal{P}_X = \{\mathcal{S}_1, \ldots, \mathcal{S}_N\} \cup \{-\mathcal{S}_1, \ldots, -\mathcal{S}_N\}$ with $\mathcal{S}_i$ as in* (5.21a) *be given. Suppose there exist matrices U_i, $W_i \in \mathbb{S}^{3\times 3}_{\geq 0}$ for all $i \in \mathcal{N}$, and $K_j \in \mathbb{S}^{2\times 2}_{\geq 0}$, $L_j \in \mathbb{R}^q$ for all $j \in \mathcal{M}$, a matrix $\Phi \in \mathbb{S}^{q\times q}$, and constant $\gamma > 0$, such that $P_i = F_i^\top \Phi F_i$, $i \in \mathcal{N}$ satisfies the following LMI conditions:*

$$P_i - S_i^\top W_i S_i \succ 0, \quad i \in \mathcal{N}, \tag{5.27}$$

$$\begin{bmatrix} A_1^\top P_i + P_i A_1 + S_i^\top U_i S_i & P_i B & C^\top \\ B^\top P_i & -\gamma I & D^\top \\ C & D & -\gamma I \end{bmatrix} \prec 0, \quad i \in \mathcal{N}, \tag{5.28}$$

$$\begin{bmatrix} \mathcal{A}_2^\top P_j + P_j \mathcal{A}_2 + T_j^\top K_j T_j + Q_j & P_j B & C^\top \\ B^\top P_j & -\gamma I & D^\top \\ C & D & -\gamma I \end{bmatrix} \prec 0, \quad j \in \mathcal{M}, \tag{5.29}$$

in which $Q_j = L_j H + (L_j H)^\top$, and $\mathcal{A}_2 = A_2 + \mu b H$, with $b = [0\ 1]^\top$ and $\mu \in \mathbb{R}$ a fixed number, and where S_i, T_j, H, and F_i are given in (5.22) *and* (5.23). *Then, the closed-loop system in* (5.5) *is input-to-state stable (ISS) and has a finite $\mathcal{L}_2$-gain from w to q smaller than or equal to γ.*

Proof The proof can be found in Sect. A.4.4. □

Remark 5.1 The motivation for using the extended system matrix $\mathcal{A}_2 = A_2 + \mu b H$ rather than A_2 in this case, comes from numerical considerations. Since A_2 results from explicit differentiation of the algebraic constraint in gain mode, this matrix is singular. Consequently, the LMI problem in (5.27)–(5.29) may become ill-conditioned, which potentially leads to numerical issues; note that $\mathcal{A}_2$ being singular corresponds to $\mu = 0$. By appending A_2 with an additional matrix $\mu b H$, the resulting matrix $\mathcal{A}_2$ becomes non-singular, thereby potentially leading to numerically more favourable conditions; see also the related discussion in Remark 2.2.

Note that the gain mode dynamics remain unchanged, that is, $A_2x = \mathcal{A}_2x$ for all $x \in \mathcal{X}_2$. A suitable choice for μ in this case may be one that minimizes the condition number of $\mathcal{A}_2$.

Inspired by the results for generic PWL systems in [16, 26], and reset control systems in, e.g., [1, 22], the following theorem presents conditions for computing an upper-bound on the $\mathcal{H}_2$-norm of the closed-loop system in (5.5).

Theorem 5.2 *Consider the closed-loop system in* (5.5) *with* $w = 0$ *and an initial condition* $x(0) = x_0 \in \mathcal{X}$. *Let a polyhedral partition* $\mathcal{P}_\mathcal{X} = \{\mathcal{S}_1, \ldots, \mathcal{S}_N\} \cup \{-\mathcal{S}_1, \ldots, -\mathcal{S}_N\}$ *with* $\mathcal{S}_i$ *as in* (5.21a) *be given. Suppose there exist matrices* U_i, $W_i \in \mathbb{S}_{\geq 0}^{3\times 3}$ *for all* $i \in \mathcal{N}$, *and* $K_j \in \mathbb{S}_{\geq 0}^{2\times 2}$, $L_j \in \mathbb{R}^q$, *for all* $j \in \mathcal{M}$, *a matrix* $\Phi \in \mathbb{S}^{q\times q}$, *and a constant* $\gamma > 0$, *such that* $P_i = F_i^\top \Phi F_i$, $i \in \mathcal{N}$ *satisfies the following LMI conditions:*

$$P_i - S_i^\top W_i S_i \succ 0, \quad i \in \mathcal{N}, \tag{5.30}$$

$$A_1^\top P_i + P_i A_1 + S_i^\top U_i S_i + C^\top C \prec 0, \quad i \in \mathcal{N}, \tag{5.31}$$

$$\mathcal{A}_2^\top P_j + P_j \mathcal{A}_2 + T_j^\top K_j T_j + Q_j + C^\top C \prec 0, \quad j \in \mathcal{M}, \tag{5.32}$$

$$\gamma^2 - x_0^\top P_r x_0 \geq 0, \quad r \in \mathcal{J}(x_0) \tag{5.33}$$

in which $Q_j = L_j H + (L_j H)^\top$, $\mathcal{A}_2 = A_2 + \mu b H$, *with* $b = [0\ 1]^\top$ *and* $\mu \in \mathbb{R}$ *a fixed number,* S_i, T_i, H, *and* F_i *are defined in* (5.22) *and* (5.23), *and where* $\mathcal{J}(x_0) := \{i \in \mathcal{N} \mid x_0 \in \mathcal{S}_i \cup -\mathcal{S}_i\}$. *Then, for initial condition* $x_0 \in \mathcal{X}$, *the closed-loop system in* (5.5) *has an* $\mathcal{H}_2$*-norm smaller than or equal to* γ.

Proof Taking similar steps as in the proof of Theorem 5.1, it follows that conditions (5.30)–(5.32) imply that the function $V(x) = x^\top P_i x$ if $x \in \mathcal{S}_i \cup -\mathcal{S}_i$, $i \in \mathcal{N}$ satisfies $V(x(t)) - V(x(0)) \leq -\|q\|^2$. Using $V(x(t)) \geq 0$, $x(0) = x_0$, and (5.33) yields $\|q\|_2 \leq \gamma$, thereby completing the proof. □

The following result shows that, similar to the quadratic case in Chap. 4 (see Corollary 4.1), the conditions in Theorems 5.1 and 5.2 resulting from the use of a PWQ function provide stronger robust (with respect to arbitrary small state perturbations) stability and performance guarantees for the closed-loop system in (5.5).

Corollary 5.1 *If the conditions in Theorems 5.1 and 5.2 are satisfied, then the Krasovskii regularization of* (5.5) *(see (2.42) in Sect. 2.3.4) is ISS and has a finite* $\mathcal{L}_2$*-gain /* $\mathcal{H}_2$*-norm smaller than or equal to* γ.

Proof Note that by construction, in each cell $\mathcal{T}_j \subset \mathcal{S}_j$, $j \in \mathcal{M}$ with $\mathcal{M} \subset \mathcal{N}$ a "common" P_j is employed for both the integrator mode and gain mode. As such, it readily follows that for all $x \in \mathcal{T}_j \cup -\mathcal{T}_j$, $j \in \mathcal{M}$

$$x^\top \nabla P_j \left(\overline{\text{co}}(A_1 x, A_2 x) + Bw\right) \leq s(w, q, x). \tag{5.34}$$

The remaining issue to worry about is the point where $Ex = (z, u, \dot{z})^\top = 0$, since in that point all piecewise functions come together, and (5.34) needs to hold for all $j \in \mathcal{N}$. Note, however, that for $Ex = 0$ one has $A_1 x = A_2 x$. Since $x^\top \nabla P_i (A_1 x + Bw) \leq s(w, q, x)$, for all $x \in \mathcal{S}_i \cup -\mathcal{S}_i$, $i \in \mathcal{N}$, which contains the point $Ex = 0$, the result follows. □

5.4.2 On (in)Feasibility of the LMI Conditions

When the LMIs in Theorem 5.1 or Theorem 5.2 turn out to be infeasible for a certain partitioning, this partitioning can be iteratively refined to increase the possibility of finding a solution, if existing. In this regard, it might be useful to know in advance under which conditions and for which partitioning the LMIs will not be feasible, such that these cases do not have to be considered. In general, this is difficult to assess since the LMI conditions provide only sufficient conditions for stability. However, some insights regarding (in)feasibility of the LMIs can be obtained when one or both submodes of the HIGS-controlled system are unstable. This is formalized in the following proposition.

Proposition 5.4 *Let a simplicial partitioning $\mathcal{P}_\mathcal{F}$ be given. The LMI conditions in Theorems 5.1 and 5.2 do not admit a feasible solution for that specific partitioning $\mathcal{P}_\mathcal{F}$ if one (or both) of the following conditions holds:*

C1. *The matrix A_1 given in* (5.8) *has eigenvalues $\lambda \in \mathbb{C}$ with $\text{Re}\{\lambda\} > 0$ that satisfy for some $i \in \mathcal{N}$*

$$\left\{ \begin{bmatrix} \text{Re}\{\lambda\} \\ \omega_h \\ \text{Re}\{\lambda\}^2 - \text{Im}\{\lambda\}^2 \end{bmatrix}, \begin{bmatrix} \text{Im}\{\lambda\} \\ 0 \\ 2\text{Re}\{\lambda\}\,\text{Im}\{\lambda\} \end{bmatrix} \right\} \subset \hat{\mathcal{S}}_i \cup -\hat{\mathcal{S}}_i, \tag{5.35}$$

with $\hat{\mathcal{S}}_i$ defined in (5.13).

C2. *The matrix $A_g - k_h B_g C_g$ with A_g, B_g, C_g given in* (5.1) *has eigenvalues $\lambda \in \mathbb{C}$ with* $\text{Re}\{\lambda\} > 0$ *that satisfy for some $j \in \mathcal{M}$*

$$\left\{ \begin{bmatrix} 1 \\ k_h \\ \text{Re}\{\lambda\} \end{bmatrix}, \begin{bmatrix} 0 \\ 0 \\ \text{Im}\{\lambda\} \end{bmatrix} \right\} \subset \hat{\mathcal{T}}_j \cup -\hat{\mathcal{T}}_j, \tag{5.36}$$

with $\hat{\mathcal{T}}_j$ defined in (5.15).

Proof The proof is given in Sect. A.4.5. □

The result in Proposition 5.4 shows the benefits of considering a partitioning in the full $(z, u, \dot{z})$-space in contrast to, e.g., a partitioning in the (z, u)-space. Namely, when (5.35) and/or (5.36) are true, one can refine the partitioning to render these conditions false, whereas for a simpler partitioning and refinement strategy this may not be possible due to limited flexibility. For example, in case the gain mode region is not partitioned (see for instance the partitioning strategy in [29]), one has $\hat{\mathcal{T}} \cup -\hat{\mathcal{T}} = \mathcal{F}_2$ such that (5.36) is always satisfied for any eigenvalue λ of the matrix $A_g - k_h B_g C_g$ that satisfies $0 \le \mathrm{Re}\{\lambda\} \le \frac{\omega_h}{k_h}$ and $\mathrm{Im}\{\lambda\} \in \mathbb{R}$. When the gain mode region is partitioned, this is not necessarily true. With these conditions one may test upfront if a simpler, computationally less demanding partitioning could potentially work, or a more involved three-dimensional partitioning should be considered. Moreover, these conditions give direct insights into some of the necessary properties that a partition refinement should admit. The application and relevance of Proposition 5.4 will be illustrated in Sect. 5.5.

Closer inspection of the results in Proposition 5.4 reveals that, under some simple algebraic conditions on the eigenvalues related to the system matrices in integrator mode and gain mode, there exists no partitioning for which the LMIs are feasible. These conditions are summarized in the following corollary.

Corollary 5.2 *There is no simplicial symmetric partitioning $\mathcal{P}_{\mathcal{F}}$ of $\mathcal{F}$ such that the LMI conditions in Theorems 5.1 and 5.2 admit a feasible solution, if one or both of the following conditions hold:*

(i) there is a real eigenvalue λ of A_1 that satisfies $\lambda > \frac{\omega_h}{k_h}$;
(ii) there is a real eigenvalue λ of $A_g - k_h B_g C_g$ that satisfies $0 \le \lambda < \frac{\omega_h}{k_h}$.

Proof Consider the vectors

$$v_1 = \begin{bmatrix} \mathrm{Re}\{\lambda\} \\ \omega_h \\ \mathrm{Re}\{\lambda\}^2 - \mathrm{Im}\{\lambda\}^2 \end{bmatrix}, \quad \text{and} \quad v_2 = \begin{bmatrix} 1 \\ k_h \\ \mathrm{Re}\{\lambda\} \end{bmatrix}. \tag{5.37}$$

Suppose (i) holds. Since $\mathrm{Im}\{\lambda\} = 0$ by assumption, it follows that $v_1 \in \mathcal{F}_1$. As $\hat{\mathcal{S}}_i \cup -\hat{\mathcal{S}}_i \subset \mathcal{F}_1$ for all $i \in \mathcal{N}$, there always exists a region $\hat{\mathcal{S}}_i \cup -\hat{\mathcal{S}}_i$ such that $v_1 \in \hat{\mathcal{S}}_i \cup -\hat{\mathcal{S}}_i$. Similarly, assume (ii) holds such that $v_2 \in \mathcal{F}_2$. Since $\mathrm{Im}\{\lambda\} = 0$ and $\hat{\mathcal{T}}_j \cup -\hat{\mathcal{T}}_j \subset \mathcal{F}_2$ for all $j \in \mathcal{M}$, there exists a region $\hat{\mathcal{T}}_j \cup -\hat{\mathcal{T}}_j$ such that $v_2 \in \hat{\mathcal{T}}_j \cup -\hat{\mathcal{T}}_j$. Consequently, condition (5.35) and/or (5.36) are satisfied for all partitions, and thus the LMIs are not feasible. □

Interestingly, this geometric result can directly be interpreted in terms of unstable closed-loop system behaviour. Indeed, the vectors v_1 and v_2 in (5.37) define a possible direction along which the $(z, u, \dot{z})$-trajectory can evolve in the regions $\mathcal{F}_1$ or $\mathcal{F}_2$. When condition (i) in Corollary 5.2 is satisfied, the z-trajectory evolves at a rate of at least $1/k_h$ times faster than the u-trajectory, i.e., $0 \le du/dz < k_h$. Consequently, the sector boundary $u = k_h z$ cannot be reached such that switching, as a potentially stabilizing mechanism, is not initiated. A similar interpretation is given for the gain mode. When condition (ii) is satisfied, the $\dot{z}$-trajectory evolves at least ω_h/k_h times slower than the z-trajectory. As such, $0 \leq d\dot{z}/dz < \omega_h/k_h$ and the switching boundaries $\omega_h z = k_h \dot{z}$ or $z = 0$ cannot be reached. Note that a condition on the magnitude of the eigenvalues is thus a consequence of the fact that trajectories of HIGS evolve in a sector. The above observations can be translated into *necessary* conditions for input-to-state stability as formalized in the following theorem.

Theorem 5.3 *A necessary condition for the closed-loop system in* (5.5) *to be input-to-state stable is that none of the conditions in Corollary 5.2 hold true.*

Proof If the conditions in Corollary 5.2 are satisfied, then one (or both) of the sub-modes is unstable, and according to Corollary 5.2 (see also the proof of Proposition 5.4) it follows that the eigenvector $v \in \mathbb{R}^m$ corresponding to the unstable real eigenvalue satisfies $Ev = \alpha_1 v_1 \in \mathcal{F}_1$ or $E\Theta v = \alpha_2 v_2 \in \mathcal{F}_2$, with v_1, v_2 given in (5.37), $\alpha_1, \alpha_2 \in \mathbb{R}$, and $\Theta = [I, -k_h C_g^\top]^\top$. When starting in the integrator mode with initial condition $x(0) = v$ and $w = 0$, the corresponding trajectory evolves according to $x(t) = e^{\lambda t} v$. For the integrator mode one finds for all $t \in \mathbb{R}_{\geq 0}$

$$(z(t), u(t), \dot{z}(t))^\top = e^{\lambda t} Ev \in \mathcal{F}_1. \tag{5.38}$$

Similarly, starting with initial condition $x(0) = \Theta v$ in gain mode yields $x(t) = e^{\lambda t}\Theta v$ such that for $t \in \mathbb{R}_{\geq 0}$

$$(z(t), u(t), \dot{z}(t))^\top = e^{\lambda t} E\Theta v \in \mathcal{F}_2. \tag{5.39}$$

The $(z, u, \dot{z})$-trajectory moves along the line spanned by $Ev \in \mathcal{F}_1$ or $E\Theta v \in \mathcal{F}_2$. The system cannot switch: it remains in the unstable linear mode and thus there is a trajectory that diverges exponentially. This directly violates the property of an ISS system that for zero inputs, the origin is asymptotically stable for any initial condition $x(0) \in \mathcal{X}$. □

Remark 5.2 For planar HIGS-controlled systems, being the (well-posed) interconnection of HIGS with a one-dimensional linear plant, the conditions in Theorem 5.3 can be shown to be necessary and sufficient, see also a related result for reset systems in [22, Theorem 3] and linear complementary systems in [6, Theorem III.3].

5.5 Numerical Examples

In this section, applicability of the time-domain tools for stability and performance analysis is demonstrated by means of two numerical examples. In the first example, the potential for reducing conservatism in the analysis when using PWQ functions is illustrated for the mass–spring–damper system discussed previously in Chap. 4. In the second example, stability and performance of a HIGS-based control design for a fourth-order motion system is studied in detail. For this example, the relevance of Proposition 5.4 is highlighted in particular.

5.5.1 Mass–Spring–Damper System Revisited

Recall the example presented earlier in Sect. 4.3, in which a mass–spring–damper system, given by

$$P(s) = \frac{1}{ms^2 + bs + k}, \tag{5.40}$$

with $m = 1$ kg, $b = 0.0564$ Ns/m, and $k = 1$ N/m, is placed in feedback with the series interconnection of HIGS and an LTI filter, the latter given by

$$C(s) = k_p \left(\frac{s + \omega_z}{s + \omega_p} \right), \tag{5.41}$$

where $k_p = 1.4$ N/m, $\omega_z = 5$ rad/s, and $\omega_p = 6.95$ rad/s. Note that this example serves the purpose of demonstrating the potential benefits when using a state space partition in a stability and performance analysis, and is not designed to achieve the best performance in $\mathcal{L}_2$ or $\mathcal{H}_2$ sense.

Stability Analysis

For assessing ISS of the closed-loop system, the LMI conditions in Theorem 5.1 are solved by means of the MATLAB® toolbox `YALMIP` [18] together with the external solver `MOSEK` [3] for different values of ω_h and k_h, and using different partitioning strategies. In particular, a comparison is made between (i) using no partitioning (leading to a single quadratic Lyapunov function as in Chap. 4), (ii) a two-dimensional partitioning in which the $[0, k_h]$-sector in the (z, u)-space is iteratively partitioned into a maximum number of 1200 smaller sub-sectors, and (iii) a three-dimensional partitioning in which the $(z, u, \dot{z})$-space is iteratively partitioned into at most 1200 three-dimensional subregions. The results are presented in Fig. 5.6.

From Fig. 5.6a it can be seen that, compared to the simulation-based stable region (grey) obtained from applying a unit step input to the system, a considerable degree of conservatism remains in concluding on ISS of the closed-loop system when using an LMI-based approach without partitioning (black dots). This is a result from considering a common (smooth) quadratic ISS-Lyapunov function. In Fig. 5.6b, it is shown that a two-dimensional partitioning of the $[0, k_h]$-sector (see also the strategy discussed in [29]) is able to partly alleviate this problem, resulting in a significantly larger set of parameters for which ISS can be guaranteed. For parameter pairs close to the edge of the stable region, however, also this approach is consistently unable to provide stability guarantees. Finally, Fig. 5.6c illustrates the potential of a three-dimensional partitioning in terms of reducing conservatism. The resulting range of

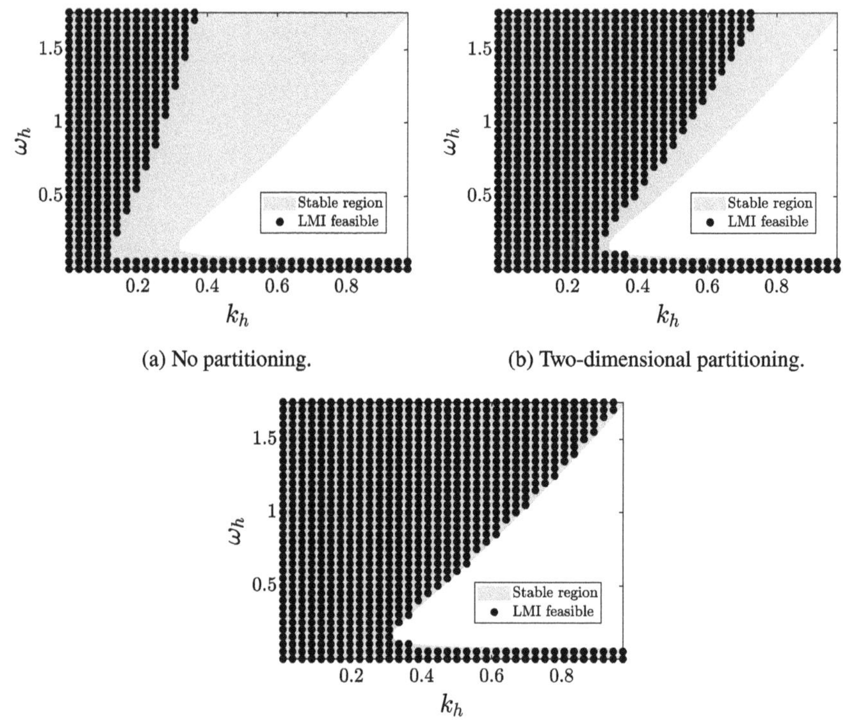

(a) No partitioning.

(b) Two-dimensional partitioning.

(c) Three-dimensional partitioning.

Fig. 5.6 Comparison between solving the LMI conditions in Theorem 5.1 using different partitioning strategies for different parameter values of ω_h and k_h. See also Fig. 4.5 for a similar analysis but on the basis of frequency-domain conditions

(k_h, ω_h)-values for which closed-loop stability can be concluded on the basis of the LMI conditions closely coincides with the stable region found from time-series simulations. The results seem to closely resemble what would be expected from a necessary condition for closed-loop stability of the nonlinear feedback system.

Performance Analysis: $\mathcal{L}_2$-gain

An upper-bound on the $\mathcal{L}_2$-gain of the closed-loop system from w to y (recall Fig. 4.4) with $k_h = 0.23$ and for various values of ω_h is computed by minimizing γ in Theorem 5.1 under the LMI constraints (5.27)–(5.29), and with different partitioning strategies. The results from solving the LMIs are shown in Fig. 5.7.

In addition, an estimate obtained from simulated experiments is provided in this figure. For this purpose, a truncated sinusoid, defined by

$$w(t) = \begin{cases} \sin(\omega t) & \text{for all } t \in [0, T], \\ 0 & \text{otherwise}, \end{cases} \tag{5.42}$$

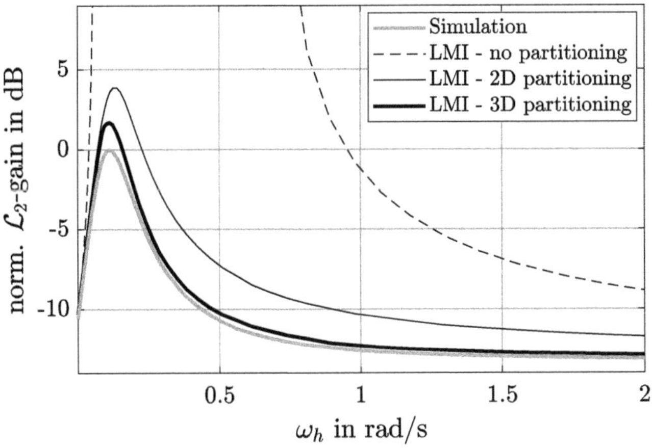

Fig. 5.7 Estimated $\mathcal{L}_2$-gain (normalized) obtained by means of (i) Theorem 5.1 with different partitioning strategies, and (ii) time-domain simulations

with $T = 100 \cdot 2\pi/\omega$ and $\omega > 0$ is applied to the closed-loop system at various frequencies. The simulation time is set to $t_s = 10T$ seconds to ensure the response has sufficiently settled. A lower-bound on the $\mathcal{L}_2$-gain is then estimated by taking the maximum ratio of the $\mathcal{L}_2$-norms of the input w and output $q = y$.

The result in Fig. 5.7 clearly illustrates the benefits of a state space partition as the estimates become significantly tighter. Note that for all values of ω_h a small discrepancy between the estimates resulting from the LMIs with a three-dimensional partitioning and the time-domain simulations remains. This may be caused by (i) numerical artefacts, (ii) remaining conservatism in the analysis due to a finite partitioning, and (iii) the chosen class of input signals, which may only provide a lower-bound for the true $\mathcal{L}_2$-gain, that is, $\gamma_{\text{sim}} \leq \gamma_{\text{true}} \leq \gamma_{\text{LMI}}$.

Performance Analysis: $\mathcal{H}_2$-norm

An upper-bound on the $\mathcal{H}_2$-norm of the closed-loop system with $w = 0$, and when subject to an initial condition $x(0)$ that corresponds to an initial state $y(0) = 1, \dot{y}(0) = 0$ of the simply supported mass, and initial controller states set to zero, is computed for $k_h = 0.23$ and various ω_h. Specifically, γ in Theorem 5.2 is minimized under the LMI constraints (5.27)–(5.29), and with different partitionings. The results are shown in Fig. 5.8, along with the true $\mathcal{H}_2$-norm obtained from time-series simulations.

Again, a clear benefit can be seen from using a three-dimensional partitioning strategy as the estimate of the $\mathcal{H}_2$-norm closely corresponds to the value obtained by simulations, emphasizing the accuracy of the LMIs.

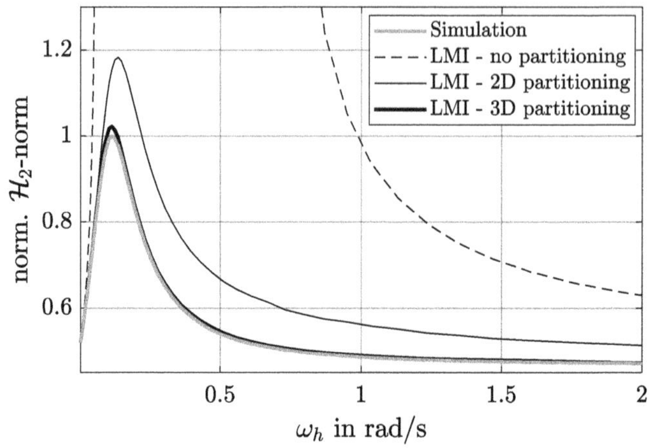

Fig. 5.8 Estimated $\mathcal{H}_2$-norm (normalized) obtained by means of Theorem 5.2 with different partitioning strategies, and the true $\mathcal{H}_2$-norm obtained through time-domain simulations

5.5.2 Fourth-Order Motion System

For the second example, consider a typical motion system represented by the fourth-order transfer function

$$P(s) = \frac{1}{ms^2} - \frac{1}{m(s^2 + 2\beta_0\omega_0 s + \omega_0^2)}, \tag{5.43}$$

which is a combination of rigid-body and non-rigid-body dynamics, and where $m = 18$ kg, $\omega_0 = 1200 \cdot 2\pi$ rad/s, and $\beta_0 = 0.03$. This system is placed in feedback with a controller C_{fb} being the series interconnection of a linear PID-filter $C_{pid}(s)$, a second-order low-pass filter $C_{lp}(s)$, and a notch filter $C_n(s)$, which are constructed as

$$C_{pid}(s) = k_p\left(1 + \frac{\omega_i}{s} + \frac{s}{\omega_d}\right), \tag{5.44}$$

$$C_{lp}(s) = \frac{\omega_{lp}^2}{s^2 + 2\beta\omega_{lp}s + \omega_{lp}^2}, \tag{5.45}$$

$$C_n(s) = \frac{\omega_p^2}{\omega_z^2} \cdot \frac{s^2 + 2\beta_z\omega_z s + \omega_z^2}{s^2 + 2\beta_p\omega_p s + \omega_p^2}. \tag{5.46}$$

To potentially improve transient and steady-state performance, HIGS is added in series with an additional output filter $C_z(s) = \frac{s+\omega_c}{\omega_c}$ with $\omega_c = \frac{\omega_h}{k_h}|1 + \frac{4j}{\pi}|$, which

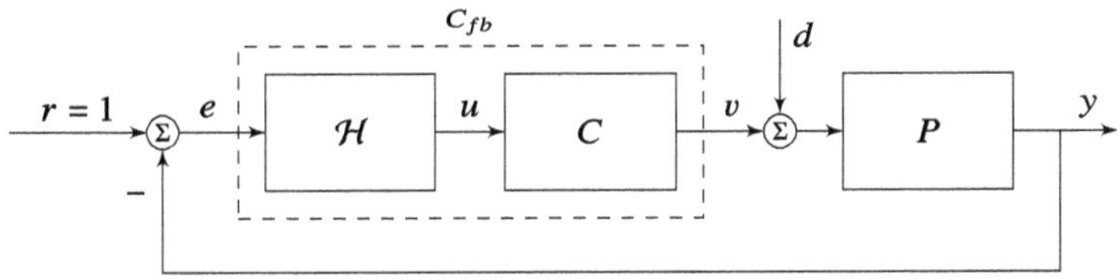

Fig. 5.9 Schematic representation of the closed-loop system

effectively renders the combination to behave like a gain with phase lead (see also Chap. 3). The linear portion of the controller is then given by

$$C(s) := C_z(s)C_{lp}(s)C_{pid}(s)C_n(s). \tag{5.47}$$

The system is subject to a square integrable input disturbance d, and a unit step input $r(t) = 1$ for $t \in \mathbb{R}_{\geq 0}$, $r(t) = 0$ otherwise. The specific closed-loop configuration is depicted in Fig. 5.9, and satisfies Assumption 2.5.

The HIGS-based controller, i.e., the combination of (5.47) and HIGS is designed by means of a loop-shaping procedure [10, 27, 28] in which HIGS is replaced by its describing function $\mathcal{D}(j\omega)$ (see Sect. 2.2.3). The aim is to maximize the bandwidth, i.e., the frequency at which the quasi-linear open-loop frequency-response function $L(j\omega) := P(j\omega)C(j\omega)\mathcal{D}(j\omega)$ crosses 0 dB for the first time, while satisfying a peaking constraint of 8 dB on the quasi-linear sensitivity function $S(j\omega) := (1 + L(j\omega))^{-1}$. The phase benefits associated with HIGS may allow for an increase in bandwidth as compared to an LTI design. The loop-shaping design procedure exploited here will be discussed in detail in Chap. 7.

On the basis of this loop-shaping approach, the following parameter values are obtained: $k_p = 4.816 \cdot 10^7$ N/m, $\omega_i = 150 \cdot 2\pi$ rad/s, $\omega_d = 240 \cdot 2\pi$ rad/s, $\omega_{lp} = 1400 \cdot 2\pi$ rad/s, $\beta = 0.7$, $\omega_z = 1160 \cdot 2\pi$ rad/s, $\beta_z = 0.01$, $\omega_p = 850 \cdot 2\pi$, $\beta_p = 0.3$, $\omega_h = 300 \cdot 2\pi$ rad/s, $k_h = 1$, and the bandwidth is 305 Hz. A Nyquist-like plot of the describing function $L(j\omega)$, together with the open-loop characteristics of the linear gain- and integrator mode is shown in Fig. 5.10.

Stability Analysis

For verifying input-to-state stability of the resulting design, the conditions in Theorem 5.1 are checked. However, before doing so, note from Fig. 5.10 that the individual LTI integrator mode subsystem and gain mode subsystem are unstable. In fact, it turns out that one of the complex eigenvalue pairs of the system matrix in gain mode is given by $\lambda = 52.2 \pm 5102j$, and satisfies

$$0 \leq \mathrm{Re}\{\lambda\} = 52.2 \leq \omega_h/k_h = 300 \cdot 2\pi.$$

According to Proposition 5.4 the closed-loop system does not admit a common quadratic Lyapunov function, but also there does not exist a PWQ Lyapunov function with *any* two-dimensional partitioning, i.e., a partitioning in which the gain mode

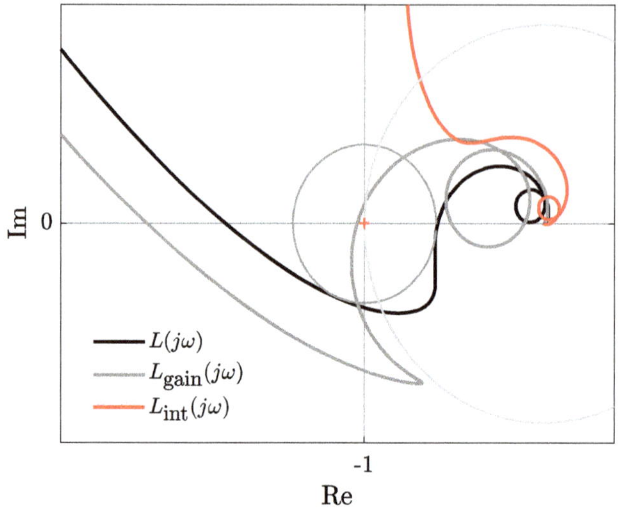

Fig. 5.10 Nyquist plot of the quasi-linear open-loop frequency-response function $L(j\omega)$ (black) together with the associated linear gain mode characteristics (grey) and linear integrator mode characteristics (red)

region is not partitioned [29]. Therefore, when existing, a PWQ Lyapunov function can only be found with a three-dimensional partitioning (or more). A feasible solution to the LMIs in Theorem 5.1 is found with a division of the region $\hat{\mathcal{S}}$ into 28 simplicial cones.

The closed-loop system is simulated with $r = d = 0$, and random initial conditions for the plant and LTI part of the controller, whereas the initial state of HIGS is set to $x_h(0) = 0$. The corresponding evolution of the Lyapunov function along trajectories of the system is shown in Fig. 5.11. As can be seen for a specific solution, this function is positive, continuous, and monotonically decreasing.

Performance Analysis: $\mathcal{L}_2$-gain

Performance in terms of the $\mathcal{L}_2$-gain is studied using Theorem 5.1. Similar as before, γ is minimized under the LMI constraints (5.27)–(5.29). The result for an increasing number of subdivisions N of the three-dimensional region $\mathcal{F}$ is given in Table 5.1, together with an estimate for the lower-bound obtained from simulations with the truncated sinusoid in (5.42) as an input disturbance d at different frequencies ω. The maximum for this lower-bound was found at an input frequency of $\omega = 229.78 \cdot 2\pi$ rad/s.

The results in Table 5.1 show the benefit of a partition refinement for improving the accuracy of the LMI approximation. Remark that a gap between the LMI approximation and the simulation result remains due to the fact that the chosen input signal may not necessarily reflect the worst-case disturbance for the closed-loop system, numerical artefacts, and remaining conservatism.

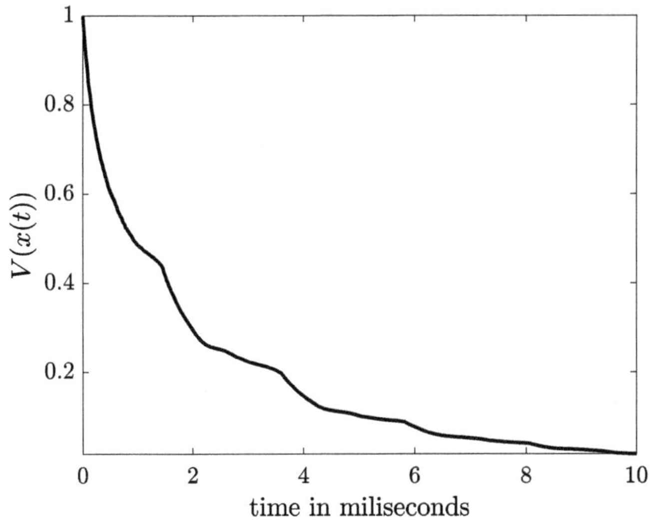

Fig. 5.11 Time-evolution of the PWQ Lyapunov function (normalized) along the trajectories of the system

Table 5.1 $\mathcal{L}_2$-gain estimation

	$N = 70$	$N = 100$	$N = 250$	$N = 700$	$N = 1200$
$\gamma_{\text{sim}}(\cdot 10^7)$	1.24	1.24	1.24	1.24	1.24
$\gamma_{\text{LMI}}(\cdot 10^7)$	4.59	3.17	2.25	1.93	1.73
Difference	270%	155%	81%	56%	39%

Performance Analysis: $\mathcal{H}_2$-norm

For a final exposition of the tools presented in this chapter, an estimate for the $\mathcal{H}_2$-norm of the error response when the system is subject to a unit step input is computed, i.e., $q = e$ in this setting. For this purpose, a state-transformation is applied as follows. When being in a state of equilibrium with $(z, u) = (0, 0)$ the system must be in integrator mode. The equilibrium state is determined as $x^* = -A_1^{-1}B$, where invertibility of A_1 follows from the absence of zero eigenvalues. Remark that due to the presence of an integrator in C_{fb}, the only possibility for the tracking error $e = r - y$ is to be zero when the system is in equilibrium. Consider the transformation $\xi := x - x^*$ for which simple calculations show that the dynamics of the transformed system read

$$\dot{\xi} = A_i\xi, \quad \text{when} \quad \xi \in \mathcal{X}_i, \quad i = \{1, 2\},$$

with initial condition $\xi(0) = -A_1^{-1}B$. The LMI conditions in Theorem 5.2 are solved for this transformed autonomous system with an increasing number of subdivisions N of $\mathcal{F}$. The results are provided in Table 5.2, along with the true $\mathcal{H}_2$-norm obtained from time-series simulations. Note that as an alternative for computing the $\mathcal{H}_2$-norm,

Table 5.2 $\mathcal{H}_2$-norm estimation

	$N = 70$	$N = 100$	$N = 250$	$N = 700$	$N = 1200$
γ_{sim}	0.041	0.041	0.041	0.041	0.041
γ_{LMI}	0.101	0.075	0.061	0.058	0.056
Difference	146%	83%	49%	42%	36%

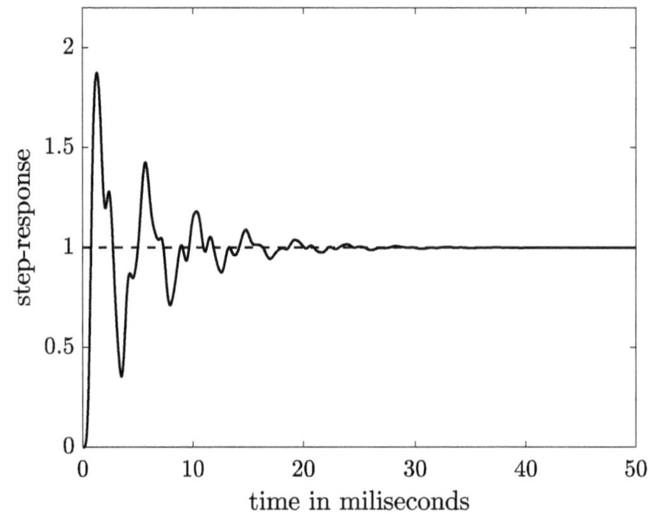

Fig. 5.12 Step response of the closed-loop system

one could have approximated the unit step input by an LTI input filter $W_r(s) = \frac{1}{s+\epsilon}$ with small $\epsilon > 0$ and initial state $x_r(0) = 1$, see also [1] for such an approach. A step response of the closed-loop system is shown in Fig. 5.12.

Although refinement of the partitioning shows a significant benefit for improving an estimation of the $\mathcal{H}_2$-norm, a discrepancy of 36% still remains. This difference may partly be due to remaining conservatism in the analysis. To potentially refine the results, one can use a different partitioning or search for, e.g., alternative (piecewise) polynomial storage functions [7, 11]. It is furthermore expected that the increased computational complexity due to an increased system dimension may lead to numerical problems, and, as such, contributes to this gap. Also note that for the previous (numerically simpler) example, the conditions lead to a very tight estimate of the true $\mathcal{H}_2$-norm. Moreover, one should keep in mind that for the current example a feasible solution to the LMIs does not even exist at all when using either no partitioning or a two-dimensional partitioning, making the conditions still relevant.

Remark 5.3 To improve numerical conditioning of the LMIs for this example, a Gramian-based balancing of the closed-loop state space realization has been performed [17, 20]. To this end, a similarity transformation $x_b = Tx$ is computed on the basis of the state space description in integrator mode, which is subsequently applied to both the integrator mode and gain mode state space models for constructing a balanced closed-loop model. In addition, input/output scaling has been used. Specifically, the input matrix B in (5.5) has been multiplied with a factor of 10^4 for lifting the $\mathcal{L}_2$-gain, thereby leading to more favourable numerical conditions. Output scaling has been applied to $\dot{z}$ such that z, u and $\dot{z}$ are approximately of the same order of magnitude.

5.6 Summary

This chapter presents time-domain conditions for stability and performance of HIGS-based control systems. For potentially reducing conservatism in the analysis and obtain tight performance estimates, the conditions exploit flexible piecewise quadratic functions defined over a tailored and a numerically robust state space partition. The latter is constructed by dividing the full region that determines mode switching of HIGS into smaller, conic regions, each to which a quadratic function is assigned. On the basis of this partition with piecewise quadratic functions, conditions for stability and performance are formulated as a convex optimization problem in terms of numerically verifiable LMIs. For achieving more insights regarding (in)feasibility of the LMIs, sufficient conditions are given under which the LMIs cannot be solved, guiding partition refinements which are important for obtaining accurate stability and performance estimates. The effectiveness of the presented tools (along with some shortcomings) has been demonstrated on numerical control examples.

Despite the $\mathcal{L}_2$-gain considered in this chapter being a valuable measure for (steady-state) time-domain performance of HIGS-based control systems, it may be conservative as it applies to a broad class of input signals and may not always reflect the actual performance objective. However, one often has more knowledge about external signals acting on the system. Therefore, it is of interest to seek for alternative ways in quantifying (steady-state) time-domain performance of systems containing HIGS. In the next chapter, the notion of *convergent systems* is adopted, which opens up the possibility for accurately characterizing steady-state performance in both time- and frequency-domain.

References

1. Aangenent, W.H.T.M., Witvoet, G., Heemels, W.P.M.H., Molengraft, M.J.G., Steinbuch, M.: Performance analysis of reset control systems. Int. J. Robust Nonlinear Control **20**, 1213–1233 (2010)
2. Ambrosino, R., Garone, E.: Piecewise quadratic Lyapunov functions over conical partitions for robust stability analysis. Int. J. Robust Nonlinear Control **25**(14), 2348–2361 (2015)
3. Andersen, E., Roos, C., Terlaky, T.: On implementing a primal-dual interior-point method for conic quadratic programming. Math. Program. **95**, 249–277 (2003)

4. Andersen, S., Giesl, P., Hafstein, S.: Common lyapunov functions for switched linear systems: Linear programming-based approach. IEEE Control Syst. Lett. **7**, 901–906 (2023)
5. Branicky, M.: Multiple Lyapunov functions and other analysis tools for switched and hybrid systems. IEEE Trans. Autom. Control **43**, 475–482 (1998)
6. Camlibel, M., Heemels, W., Schumacher, J.: Stability and controllability of planar bimodal linear complementarity systems. In: 42nd IEEE International Conference on Decision and Control, vol. 2, pp. 1651–1656 (2003)
7. Chesi, G.: LMI techniques for optimization over polynomials in control: a survey. IEEE Trans. Autom. Control **55**(11), 2500–2510 (2010)
8. Concini, C.D., Procesi, C.: Topics in Hyperplane Arrangements, Polytopes and Box-Splines. Springer, New York (2010)
9. Deenen, D.A., Sharif, B., van den Eijnden, S., Nijmeijer, H., Heemels, M., Heertjes, M.: Projection-based integrators for improved motion control: formalization, well-posedness and stability of hybrid integrator-gain systems. Automatica **133**, 109830 (2021)
10. Franklin, G., Powell, J., Emami-Naeini, A.: Feedback Control of Dynamic Systems, 5th edn. Prentice Hall (2005)
11. Garulli, A., Masi, A., Valmorbida, G., Zaccarian, L.: Global stability and finite L2m-gain of saturated uncertain systems via piecewise polynomial Lyapunov functions. IEEE Trans. Autom. Control **58**(1), 242–246 (2013)
12. Giesl, P., Hafstein, S.: Implementation of a fan-like triangulation for the CPA method to compute Lyapunov functions. In: 2014 American Control Conference, pp. 2989–2994 (2014)
13. Hassibi, A., Boyd, S.: Quadratic stabilization and control of piecewise-linear systems. In: Proceedings of the 1998 American Control Conference, vol. 6, pp. 3659–3664 (1998)
14. Iervolino, R., Vasca, F., Iannelli, L.: Cone-copositive piecewise quadratic Lyapunov functions for conewise linear systems. IEEE Trans. Autom. Control **60**(11), 3077–3082 (2015)
15. Johansson, M.: Piecewise linear control systems - a computational approach. Lecture Notes in Control and Information Sciences, vol. 284 (2002)
16. Johansson, M., Rantzer, A.: Computation of piecewise quadratic Lyapunov functions for hybrid systems. IEEE Trans. Autom. Control **43**(4), 555–559 (1998)
17. Laub, A., Heath, M., Paige, C., Ward, R.: Computation of system balancing transformations and other applications of simultaneous diagonalization algorithms. IEEE Trans. Autom. Control **32**(2), 115–122 (1987)
18. Löfberg, J.: Yalmip : a toolbox for modeling and optimization in MATLAB®. In: 2004 IEEE International Conference on Robotics and Automation, pp. 284–289 (2004)
19. Loquen, T., Nešić, D., Prieur, C., Tarbouriech, S., Teel, A., Zaccarian, L.: Piecewise quadratic Lyapunov functions for linear control systems with first order reset elements. IFAC Proc. Vol. **43**, 807–812 (2010)
20. Moore, B.: Principal component analysis in linear systems: controllability, observability, and model reduction. IEEE Trans. Autom. Control **26**(1), 17–32 (1981)
21. Nešić, D., Zaccarian, L., Teel, A.: Stability properties of reset systems. Automatica **44**(8), 2019–2026 (2008)
22. Nešić, D., Teel, A.R., Zaccarian, L.: Stability and performance of SISO control systems with first-order reset elements. IEEE Trans. Autom. Control **56**(11), 2567–2582 (2011)
23. Oehlerking, J.: Decomposition of stability proofs for hybrid systems. PhD thesis, Carl von Ossietzky Universitat (2011)
24. Palacios Roman, J.J., Hafstein, S.F., van den Eijnden, S.J., Heertjes, M.F., Heemels, W.: Stability analysis of hybrid integrator-gain systems using linear programming. IFAC-PapersOnLine **58**(11), 19–24, 8th IFAC Conference on Analysis and Design of Hybrid Systems ADHS 2024 (2024)
25. Pettersson, S., Lennartson, B.: Stability and robustness for hybrid systems. In: Proceedings of 35th IEEE Conference on Decision and Control, vol. 2, pp. 1202–1207 (1996)
26. Rantzer, A., Johansson, M.: Piecewise linear quadratic optimal control. IEEE Trans. Autom. Control **45**(4), 629–637 (2000)

27. Skogestad, S., Postlethwaite, I.: Multivariable Feedback Control: Analysis and Design, 2nd edn. Wiley and Sons (2010)
28. Steinbuch, M., Norg, M.: Advanced motion control: an industrial perspective. Eur. J. Control **4**(4), 278–293 (1998)
29. van den Eijnden, S., Heertjes, M., Nijmeijer, H.: Robust stability and nonlinear loop-shaping design for hybrid integrator-gain-based control systems. In: 2019 American Control Conference (ACC), pp. 3063–3068 (2019)
30. van den Eijnden, S., Heemels, W., Nijmeijer, H., Heertjes, M.: Stability and performance analysis of hybrid integrator-gain systems: a linear matrix inequality approach. Nonlinear Anal. Hybrid Syst. **45**, 101192 (2022)
31. Xie, L., Shishkin, S., Fu, M.: Piecewise Lyapunov functions for robust stability of linear time-varying systems. Syst. Control Lett. **31**(3), 165–171 (1997)
32. Yfoulis, C.A., Shorten, R.: A numerical technique for the stability analysis of linear switched systems. Int. J. Control **77**(11), 1019–1039 (2004)
33. Zaccarian, L., Nešić, D., Teel, A.R.: Analytical and numerical Lyapunov functions for SISO linear control systems with first-order reset elements. Int. J. Robust Nonlinear Control **21**, 1134–1158 (2011)

Chapter 6
Performance Analysis: A Convergent Dynamics Approach

6.1 Introduction

Steady-state performance of a feedback control system indicates, in a certain sense, the sensitivity of its steady-state response to external inputs. In the previous chapter, performance of HIGS-controlled systems has been characterized through the notion of the $\mathcal{L}_2$-gain, which is often interpreted as a measure for the system's worst-case energy transmission from an input to an output. Although such an approach is valid for a broad class of input signals and provides a worst-case performance metric, it does not take into account additional knowledge regarding the nature of the inputs, and in that sense can be conservative when dealing with specific inputs, such as periodic inputs. To characterize the steady-state performance of control systems containing HIGS in a more accurate manner, this chapter pursues an alternative approach that exploits the notion of *convergent systems*.

Convergent systems refer to nonlinear systems having the property that, when excited by an arbitrary bounded input, there exists a *unique* time-varying solution (related to the input) that is bounded on the whole time axis and is globally asymptotically stable. Regardless of the initial conditions of the system, all other solutions converge to this steady-state solution. Contrary to asymptotically stable linear systems, this property does not hold for nonlinear systems in general. However, proving the (uniform) convergence property leads to the guarantee of having a unique and bounded steady-state response. This property opens up the possibility for characterizing performance in an accurate manner using a variety of numerical tools such as forward integration, periodic solvers [3], and mixed time–frequency (MTF) algorithms [28], and facilitates a means to construct "nonlinear" Bode plots [1, 26, 37] and quantify the accuracy of approximate methods such as the describing function [27, 29]. Besides, (input-to-state) convergent systems have predictable and robust behaviour, in the sense that their (steady-state) solutions are insensitive to initial conditions, and small changes in inputs only lead to small changes in states and outputs. Notions closely related to convergence, i.e., describing the property of solutions converging

S. van den Eijnden et al., *Hybrid Integrator-Gain Systems*, Lecture Notes in Control and Information Sciences 499, https://doi.org/10.1007/978-3-032-18101-5_6

to each other, are known in the literature as incremental stability and contraction, see, e.g., [2, 6, 11, 12, 21, 22, 33, 42] and the references therein.

Motivated by the desirable properties and tools related to convergence, this chapter investigates the relevant convergence property for HIGS-controlled systems. For systems containing a scalar, memoryless nonlinearity, conditions for convergence typically hinge upon the nonlinearity satisfying an incremental sector condition [14, 25, 31]. HIGS, however, inherently violates such incremental sector condition due to its underlying dynamics in combination with the discontinuous vector field. Moreover, discontinuous vector fields can significantly complicate a convergence analysis as is also highlighted in [24], where conditions for convergence of piecewise affine (PWA) systems with discontinuous right-hand sides are formulated in terms of linear matrix inequalities (LMIs). In that work, discontinuities may only occur due to the affine (constant) terms in the dynamics. For HIGS, however, discontinuities in the dynamics also result from other terms, so that the results in [24] do not translate to the setting considered here. In [23, 42, 43] incremental piecewise quadratic/polynomial Lyapunov functions are used for deriving incremental stability properties of classes of PWA systems. These results, however, do not exploit knowledge regarding (incremental) properties of the closed-loop system dynamics, possibly making them conservative in the current context.

In this chapter, a piecewise quadratic approach is pursued as well, which takes into account specific incremental properties of the vector field of HIGS. In particular, by partitioning HIGS' incremental input–output space, regional properties are identified that can be conveniently exploited for constructing appropriate piecewise incremental Lyapunov functions. Essentially, these functions result from "connecting" separate incremental storage functions (each linked to different incremental properties) in a suitable manner. It turns out that the strong integrator dynamics in HIGS pose an additional challenge for deriving useful conditions for (input-to-state) convergence of HIGS-controlled systems. For that reason, in this chapter, a small modification to the vector field in integrator mode is made to allow for, e.g., weak integrator characteristics. Under certain assumptions on the modified vector field, conditions for convergence are established, which are formulated in terms of numerically tractable LMIs. A frequency-domain interpretation of the LMIs, which allows for a graphical verification of the conditions that may yield valuable tuning insights, is presented as well. The conditions are extended to provide rigorous error bounds on steady-state performance predictions based on the describing function approximation. Results on incremental stability of discrete-time HIGS-based control systems and variants can be found in [4].

The organization of this chapter is as follows. In Sect. 6.2 the system set-up and a modification to the original HIGS description are discussed. In Sect. 6.3, the incremental form of the closed-loop dynamics is formulated, and local properties are identified that play an essential role in the analysis. Time- and frequency-domain conditions for convergence are presented in Sects. 6.4 and 6.5. An extension toward steady-state performance estimation is made in Sect. 6.6. Application of the presented tools is demonstrated on a numerical example in Sect. 6.7. The main findings in this chapter are summarized in Sect. 6.8.

6.2 System Setting and Problem Formulation

In this chapter, a modified version of HIGS is considered in which the vector field in integrator mode is allowed to include, e.g., weak integrator characteristics. This section starts with discussing this modification, and subsequently presents the closed-loop system configuration and problem formulation.

6.2.1 An Extended Hybrid Integrator Design

To generalize (and relax) the strong integrator dynamics of HIGS, below a modified hybrid integrator design is presented in which the vector field in integrator mode is allowed to be nonlinear. In particular, consider the discontinuous dynamical system

$$\mathcal{H}: \begin{cases} \dot{x}_h(t) = f_h(x_h(t), z(t)), & \text{if } (z(t), u(t), \dot{z}(t)) \in \hat{\mathcal{F}}_1, & (6.1\text{a}) \\ x_h(t) = k_h z(t), & \text{if } (z(t), u(t), \dot{z}(t)) \in \hat{\mathcal{F}}_2, & (6.1\text{b}) \\ u(t) \ = x_h(t), & & (6.1\text{c}) \end{cases}$$

where $x_h(t) \in \mathbb{R}$ denotes the state, $z(t) \in \mathbb{R}$ is the input, $\dot{z}(t)$ denotes the time-derivative of z at time t, $u(t) \in \mathbb{R}$ is the output, and $f_h : \mathbb{R} \times \mathbb{R} \to \mathbb{R}$ is a (non)linear function. On the basis of a similar reasoning as discussed in Sect. 2.2, the sets $\hat{\mathcal{F}}_1$ and $\hat{\mathcal{F}}_2$ dictate the active dynamics of (6.1), and are defined as

$$\hat{\mathcal{F}}_1 := \left\{(z, u, \dot{z}) \in \mathbb{R}^3 \mid (z, u, \dot{z}) \in \mathcal{F} \setminus \hat{\mathcal{F}}_2\right\}, \tag{6.2a}$$

$$\hat{\mathcal{F}}_2 := \left\{(z, u, \dot{z}) \in \mathbb{R}^3 \mid (z, u, \dot{z}) \in \mathcal{F} \vee u = k_h z \wedge f_h(x_h, z)z > k_h \dot{z} z\right\}, \tag{6.2b}$$

with $\mathcal{F}$ the $[0, k_h]$-sector given by

$$\mathcal{F} = \left\{(z, u, \dot{z}) \in \mathbb{R}^3 \mid zu \geq \frac{1}{k_h} u^2\right\}. \tag{6.3}$$

Concerning the function f_h in (6.1), the following assumptions are made.

Assumption 6.1 The function f_h in (6.1) satisfies $f_h(0, 0) = 0$ and $f_h(0, z)z \geq 0$.

Assumption 6.2 There exist constants $c_1, c_2 \in \mathbb{R}$ such that the function f_h in (6.1) satisfies for all $x_h', z', x_h'', z'' \in \mathbb{R}$

$$\left((f_h(x_h', z') - f_h(x_h'', z'')\right) \delta x_h \leq c_1 \delta x_h^2 + c_2 \delta x_h \delta z, \tag{6.4}$$

where $\delta x_h := x_h' - x_h''$, and $\delta z := z' - z''$.

Assumption 6.1 ensures $(x_h, z) = (0, 0)$ to be an equilibrium point of (6.1) for zero input, and ensures for $x_h = 0$, $z \neq 0$ that the vector field in integrator mode points toward the interior of $\mathcal{F}$. The latter is important for guaranteeing that trajectories of (6.1) cannot escape the $[0, k_h]$-sector through the line $(z, u) = (z, 0)$. Assumption 6.2 can be seen as a monotonicity type of condition, and essentially puts a linear growth restriction on f_h. For functions of the form $f_h(x_h, z) = g(x_h) + \omega_h z$, with $\omega_h \in \mathbb{R}_{\geq 0}$, $g(\cdot)$ a globally Lipschitz continuous function, and $g(0) = 0$, Assumption 6.1 and Assumption 6.2 are always satisfied. A simple choice for $g(\cdot)$ includes (but is not limited to) $g(x_h) = \alpha x_h$, $\alpha \in \mathbb{R}$, leading to first-order (weak integrator) dynamics in integrator mode. This choice strongly links to a first-order reset element (FORE) as discussed in, e.g., [7, 18]. In fact, when f_h is linear, $\mathcal{H}$ in (6.1) has comparable properties as the FORE, but with the distinct advantage that the generated control outputs are continuous. For $g(x_h) = 0$, one recovers $f_h(x_h, z) = \omega_h z$, i.e., HIGS as considered in the previous chapters.

6.2.2 *Closed-Loop System Description*

Consider the feedback interconnection in Fig. 6.1. The LTI portion of the closed-loop system in Fig. 6.1 is described in state space by

$$G : \begin{cases} \dot{x}_g(t) &= A_g x_g(t) + B_g u(t) + B_w w(t), \\ y(t) &= C_g x_g(t), \\ q(t) &= C_q x_g(t) + D_q u(t) + D_w w(t), \end{cases} \tag{6.5}$$

with states $x_g(t) \in \mathbb{R}^m$, external inputs $w(t) \in \mathbb{R}^p$, controlled output $y(t) \in \mathbb{R}$, control input $u(t) \in \mathbb{R}$, and performance output $q(t) \in \mathbb{R}^r$, all at time $t \in \mathbb{R}_{\geq 0}$. As in previous chapters, w is assumed to be piecewise continuous, i.e., $w \in PC$. The system in (6.5) is assumed to be a minimal representation, and the transfer functions from u to y, and w to y are denoted by

$$G_{yu}(s) = C_g \left(sI - A_g\right)^{-1} B_g, \tag{6.6a}$$

$$G_{yw}(s) = C_g \left(sI - A_g\right)^{-1} B_w. \tag{6.6b}$$

Recall from Assumption 2.5 that the transfer functions $G_{yu}(s)$ and $G_{yw}(s)$ in (6.6) have a relative degree of at least two, such that $C_g B_g = 0$, and $C_g B_w = 0$.

Due to the piecewise nonlinear nature of $\mathcal{H}$ in (6.1), the closed-loop system resulting from the negative feedback interconnection of G in (6.5) (with $z = -y = -C_g x_g$) and $\mathcal{H}$ in (6.1) admits the state space description

$$\Sigma : \begin{cases} \dot{x}(t) = A_i x(t) + Bw(t) + b f_i(x(t)) \quad \text{if } x(t) \in \hat{\mathcal{X}}_i,\ i \in \{1, 2\}, \\ q(t) = Cx(t) + Dw(t), \end{cases} \tag{6.7}$$

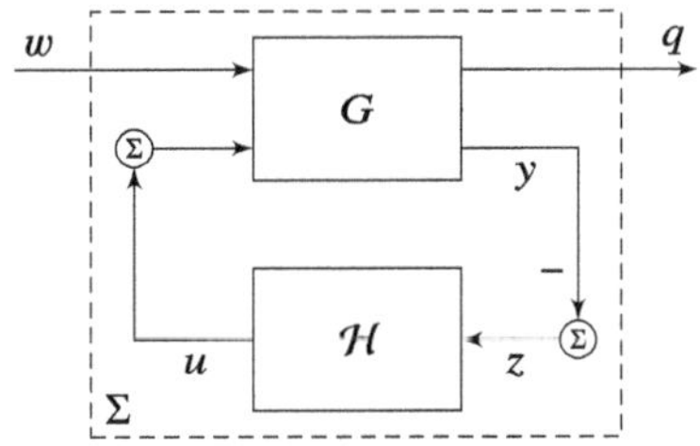

Fig. 6.1 Feedback interconnection of a MIMO LTI system G and a SISO dynamic nonlinearity $\mathcal{H}$ with inputs w and outputs q

with states $x(t) = [x_g(t)^\top, x_h(t)]^\top \in \mathbb{R}^n, n = m+1$, and where (with some abuse of notation) $f_1(x) = f_h(x_h, -C_g x_g) + \omega_h C_g x_g$, and $f_2(x) = 0$. The sets $\hat{\mathcal{X}}_i, i \in \{1,2\}$, in (6.7) are given by

$$\hat{\mathcal{X}}_i = \left\{ x \in \mathbb{R}^n \mid Ex \in \hat{\mathcal{F}}_i \right\}, \tag{6.8}$$

in which the matrix E is such that $Ex = (z, u, \dot{z})^\top$, i.e.,

$$E^\top = \begin{bmatrix} -C_g^\top & 0 & -(C_g A_g)^\top \\ 0 & 1 & 0 \end{bmatrix}. \tag{6.9}$$

Note that $\hat{\mathcal{X}}_1 \cup \hat{\mathcal{X}}_2 = \hat{\mathcal{X}} := \left\{ x \in \mathbb{R}^n \mid Ex \in \hat{\mathcal{F}}_1 \cup \hat{\mathcal{F}}_2 \right\}$. The mode-dependent system matrices are given by

$$A_1 = \begin{bmatrix} A_g & B_g \\ -\omega_h C_g & 0 \end{bmatrix}, \quad \text{and} \quad A_2 = \begin{bmatrix} A_g & B_g \\ -k_h C_g A_g & 0 \end{bmatrix}, \tag{6.10}$$

and, furthermore, $b = [0\ 1]^\top$, $B = [B_w^\top\ 0]^\top$, $C = [C_q\ D_q]$, and $D = D_w$.

A few words on well-posedness properties of the nonlinear closed-loop system in (6.7) are in order. Clearly, for the system with $f_h(x_h, z) = \omega_h z$, i.e., the original description of HIGS, global existence of solutions is formally guaranteed in Chap. 2 for all inputs $w \in PC$. Under Assumption 6.2, which puts a linear growth restriction on f_h, these results can be extended, thus guaranteeing global existence of solutions to the more general description in (6.7) as well; see also [16, Corollary 1] for further details.

Remark 6.1 For assessing input-to-state stability or $\mathcal{L}_2/\mathcal{H}_2$-performance properties of the closed-loop system in (6.7), one may extend the tools developed in Chaps. 4 and 5 accordingly. In case $f_h(x_h, z) = -\alpha x_h + \omega_h z$, i.e., the underlying dynamics are linear, such an extension is straightforward. On the other hand, when f_h is nonlinear, application of the results may be more involved. A possible approach to leverage the previously introduced tools in this case is to view (6.7) as a "perturbed" version of the piecewise linear system considered in the previous chapters, where the perturbation is specified by $f_1(x) = f_h(C_u x, C_z x) - \omega_h C_z x$, with $C_u x = x_h$ and $C_z x = z$. By assuming that $|f_1(x)| \leq \epsilon |x|$ for some $\epsilon \geq 0$, one can derive stability and performance conditions for the true piecewise nonlinear system by combining tools from

Chaps. 4 and 5 for the nominal piecewise linear system with robust stability type of arguments (e.g., small-gain conditions), see also [20, Chap. 9] and related ideas in [30, Sect. VII]. In dealing with the (possibly non-convex) set $\hat{\mathcal{X}}_2$, conditions related to the gain mode dynamics in (6.7) should be posed for the extended set $\tilde{\mathcal{X}}_2$, defined by $\tilde{\mathcal{X}}_2 = \left\{x \in \mathbb{R}^n \mid Ex \in \tilde{\mathcal{F}}_2\right\}$, in which

$$\tilde{\mathcal{F}}_2 := \left\{(z, u, \dot{z}) \in \mathbb{R}^3 \mid (z, u, \dot{z}) \in \mathcal{F} \vee u = k_h z \wedge (k_h c_1 + c_2)z^2 > k_h \dot{z} z\right\}, \tag{6.11}$$

allowing for formulating LMI conditions. This extended set results from applying Assumption 6.2 to the inequality $f_h(x_h, z)z > k_h \dot{z} z$ in gain mode. Clearly $\hat{\mathcal{F}}_2 \subseteq \tilde{\mathcal{F}}_2$, so that when conditions hold for all $x \in \tilde{\mathcal{X}}_2$, these also hold for all $x \in \hat{\mathcal{X}}_2$. For the nonlinear system in (6.7), the results presented in earlier chapters are thus still highly relevant, particularly when the characteristics of (6.1) do not deviate "too much" from the characteristics of HIGS as in earlier chapters.

6.2.3 Problem Formulation

The aim in this chapter is to study the behaviour of (steady-state) solutions to the closed-loop system in (6.7). Of particular interest is the stability of solutions with respect to each other. This property can be characterized through the notions of incremental (input-to-state) stability (δISS), and (input-to-state) convergence (ISC).

Definition 6.1 ([2]) The closed-loop system in (6.7) is said to be incrementally asymptotically stable (δAS), if there exists a $\mathcal{KL}$-function β such that for any bounded input $w \in PC$, and initial conditions $x_0', x_0'' \in \hat{\mathcal{X}}$, all corresponding solutions to (6.7) satisfy

$$\|x(t, x_0', w) - x(t, x_0'', w)\| \leq \beta\left(\|x_0' - x_0''\|, t\right), \tag{6.12}$$

for all times $t \in \mathbb{R}_{\geq 0}$. If the function β can be taken of the form

$$\beta(\|x_0' - x_0''\|, t) = re^{-kt}\|x_0' - x_0''\| \tag{6.13}$$

for some $r, k > 0$, the system is called incrementally exponentially stable (δES).

Definition 6.2 ([2]) The closed-loop system in (6.7) is said to be incrementally input-to-state stable (δISS), if there exist a $\mathcal{KL}$-function β and a $\mathcal{K}$-function γ such that for any initial conditions $x_0', x_0'' \in \hat{\mathcal{X}}$ and any bounded inputs $w', w'' \in PC$, the corresponding solutions to (6.7) satisfy

$$\|x(t, x_0', w') - x(t, x_0'', w'')\| \leq \beta\left(\|x_0' - x_0''\|, t\right) + \gamma\left(\sup_{0 \leq \tau \leq t} \|w'(\tau) - w''(\tau)\|\right), \tag{6.14}$$

for all times $t \in \mathbb{R}_{\geq 0}$.

Note that from the definitions of δAS, δES, and δISS, it is immediately clear that solutions x to (6.7) given $w \in PC$ and initial conditions $x(0) = x_0 \in \hat{\mathcal{X}}$ are unique.

Definition 6.3 ([10, 25]) The closed-loop system in (6.7) is said to be uniformly convergent (UC), if:

1. All solutions x of (6.7) corresponding to initial condition $x(t_0) = x_0$ with $x_0 \in \hat{\mathcal{X}}$, $t_0 \in \mathbb{R}$, and bounded input $w \in PC$ are defined for all times $t \in \mathbb{R}_{\geq t_0}$.
2. There exists a solution $\bar{x}_w : \mathbb{R} \to \mathbb{R}^n$ of (6.7), called a steady-state solution, depending on the input w and defined and bounded for all $t \in \mathbb{R}$.
3. The steady-state solution $\bar{x}_w$ is uniformly asymptotically stable, i.e., there exists a $\mathcal{KL}$-function β such that for every initial condition $x(t_0) = x_0 \in \hat{\mathcal{X}}$, $t_0 \in \mathbb{R}$, and every bounded input $w \in PC$, all solutions $x : \mathbb{R}_{\geq t_0} \to \mathbb{R}^n$ to (6.7) as in item 1 above satisfy
$$\|x(t) - \bar{x}_w(t)\| \leq \beta(\|x_0 - \bar{x}_w(t_0)\|, t - t_0), \tag{6.15}$$
for all times $t \in \mathbb{R}_{\geq t_0}$.

Definition 6.4 ([10, 25]) The closed-loop system (6.7) is said to be input-to-state convergent (ISC), if it is UC, and additionally there exist a $\mathcal{KL}$-function β and a $\mathcal{K}$-function γ such that all solutions $x : \mathbb{R}_{\geq t_0} \to \mathbb{R}^n$ of the system (6.7) corresponding to bounded inputs $w' \in PC$ and all steady-state solutions $\bar{x}_w$ for bounded input $w \in PC$ satisfy

$$\|x(t) - \bar{x}_w(t)\| \leq \beta(\|x(t_0) - \bar{x}_w(t_0)\|, t - t_0) + \gamma\left(\sup_{t_0 \leq \tau \leq t} \|w'(\tau) - w(\tau)\|\right), \tag{6.16}$$

for all times $t \in \mathbb{R}_{\geq t_0}$.

In general, the properties of incremental (input-to-state) stability and uniform (input-to-state) convergence are not equivalent. However, as shown in [32], uniform convergence and incremental stability are equivalent on compact positively invariant sets. By virtue of this fact, for proving (input-to-state) convergence of the closed-loop system in (6.7) it is then sufficient to (1) verify incremental stability to guarantee any two solutions to converge to each other, and (2) verify for bounded inputs w the existence of at least one solution $\bar{x}_w(t)$ that is bounded on $\mathbb{R}$ [10, 45]. Item (2) can be verified through the following lemma, which will be exploited in the upcoming results.

Lemma 6.1 ([45]) *Consider system (6.7) with bounded input $w \in PC$. Suppose that*

(i) the system (6.7) admits a unique solution $x(\cdot, t_0, x_0, w)$ on $\mathbb{R}_{\geq t_0}$ for every initial condition $x(t_0) = x_0 \in \hat{\mathcal{X}}$, $t_0 \in \mathbb{R}$ and for the input w;
(ii) these solutions depend on the initial state x_0 in a continuous manner;
(iii) there exists a compact set $D \subseteq \hat{\mathcal{X}}$, which is positively invariant with respect to system (6.7) for the bounded input $w \in PC$ in the sense that for all initial conditions $x(t_0) = x_0 \in D$, $t_0 \in \mathbb{R}$ it follows that $x(t, t_0, x_0, w) \in D$ for all $t \in \mathbb{R}_{\geq t_0}$.

Then there exists at least one solution $\bar{x}_w$ to system (6.7) for the input w that is defined and bounded for all $t \in \mathbb{R}$.

The existence of a positively invariant set in item (iii) of Lemma 6.1 is guaranteed if the system in (6.7) has an ISS-Lyapunov function (in a non-incremental sense), see Definition 2.5, and is thus ISS. The existence of an ISS-Lyapunov function, and thus the existence of a compact positively invariant set can be verified using the tools presented in Chaps. 4 and 5. The main challenge in convergence analysis therefore lies in verifying incremental stability properties as discussed in item (1) above, and which are also directly useful for guaranteeing items *(i)* and *(ii)* in Lemma 6.1.

In line with the previous chapters, the main objective in this chapter is to formulate constructive time- and frequency-domain conditions for guaranteeing incremental stability, i.e., verifying item (1) above, and with that convergence of the closed-loop system in (6.7). In doing so, the *incremental form* of (6.7) and its properties are studied first.

6.3 Incremental form and System Properties[1]

This section presents the incremental dynamics of (6.7) along with some properties that serve a key purpose in deriving useful conditions for guaranteeing convergence.

6.3.1 Incremental Dynamics

Define $\delta x(t) := x'(t) - x''(t) \in \mathbb{R}^n$ as the difference between two trajectories $x'(t) = x(t, x_0', w')$ and $x''(t) = x(t, x_0'', w'')$ generated by the closed-loop system in (6.7) subject to bounded inputs $w', w'' \in PC$, and initial conditions $x_0', x_0'' \in \hat{\mathcal{X}}$. The incremental form of (6.7) is written for the sake of subsequent analysis as a Lur'e-type system (i.e., an LTI part in feedback with a nonlinear part) as follows:

[1] Section 6.3 is partly reproduced with permission from "On Convergence of Systems with Sector-Bounded Hybrid Integrators", van den Eijnden S, Heertjes M, Nijmeijer H, Heemels M, 2022 IEEE 61st Conference on Decision and Control, 7636–7641. ©2022 IEEE.

$$\delta\Sigma : \begin{cases} \delta\dot{x}(t) = A\delta x(t) + B\delta w(t) + b\Delta(x'(t), x''(t)), \\ \delta q(t) = C\delta x(t) + D\delta w(t), \end{cases} \tag{6.17}$$

with incremental input $\delta w(t) := w'(t) - w''(t)$, incremental output $\delta q(t) := q'(t) - q''(t)$, system matrix

$$A = \begin{bmatrix} A_g & B_g \\ 0 & 0 \end{bmatrix}, \tag{6.18}$$

which captures the "common" part of the system matrices A_1 and A_2 in (6.10), and where matrices b, B, C, and D are specified after (6.10). The "trouble-making" nonlinearity in (6.17) is given by

$$\Delta(x', x'') = \begin{cases} f_h(x_h', z') - f_h(x_h'', z''), & \text{if } \left(Ex', Ex''\right) \in \hat{\mathcal{F}}_1 \times \hat{\mathcal{F}}_1, \\ k_h \delta\dot{z}, & \text{if } \left(Ex', Ex''\right) \in \hat{\mathcal{F}}_2 \times \hat{\mathcal{F}}_2, \\ f_h(x_h', z') - k_h \dot{z}'', & \text{if } \left(Ex', Ex''\right) \in \hat{\mathcal{F}}_1 \times \hat{\mathcal{F}}_2, \\ k_h \dot{z}' - f_h(x_h'', z''), & \text{if } \left(Ex', Ex''\right) \in \hat{\mathcal{F}}_2 \times \hat{\mathcal{F}}_1 \end{cases} \tag{6.19}$$

with increments $\delta x_h(t) := x_h'(t) - x_h''(t)$, and $\delta z(t) := z'(t) - z''(t)$, and where $(Ex', Ex'') = ((Ex')^\top, (Ex'')^\top)^\top$. We stress that Δ in (6.19) represents the right-hand side of the discontinuous incremental dynamics of (6.1), that is,

$$\delta\dot{x}_h = \Delta(x', x''). \tag{6.20}$$

Note that the last three lines in (6.19) result from explicit differentiation of the algebraic constraint in (6.1) in gain mode.

6.3.2 *Properties*

In order to derive conditions under which the closed-loop system in (6.7) is incrementally stable, the incremental dynamics in (6.20) are studied in more detail. The following two properties of (6.20) will be particularly useful in the upcoming analysis. These essentially show that, under Assumptions 6.1 and 6.2, the incremental dynamics in (6.20) satisfy incremental gain and incremental dissipativity properties in subregions of the incremental input–output space.

Property 6.1 Suppose that Assumptions 6.1 and 6.2 are satisfied. Then, the incremental system in (6.20) with Δ in (6.19) satisfies

$$(\delta x_h)^2 \leq k_h \delta z \delta x_h, \quad \text{for } (\delta z, \delta x_h) \in \Omega_1, \tag{6.21a}$$

$$(\delta\dot{x}_h)\delta x_h \leq c_1 \delta x_h^2 + c_2 \delta x_h \delta z, \quad \text{for } (\delta z, \delta x_h) \in \Omega_2, \tag{6.21b}$$

where

$$\Omega_1 := \left\{ (\delta z, \delta x_h) \in \mathbb{R}^2 \mid \delta z \delta x_h \geq \frac{1}{k_h} \delta x_h^2 \right\}, \tag{6.22a}$$

$$\Omega_2 := \mathbb{R}^2 \backslash \Omega_1. \tag{6.22b}$$

Proof The proof can be found in Sect. A.5.1. □

Property 6.2 Suppose that Assumptions 6.1 and 6.2 are satisfied. Then, the incremental system in (6.20) with Δ in (6.19) satisfies the inequality

$$\delta \dot{x}_h (\delta x_h - k_h \delta z) \leq (c_1 \delta x_h + c_2 \delta z)(\delta x_h - k_h \delta z), \tag{6.23}$$

for all $(\delta z, \delta x_h) \in \Omega_2$, with Ω_2 defined in (6.22).

Proof For all $(\delta z, \delta x_h) \in \Omega_2$ it holds that $\delta x_h (\delta x_h - k_h \delta z) > 0$ (this is by definition of Ω_2), and thus δx_h and $\delta x_h - k_h \delta z$ have similar signs and are both non-zero. As a consequence of Property 6.1, in particular of (6.21b) one arrives at (6.23). □

Interestingly, both (6.21b) and (6.23) resemble *one-sided Lipschitz-like* characteristics of the vector field of (6.1). The concept of a one-sided Lipschitz condition [19] (which strongly connects to the concept of monotonicity [5, 17]) plays an important role in characterizing asymptotic/incremental stability properties of non-smooth systems through quadratic Lyapunov functions. Recently, in [17], comparable properties have been exploited for deriving incremental stability of (discontinuous) projected dynamical systems (PDS), which show strong parallels with the formulation in (6.7), with the main difference being that the projection set in [17] is a convex set (see also the discussion in Sect. 2.2.4). It will become clear in the next section that Properties 6.1 and 6.2 are instrumental in formulating conditions for incremental stability of the closed-loop system in (6.7).

6.4 Time-Domain Conditions for Input-to-State Convergence

In this section, the properties derived above are exploited in formulating time-domain conditions for input-to-state convergence. Two approaches are explored: (i) an input–output approach that uses a small-gain argument, and (ii) a Lyapunov-based approach.

6.4.1 A Small-Gain Condition[2]

Before deriving explicit conditions for convergence using small-gain arguments, consider first the following observation. When regarding δz as an input to the incremental system in (6.19), one may recognize from (6.21b) in Property 6.1 that δx_h^2 could resemble a regional δISS-Lyapunov function for this system provided $c_1 < 0$. Combining this with the incremental gain property in (6.21a), in fact, suggests (6.1) to be δISS. This important insight is formalized in the next result.

Lemma 6.2 *Consider the incremental system in (6.20) with Δ specified in (6.19). Suppose that Assumptions 6.1 and 6.2 are satisfied with $c_1 < 0$ and, furthermore, suppose that δz is bounded in the sense that $\|\delta z(t)\| < \infty$ for all $t \in \mathbb{R}_{\geq 0}$. Then,*

$$\|\delta x_h(t)\| \leq e^{-\epsilon t}\|\delta x_h(0)\| + \gamma_z \left(\sup_{0\leq\tau\leq t} \|\delta z(\tau)\| \right), \tag{6.24}$$

with $\gamma_z = \max\left\{k_h, \frac{c_2}{|c_1|-\epsilon}\right\}$ and $0 < \epsilon < |c_1|$.

Proof Consider the set

$$M := \left\{(\delta z, \delta x_h) \in \mathbb{R}^2 \mid |\delta z| \leq \rho, |\delta x_h| \leq \gamma_z \rho\right\}, \quad \rho := \sup_{0\leq\tau\leq t} |\delta z(\tau)|, \tag{6.25}$$

which is visualized in Fig. 6.2 by the black rectangle.

Suppose that $(\delta z(0), \delta x_h(0)) \notin M$. Note from the fact that $-\rho \leq \delta z(0) \leq \rho$, this implies that $\|\delta x_h(t)\| > \gamma_z \rho \geq \gamma_z\|\delta z(t)\|$ for all $t \geq 0$ for which $(\delta z(t), \delta x_h(t)) \notin M$, such that $(\delta z(t), \delta x_h(t)) \in \Omega_2$, see also Fig. 6.2. Then, it immediately follows from Property 6.1 with $c_1 < 0$ that (6.21b) satisfies

$$(\delta \dot{x}_h)\delta x_h \leq c_1 \delta x_h^2 + c_2 \delta x_h \delta z \leq -\epsilon \delta x_h^2, \tag{6.26}$$

due to $\|\delta z(t)\| \leq \frac{1}{\gamma_z}\|\delta x_h(t)\| \leq \frac{|c_1|-\epsilon}{c_2}\|\delta x_h(t)\|$. By the Bellmann–Grönwall lemma, this leads to

$$\|\delta x_h(t)\| \leq e^{-\epsilon t}\|\delta x_h(0)\| \tag{6.27}$$

for all time $t \geq 0$ as long as $(\delta z(t), \delta x_h(t)) \notin M$. Then, there must exist a finite time $t' \geq 0$ for which $(\delta z(t'), \delta x_h(t')) \in M$. Once $(\delta z(t'), \delta x_h(t')) \in M$, it remains in M for all times $t \geq t'$, as it follows from (6.27) that M is a positively invariant set, and

$$\|\delta x_h(t)\| \leq \gamma_z \rho \quad \text{for all } t \geq t'. \tag{6.28}$$

---

[2] Section 6.4.1 is partly reproduced with permission from IEEE Control Systems Letters, Vol 4, van den Eijnden S, Heertjes M, Nijmeijer H, Heemels M, A small-gain approach to incremental input-to-state stability of hybrid integrator-gain systems, 2443–2448. ©2023 IEEE.

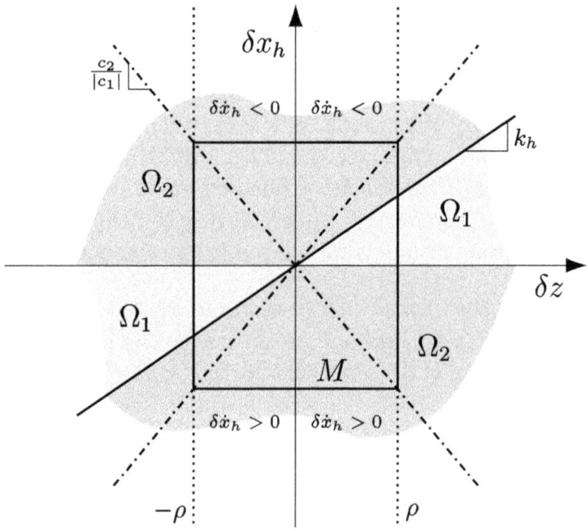

Fig. 6.2 Sketch of the mechanism underlying the δISS property of the system in (6.1) Reprinted with permission from IEEE Control Systems Letters, Vol 4, van den Eijnden S, Heertjes M, Nijmeijer H, Heemels M, A small-gain approach to incremental input-to-state stability of hybrid integrator-gain systems, 2443–2448. ©2023 IEEE

In case $(\delta z(0), \delta x_h(0)) \in M$, the above applies with $t' = 0$. Combining the inequalities in (6.27) and (6.28) leads to (6.24). A sketch of the mechanism underlying the δISS property is illustrated in Fig. 6.2.

□

The next result is a direct consequence of Lemma 6.2.

Theorem 6.3 *Consider the closed-loop system (6.7), and suppose that this system admits an ISS-Lyapunov function. Suppose that the matrix A_g in (6.10) related to the LTI part G in (6.5) is Hurwitz, and the small-gain relation*

$$\bar{\gamma}_z \gamma_u < 1 \tag{6.29}$$

is satisfied, where $\bar{\gamma}_z = \max\left\{k_h, \frac{c_2}{|c_1|}\right\}$ and $\gamma_u = \int_0^\infty |C_g e^{A_g \tau} B_g| d\tau$ with A_g, B_g, and C_g in (6.5). Then, the closed-loop system in (6.7) is input-to-state convergent.

Proof For a proof, see Sect. A.5.2. □

A few remarks on Theorem 6.3 are in order:

(1) A crucial assumption in Theorem 6.3 is the existence of an ISS-Lyapunov function. Existence of such function guarantees the existence of a forward invariant set that is needed in Lemma 6.1. Besides, it guarantees boundedness of solutions

to the non-incremental closed-loop system in (6.7), i.e., $\sup_{t\in\mathbb{R}_{\geq 0}} \|x(t)\| < \infty$. Existence of an ISS-Lyapunov function can be verified using tools for ISS as discussed in Chaps. 4 and 5.

(2) It is interesting to observe that $\gamma_u = \int_0^\infty |C_g e^{A_g\tau} B_g| dt$ corresponds to the induced $\mathcal{L}_\infty$-norm of the SISO LTI system described by the transfer function $G_{yu}(s) = C_g(sI - A_g)^{-1} B_g$ as in (6.6a). This is equivalent to the $\mathcal{L}_1$-norm of the corresponding impulse response function $g(t) = C_g e^{A_g t} B_g$ (for zero initial conditions and no disturbances), see also [36, p. 540]. As such, γ_u can be obtained from, e.g., measurement data [34].

Application of the incremental small-gain result in Theorem 6.3 will be shown later on a numerical example in Sect. 6.7, see also [40] for examples.

6.4.2 Lyapunov-Based Conditions[3]

A different approach to studying incremental stability of the closed-loop system in (6.7) is via incremental Lyapunov-based methods as presented in the next result.

Theorem 6.4 *Consider the closed-loop system in (6.7). Suppose there exist a locally Lipschitz continuous function* $V : \mathbb{R}^n \times \mathbb{R}^n \to \mathbb{R}_{\geq 0}$, *class* $\mathcal{K}_\infty$*-function* α_1, *and class* $\mathcal{K}$*-functions* α_2, α_3, *and* β *satisfying for all* $x', x'' \in \hat{\mathcal{X}}$, *and all* $w', w'' \in \mathbb{R}^p$

$$\alpha_1(\|x' - x''\|) \leq V(x', x'') \leq \alpha_2(\|x' - x''\|), \tag{6.30a}$$

$$\max_{v\in\partial V(x',x'')} \langle v, F(x', w', x'', w'')\rangle \leq -\alpha_3(\|x' - x''\|) + \beta(\|w' - w''\|), \tag{6.30b}$$

with

$$F(x', w', x'', w'') = A(x' - x'') + B(w' - w'') + b\Delta(x', x''), \tag{6.31}$$

where A, B, *and* b *are the system matrices defined around (6.17), and where* $\Delta(x', x'')$ *is given in (6.19). Then, the closed-loop system in (6.7) is input-to-state convergent.*

Proof By Theorems 2.7 and 2.8, it can be shown that conditions (6.30a) and (6.30b) imply the closed-loop system in (6.7) to be incrementally input-to-state stable (δISS), and, in turn, incrementally asymptotically stable (δAS). From the definition of δAS (Definition 6.1) it then immediately follows that a solution corresponding to an initial state $x_0 \in \hat{\mathcal{X}}$ at time $t = 0$ and input $w \in PC$ is unique, and depends continuously on these initial states.

[3] The text in Sect. 6.4.2 between Lemma 6.3 and Theorem 6.5 is partly reproduced with permission from "On Convergence of Systems with Sector-Bounded Hybrid Integrators", van den Eijnden S, Heertjes M, Nijmeijer H, Heemels M, 2022 IEEE 61st Conference on Decision and Control, 7636–7641. ©2022 IEEE.

Next, take $(x(t), w(t))$ with $w \in PC$ to be an arbitrary trajectory of the system and let $w'(t) \equiv 0$, $x'(t) \equiv 0$ for all $t \in \mathbb{R}$. Due to Assumption 6.1, it follows that $f(0, 0) = 0$, and the point $(x, w) = (0, 0)$ is an equilibrium of the system (6.7). Then, using standard arguments it follows that (6.30a) and (6.30b) imply the set $D(c) := \{x \in \mathbb{R}^n \mid V(x) \leq \alpha_2 \circ \alpha^{-1} \circ \beta(c)\}$ to be a compact positively invariant set for every input $w \in PC$ satisfying $|w(t)| \leq c$ for all $t \in \mathbb{R}$. Here, $\circ$ denotes the composition of functions. By Lemma 6.1 there exists a solution $\bar{x}_w$ that is defined and bounded for all $t \in \mathbb{R}$. From the fact that the system is δAS, it follows that this solution is uniformly asymptotically stable, and, as a consequence, the system is uniformly convergent.

Furthermore, since the system is δISS, one finds by setting $x'(t) = \bar{x}_w(t)$ that (6.14) implies (6.16) to hold for some appropriate functions β and γ. Hence, the system is input-to-state convergent (ISC). This completes the proof. □

Although Theorem 6.4 provides a generic result in terms of the class of admissible incremental Lyapunov functions, finding a suitable function remains a challenge. To come up with a rationale for constructing an appropriate candidate function, the following auxiliary lemma that follows directly from the properties discussed in Sect. 6.3 will be particularly helpful.

Lemma 6.3 *Suppose Assumptions 6.1 and 6.2 are satisfied, and there exist a positive definite matrix* $Q \in \mathbb{S}^{m \times m}$ *and a number* $0 \leq \tau \leq 1$ *that satisfy*

$$\mathcal{A}^\top Q + Q\mathcal{A} \prec 0 \quad \text{and} \quad Qb = C_u^\top - \tau k_h C_z^\top, \tag{6.32}$$

with matrices

$$\mathcal{A} = \begin{bmatrix} A_g & B_g \\ -c_2 C_g & c_1 \end{bmatrix}, \qquad b = C_u^\top = \begin{bmatrix} 0_m \\ 1 \end{bmatrix}, \tag{6.33}$$

and $C_z = [-C_g, 0]$. *Then, for all* $(\delta z, \delta x_h) \in \Omega_2$ *with* Ω_2 *defined in (6.22), the incremental dynamics in (6.17) satisfy*

$$\left\langle \frac{\partial W(\delta x)}{\partial \delta x}, F(x', w', x'', w'') \right\rangle \leq -\nu \|\delta x\|^2 + \gamma \|\delta w\|^2, \tag{6.34}$$

for some $\nu, \gamma > 0$, *and where* $W(\delta x) = \delta x^\top Q \delta x$, *and* $F(x', w', x'', w'')$ *given in (6.31).*

Proof The incremental dynamics in (6.17) can be written as the "perturbed" LTI system

$$\delta \dot{x}(t) = \mathcal{A}\delta x(t) + b\phi(t) + B\delta w(t), \tag{6.35}$$

where $\mathcal{A}$ is given in (6.33), and $\phi(t) = \delta\dot{x}_h(t) - c_1\delta x_h(t) - c_2\delta z(t)$. By the equality in (6.32) together with Properties 6.1 and 6.2, one finds for all $(\delta z, \delta x_h) \in \Omega_2$

$$\begin{aligned}\delta x^\top Q b\phi &= \phi(C_u - \tau k_h C_z)\delta x = \phi(\delta x_h - \tau k_h \delta z)\\ &= (\delta\dot{x}_h - c_1\delta x_h - c_2\delta z)\left((1-\tau)\delta x_h + \tau(\delta x_h - k_h\delta z)\right) \leq 0.\end{aligned} \tag{6.36}$$

Combining this with the assumption that $\delta x^\top(\mathcal{A}^\top Q + Q\mathcal{A})\delta x \leq -\epsilon\|\delta x\|^2$ for some sufficiently small $\epsilon > 0$, it follows that for all $(\delta z, \delta x_h) \in \Omega_2$

$$\begin{aligned}\frac{\partial W(\delta x)}{\partial \delta x}\delta\dot{x} &= \delta x^\top(\mathcal{A}^\top Q + Q\mathcal{A})\delta x + 2\delta x^\top Q B\delta w + 2\delta x^\top Q b\phi\\ &\leq -\epsilon\|\delta x\|^2 + \|QB\|\left(\mu\|\delta x\|^2 + \frac{1}{\mu}\|\delta w\|^2\right),\end{aligned} \tag{6.37}$$

where Young's inequality is used with $\mu > 0$. Selecting $\mu < \epsilon/\|QB\|$ leads to the result.

□

The result in Lemma 6.3 forms the inspiration for constructing a specific *incremental* Lyapunov function for the closed-loop system in (6.7) as follows. A function of the form $W(\delta x) = \delta x^\top Q\delta x$ with $Q \in \mathbb{S}^{n\times n}$ satisfying (6.32) in Lemma 6.3 qualifies as a "regional" incremental Lyapunov function for the system in (6.7) when $(\delta z, \delta x_h) \in \Omega_2$, with Ω_2 defined in (6.22). Due to the equality in (6.32), Q can be written explicitly as

$$Q = \begin{bmatrix} Q_{11} & \tau k_h C_g^\top \\ \tau k_h C_g & 1 \end{bmatrix}, \tag{6.38}$$

with $Q_{11} \in \mathbb{S}^{m\times m}$. The function W then evaluates to $W(\delta x) = \delta x_g^\top Q_{11}\delta x_g + v(\delta x_h, \delta z)$ with $v(\delta x_h, \delta z) := \delta x_h(\delta x_h - 2\tau k_h\delta z)$. For all $(\delta z, \delta x_h)$ in the region Ω_1, however, this specific function W may not qualify as an incremental Lyapunov function as the property in (6.34) on the time-derivative of W is not guaranteed to hold, and thus condition (6.30b) in Theorem 6.4 is not guaranteed to hold. In fact, it can even be shown that there are cases for which (6.34) is strictly violated for some $(\delta z, \delta x_h) \in \Omega_1$.

On the other hand, for all $(\delta z, \delta x_h) \in \Omega_1$ the incremental sector condition in (6.21a) holds true. Then, a natural incremental Lyapunov function candidate for this region is one that stems from the circle criterion and is given by $U(\delta x_g) = \delta x_g^\top M x_g$ with $M \in \mathbb{S}^{m\times m}$ a positive definite matrix that satisfies the corresponding circle criterion conditions (see, e.g., the conditions posed in Theorem 4.2).

The question that arises at this point is how to "patch" the functions W and U together over the boundaries shared by Ω_1 and Ω_2 to obtain a continuous function, and not losing the incremental decrease condition. A particular idea is to fix $\tau = 1/2$ and set $M = Q_{11}$ a priori. Then, for all $(\delta x_h, \delta z) \in \Omega_1 \cap \Omega_2$ one finds $v(\delta x_h, \delta z) = 0$ and $W(\delta x) = U(\delta x_g)$. Setting $\tau = 1/2$, however, is not necessary and may therefore introduce conservatism in the analysis.

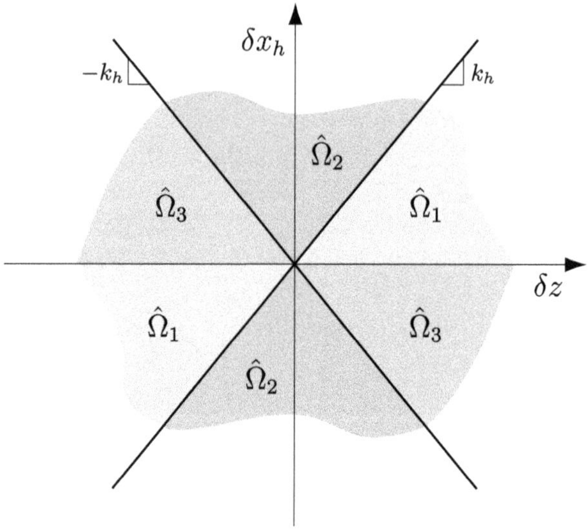

Fig. 6.3 Partitioning of the incremental input–output space of (6.19) induced by the choice for the piecewise quadratic function V in (6.39). Here, $\hat{\Omega}_1 = \Omega_1$ and $\hat{\Omega}_2 \cup \hat{\Omega}_3 = \Omega_2$

To circumvent this, observe that when keeping the choice for τ free, $v(\delta x_h, \delta z)$ can be rewritten as $v(\delta x_h, \delta z) = (1 - 2\tau)\delta x_h^2 + 2\tau \delta x_h(\delta x_h - k_h \delta z)$. While the second term vanishes at the boundaries of Ω_1 and Ω_2, i.e., where $\delta x_h = 0$ or $\delta x_h = k_h \delta z$, the first term only vanishes at $\delta x_h = 0$. This would lead to a discontinuity between the functions W and U as considered above. To account for this term, a further partitioning of the incremental input–output space of (6.19) as depicted in Fig. 6.3 is proposed. Based on this partitioning and the reasoning above, a candidate piecewise quadratic incremental Lyapunov function of the form

$$V(\delta x) = \delta x_g^\top M \delta x_g + k \max\left\{\delta x_h^2, k_h^2 \delta z^2\right\} + \max\left\{0, \delta x_h(\delta x_h - k_h \delta z)\right\}, \quad (6.39)$$

with $k > -\frac{1}{2}$ is proposed. The max functions in (6.39) correspond to "transitions" from one region to another and are active in their corresponding regions. By construction, V in (6.39) is a locally Lipschitz continuous piecewise quadratic function.

Based on the ideas outlined above, the next theorem presents sufficient conditions for the function V in (6.39) to classify as an incremental Lyapunov function through which input-to-state convergence of the closed-loop system in (6.7) is guaranteed.

Theorem 6.5 *Consider the closed-loop system in (6.7), and suppose that Assumptions 6.1 and 6.2 are satisfied. Suppose there exist constants* $k > -\frac{1}{2}$, $\tau_1, \tau_2 \geq 0$ *and a matrix* $M \in \mathbb{S}^{m \times m}$ *that satisfy the LMI conditions*

$$M + k k_h^2 C_g^\top C_g \succ 0, \quad (6.40a)$$

and

$$\begin{bmatrix} A_g^\top M + MA_g & MB_g \\ B_g^\top M & 0 \end{bmatrix} + k \begin{bmatrix} 0 & -c_2 C_g^\top \\ -c_2 C_g & 2c_1 \end{bmatrix} + T \prec 0, \tag{6.40b}$$

$$\begin{bmatrix} A_g^\top M + MA_g & MB_g \\ B_g^\top M & 0 \end{bmatrix} + k \begin{bmatrix} k_h^2 (A_g^\top C_g^\top C_g + C_g^\top C_g A_g) & 0 \\ 0 & 0 \end{bmatrix} + \tau_1 S_1 \prec 0, \tag{6.40c}$$

$$\begin{bmatrix} A_g^\top M + MA_g & MB_g \\ B_g^\top M & 0 \end{bmatrix} + k \begin{bmatrix} k_h^2 (A_g^\top C_g^\top C_g + C_g^\top C_g A_g) & 0 \\ 0 & 0 \end{bmatrix} + \tau_2 S_2 + T \prec 0, \tag{6.40d}$$

where

$$T = \begin{bmatrix} -k_h c_2 C_g^\top C_g & \left(\frac{k_h c_1}{2} - c_2\right) C_g^\top + \frac{k_h}{2} A_g^\top C_g^\top \\ C_g \left(\frac{k_h c_1}{2} - c_2\right) + \frac{k_h}{2} C_g A_g & 2c_1 \end{bmatrix}, \tag{6.41a}$$

and

$$S_1 = \begin{bmatrix} 0 & -k_h C_g^\top \\ -k_h C_g & -2 \end{bmatrix}, \quad S_2 = \begin{bmatrix} 0 & k_h C_g^\top \\ k_h C_g & -2 \end{bmatrix}. \tag{6.41b}$$

Then, the closed-loop system in (6.7) is input-to-state convergent.

Proof The proof is provided in Sect. A.5.3. □

Remark 6.2 Necessary conditions for the LMIs in (6.40) to be feasible are $\tau_1 > 0$, and $c_1 < 0$. The choice $k = 0$ makes the LMI in (6.40d) redundant since satisfying (6.40b) in that case ensures that (6.40d) is satisfied for $\tau_2 = 0$. Note that this case just results in a piecewise function defined in the original partitioning of the incremental region, i.e., in Ω_1 and Ω_2, and gives the same conditions as in [39].

Remark 6.3 The above results show a strong relation with recently presented results for so-called oblique projected dynamical systems (oPDS) with state constraints [17]. In particular, the conditions presented in Lemma 6.3 correspond to typical conditions for δISS of an oPDS with states that are confined to a closed, convex constraint set (see [17, Theorem 2]). In the context of HIGS, it was shown in [35] that, under some additional conditions, a HIGS-controlled system as in (6.7) can be reformulated as a switched oPDS. In that case, the positive definite matrix associated with state projections onto a constraint set coincides with the matrix Q in (6.38) with arbitrary Q_{11} and $\tau = 1$. However, as remarked earlier, the constraint set for HIGS-controlled systems is a non-convex set that is given by the union of two convex polyhedral cones, so that the results in [17] cannot be applied for showing δISS of HIGS-controlled systems. In this regard, Theorem 6.5 provides a possible approach for overcoming this hurdle, and may further help in studying δISS properties of more generic oPDS with non-convex constraint sets.

6.4.3 Uniqueness of Solutions

The notions of incremental stability and uniqueness of solutions have some important connections. Clearly, uniqueness of solutions (given an initial condition and an input function) is implied by the incremental stability property, and thus when the conditions of Theorem 6.5 are satisfied, it can be guaranteed that given a consistent initial condition $x(0) \in \hat{\mathcal{X}}$ the discontinuous dynamical system (6.7) has unique solutions. The opposite, however, may not be true. That is, given a consistent initial condition $x(0) \in \hat{\mathcal{X}}$ solutions to (6.7) may be unique, but the system may not be incrementally stable. In that sense, the conditions in Theorem 6.5 may be too restrictive for proving uniqueness of solutions. Indeed, uniqueness of solutions can still be proven by using relaxed versions of the LMI conditions in (6.40b)–(6.40d) by replacing "$\prec 0$" with "$\prec \lambda I$", $\lambda \in \mathbb{R}$, thus allowing the left-hand side to be positive definite. In fact, it can be shown that for sufficiently large λ, there always exists a solution to (6.40), leading to the following uniqueness result.

Theorem 6.6 *Suppose Assumptions 6.1 and 6.2 are satisfied. Solutions to the discontinuous differential equation in (6.7) are unique in the sense that given an initial condition $x(0) \in \hat{\mathcal{X}}$, and input $w \in PC$ there is exactly one corresponding solution.*

Proof Consider the LMI conditions in Theorem 6.5 with $k = \tau_1 = \tau_2 = 0$ and "$\prec 0$" replaced with "$\prec \lambda I$", where $\lambda \in \mathbb{R}$. Then, it follows along similar lines as the proof of Theorem 6.5 that for *any* matrix $M = M^\top \succ 0$, there always exist $\alpha_1, \alpha_2 > 0$, and a sufficiently large $\lambda > 0$ such that the LMIs in (6.40) (with $\prec \lambda I$) are satisfied, and, therefore, the piecewise continuous function

$$V(\delta x) = \delta x_g^\top M \delta x_g + \max\{0, \delta x_h(\delta x_h - k_h \delta z)\} \tag{6.42}$$

satisfies

$$\alpha_1 \|\delta x\|^2 \leq V(\delta x) \leq \alpha_2 \|\delta x\|^2,$$

and

$$\frac{dV(\delta x(t))}{dt} \leq \lambda V(\delta x)$$

for almost all times $t \in \mathbb{R}_{\geq 0}$. By application of Grönwall's inequality one finds $V(\delta x(t)) \leq e^{\lambda t} V(\delta x(0))$, leading to

$$\|\delta x(t)\| \leq \alpha e^{\mu t} \|\delta x(0)\| \tag{6.43}$$

for some $\alpha, \mu > 0$. Consider two trajectories $x'(t) = x(t, x'(0), w)$, and $x''(t) = x(t, x''(0), w)$ starting from the same initial condition $x'(0) = x''(0) \in \hat{\mathcal{X}}$ so that $\delta x(0) = 0$. Then, (6.43) implies $\|\delta x(t)\| = 0$ for all times $t \in \mathbb{R}_{\geq 0}$, and thus $x'(t) \equiv x''(t)$, thereby proving uniqueness of solutions. □

Remark 6.4 It is important to realize that the crux in the proof of Theorem 6.6 is the exploitation of Properties 6.1 and 6.2, and the fact that it is possible to choose *any* positive definite matrix M, as (due to linearity of the conditions in Theorem 6.5) it is always possible to find a sufficiently large $\lambda > 0$ such that the corresponding inequalities are satisfied with "$\prec \lambda I$".

To conclude: the uniqueness result in Theorem 6.6 complements the well-posedness results in Sect. 2.3.2. That is, in addition to global existence of solutions, the solutions to (6.7) are always unique for any choice of f_h satisfying Assumptions 6.1 and 6.2, including the choice $f_h(x_h, z) = \omega_h z$ resulting in HIGS as considered in the previous chapters.

6.5 Frequency-Domain Conditions for Input-to-State Convergence

When studying the set of LMIs in (6.40) in more detail, one may recognize the individual LMI conditions (6.40b)–(6.40d) to resemble certain passivity conditions. Due to the strong link between the concept of passivity and frequency-domain properties via the Kalman–Yakubovich–Popov (KYP) lemma [44], it is interesting to verify if the set of LMIs in Theorem 6.4 admit a frequency-domain interpretation. As also motivated in Chap. 4, frequency-domain conditions are highly valuable from a practical perspective as these often lead to insights regarding design and robustness aspects, and can be verified using measured frequency-response function data.

In order to make the transition from time-domain conditions toward frequency-domain conditions, the LMIs in (6.40) are simplified by a priori assuming $k = 0$ and $c_2 \geq 0$. This assumption is useful for reducing the three LMIs in (6.40b)–(6.40d) to two LMIs (see Remark 6.2). Although such a simplification may restrict the solution space of the LMIs, and in that sense comes at the cost of introducing conservatism, the associated frequency-domain conditions can still yield valuable tuning insights and may further aid the development of more advanced (frequency-domain) conditions for incremental stability and convergence.

For $k = 0$, and $c_2 \geq 0$, it is possible to verify that the LMIs in (6.40) admit a solution if there exist a matrix $M \succ 0$ and a constant $\tau > 0$ that satisfy the LMIs

$$\begin{bmatrix} A_g^\top M + M A_g & M B_g \\ B_g^\top M & 0 \end{bmatrix} + \tau \begin{bmatrix} 0 & -k_h C_g^\top \\ -k_h C_g & -2 \end{bmatrix} \prec 0, \tag{6.44a}$$

$$\begin{bmatrix} A_g^\top M + M A_g & M B_g \\ B_g^\top M & 0 \end{bmatrix} + \begin{bmatrix} 0 & (\lambda I + \frac{k_h}{2} A_g^\top)\, C_g^\top \\ C_g\,(\lambda I + \frac{k_h}{2} A_g) & 2c_1 \end{bmatrix} \prec 0, \tag{6.44b}$$

with $\lambda = \frac{k_h c_1}{2} - c_2$. Note that the upper left term $-k_h c_2 C_g^\top C_g \preceq 0$ in the matrix T in (6.41a) is left out to further simplify the conditions. The next theorem presents a necessary and sufficient condition in the frequency-domain for feasibility of the LMIs in (6.44) with $M \succ 0$.

Theorem 6.7 *The linear matrix inequalities in (6.44) admit a feasible solution if and only if the transfer function* $G_{yu}(s) = C_g(sI - A_g)^{-1}B_g$ *is stable, i.e., the matrix* A_g *is Hurwitz, and there exist constants* $\tau, \varepsilon > 0$ *that satisfy for all* $\omega \in \mathbb{R} \cup \{\infty\}$

$$\begin{bmatrix} k_h Re\left\{G_{yu}(j\omega)\right\} + 1 & \frac{1}{2\sqrt{2}}W(j\omega) \\ \frac{1}{2\sqrt{2}}W^*(j\omega) & 1 - \frac{k_1}{2}Re\left\{U(j\omega)\right\} \end{bmatrix} \succ 0, \tag{6.45}$$

with $U(j\omega) = (k_2 + k_3 j\omega)G_{yu}(j\omega)$, $W(j\omega) = \frac{1}{\sqrt{\tau}}U^*(j\omega) - k_1\sqrt{\tau}k_h G_{yu}(j\omega)$, *and*

$$k_1 = \left(\varepsilon + \frac{1}{\varepsilon}\right)\sqrt{\frac{1}{2|c_1|}} + \left(\varepsilon - \frac{1}{\varepsilon}\right)\sqrt{\frac{\tau}{2}}k_h, \tag{6.46a}$$

$$k_2 = \left(\varepsilon + \frac{1}{\varepsilon}\right)\sqrt{\frac{1}{2|c_1|}}\left(\frac{k_h c_1}{2} - c_2\right) + \left(\frac{1}{\varepsilon} - \varepsilon\right)\sqrt{\frac{\tau}{2}}k_h, \tag{6.46b}$$

$$k_3 = \left(\varepsilon + \frac{1}{\varepsilon}\right)\sqrt{\frac{1}{2|c_1|}}\frac{k_h}{2}. \tag{6.46c}$$

Proof The proof is provided in Sect. A.5.4. □

It is interesting to mention that the diagonal elements of the matrix in (6.45) pose necessary conditions for feasibility of (6.44a) and (6.44b), respectively, and correspond to circle criterion-like conditions (see also the frequency-domain conditions for stability in Theorem 4.2). The off-diagonal elements account for the coupling in the LMIs in (6.44) through the matrix M.

Clearly, the variable ε appears in an awkward manner in (6.45). For the choice $\varepsilon = 1$ the frequency-domain inequality in (6.45) simplifies significantly, but at the cost of losing necessity of the conditions. Still, one obtains sufficient conditions for convergence that can be assessed graphically and may provide some direction toward parameter tuning and robustness.

Corollary 6.1 *Suppose the transfer function* $G_{yu}(s) = C_g(sI - A_g)^{-1}B_g$ *is stable, i.e., the matrix* A_g *is Hurwitz. Then, the closed-loop system in (6.7) is input-to-state convergent if the following frequency-domain conditions are satisfied:*

(1) The transfer function $G_{yu}(s)$ *satisfies for all* $\omega \in \mathbb{R} \cup \{\infty\}$

$$1 + k_h Re\left\{G_{yu}(j\omega)\right\} > 0. \tag{6.47}$$

(2) The transfer function $G_{yu}(s)$ *satisfies for all* $\omega \in \mathbb{R} \cup \{\infty\}$

$$1 + X(j\omega) - \beta Y(j\omega) > 0, \tag{6.48}$$

in which $\beta = 1/c_1$ *with* $c_1 < 0$ *and*

$$X(j\omega) = \frac{3k_h}{2} Re\left\{G_{yu}(j\omega)\right\} + \frac{k_h^2}{4} G_{yu}(j\omega)^2, \tag{6.49a}$$

$$Y(j\omega) = c_2 Re\left\{G_{yu}(j\omega)\right\} + \frac{k_h}{2}\omega Im\left\{G_{yu}(j\omega)\right\} + \frac{k_h c_2}{2} G_{yu}(j\omega)^2. \tag{6.49b}$$

(3) There exists a $\tau > 0$ *such that for all* $\omega \in \mathbb{R} \cup \{\infty\}$, *it follows that*

$$\left\| \frac{\sqrt{\tau}}{j\omega + 2\omega_c(\tau)} \right\| > \frac{k_h}{4} \frac{\sqrt{|\beta|}\|G_{yu}(j\omega)\|}{\sqrt{1 + X(j\omega) - \beta Y(j\omega)}}, \tag{6.50}$$

with $\omega_c(\tau) = \sqrt{\left(\frac{1}{2\beta} - \frac{c_2}{k_h}\right)^2 + \tau^2}$.

Proof The proof can be found in Sect. A.5.5. □

The first condition in Corollary 6.1 corresponds to a classical circle criterion condition and can be checked graphically by inspecting the Nyquist plot of $G_{yu}(j\omega)$. The second condition corresponds to a Popov-like condition that can be verified graphically by inspecting if the (X, Y)-locus remains to the right of a line that passes through the point $(-1, 0)$ and has a slope of $\beta = 1/c_1$ (see also Sect. 4.3.3 for comparable tests). The third condition essentially implies that the weighted magnitude characteristics of $G_{yu}(j\omega)$ should be upper-bounded by the magnitude characteristics of a specific first-order low-pass filter, for which both the gain at $\omega = 0$ and cut-off frequency are determined by the choice of τ. This can be verified in a Bode magnitude plot. Note the resemblance with assessing robust performance for SISO LTI systems (see e.g., [36, Chap. 7]) where the low-pass filter may be given the interpretation of a specific weighting filter. Application of the frequency-domain conditions to a specific example is discussed in Sect. 6.7.

6.6 Steady-State Performance Estimation

The property of convergence paves the way for quantifying the accuracy of approximate, but easy-to-use tools such as the describing function method for steady-state performance analysis of the closed-loop system in (6.7). This section presents a computationally efficient method for quantifying the accuracy of a steady-state performance estimation using a describing function approximation. The presented ideas are inspired by the results in [27, 29, 38]. A different approach that exploits the notion of contractive systems can be found in [8].

6.6.1 Linear Approximation of Steady-State Solutions

Consider an LTI system of the form

$$\hat{\Sigma} : \begin{cases} \dot{\hat{x}}_g(t) &= A_g \hat{x}_g(t) + B_g \hat{u}(t) + B_w w(t) \\ \hat{y}(t) &= C_g \hat{x}_g(t) \end{cases} \tag{6.51}$$

with states $\hat{x}_g(t) \in \mathbb{R}^m$, external inputs $w(t) \in \mathbb{R}^p$ and $\hat{u}(t) \in \mathbb{R}$, and output $\hat{y}(t) \in \mathbb{R}$ at time $t \in \mathbb{R}_{\geq 0}$. The external inputs $w : \mathbb{R}_{\geq 0} \to \mathbb{R}^p$ are assumed to be periodic with period time $T > 0$. Define the signal $\hat{u} : \mathbb{R}_{\geq 0} \to \mathbb{R}$ as

$$\hat{u}(t) = \frac{1}{2\pi} \int_{-\infty}^{\infty} \hat{U}(j\omega) e^{j\omega t} d\omega \quad \text{with} \quad \hat{U}(j\omega) = -\hat{T}(j\omega) W(j\omega), \tag{6.52}$$

in which $W(j\omega)$ denotes the Fourier series of the periodic input w, and where the complex mapping $\hat{T} : \mathbb{C} \to \mathbb{C}$ is defined as

$$\hat{T}(j\omega) = \frac{\Psi(j\omega) G_{yw}(j\omega)}{1 + \Psi(j\omega) G_{yu}(j\omega)},$$

with transfer functions $G_{yw}(s) = C_g(sI - A_g)^{-1} B_w$ and $G_{yu}(s) = C_g(sI - A_g)^{-1} B_g$, and where $\Psi : \mathbb{C} \to \mathbb{C}$ denotes a harmonic linearization of the nonlinear element $\mathcal{H}$ with respect to a harmonic input. In this regard, Ψ may for example represent the sinusoidal input describing function as derived in Chap. 2, or a two-sinusoidal/incremental input describing function, see, e.g., [13]. It is assumed that $1 + \Psi(j\omega) G_{yu}(j\omega) \neq 0$ for all $\omega \in \mathbb{R} \cup \{\infty\}$. It is important to realize that by definition $\hat{u}(t) = \hat{u}(t + T)$ is periodic with the same fundamental period time T as that of the input $w(t) = w(t + T)$.

Under the assumption that A_g is Hurwitz, and since both $\hat{u}$ and w are harmonic inputs, it follows that the limit (steady-state) solution of the LTI system in (6.51), denoted by $\hat{\bar{y}}$, is unique, periodic with period time T, and its Fourier components are expressed in the frequency-domain as

$$\hat{\bar{Y}}(j\omega) = G_{yu}(j\omega) \hat{U}(j\omega) + G_{yw}(j\omega) W(j\omega) = \frac{G_{yw}(j\omega)}{1 + \Psi(j\omega) G_{yu}(j\omega)} W(j\omega). \tag{6.53}$$

Observe that (6.53) is, in fact, the response obtained through a harmonic linearization of the closed-loop system in (6.7) [20]. Note that $\hat{\bar{y}}$ and $\hat{\bar{Y}}$ are fully known, and that $\hat{U}(j\omega) = -\Psi(j\omega) \hat{\bar{Y}}(j\omega) = -\hat{T}(j\omega) W(j\omega)$.

6.6.2 Accuracy of The Harmonic Linearization

It is of interest to quantify in some way the accuracy of the linear approximation in (6.53). To this end, suppose that the closed-loop system in (6.7) is a uniformly convergent system. As such, when forced by a periodic input $w(t) = w(t + T)$, it admits a unique and periodic steady-state solution $\bar{x}$ satisfying $\bar{x}(t) = \bar{x}(t + T)$ for all $t \in \mathbb{R}_{\geq 0}$ with $\bar{x}(t) = [\bar{x}_g(t)^\top, \bar{x}_h(t)]^\top$, see also [2].

Let ξ be the difference $\xi(t) := \bar{x}_g(t) - \hat{\bar{x}}_g(t)$, where $\hat{\bar{x}}_g(t)$ denotes the unique steady-state solution of (6.51) subject to the same periodic input w, and $\zeta(t) := \bar{x}_h(t) - \hat{\bar{x}}_h(t)$, in which $\hat{\bar{x}}_h(t)$ is the unique response of (6.1) subject to the input $\hat{\bar{z}}(t) = -\hat{\bar{y}}(t)$, i.e., the open-loop response of HIGS. Then, the dynamics of the differences ξ and ζ can be written as

$$\begin{aligned} \dot{\xi}(t) &= A_g\xi(t) + B_g\zeta(t) + B_g\Lambda(t), \\ \dot{\zeta}(t) &\overset{\text{a.e.}}{=} \frac{d}{dt}\left(\bar{x}_h(t) - \hat{\bar{x}}_h(t)\right), \\ \sigma(t) &= C_g\xi(t), \end{aligned} \tag{6.54}$$

where $\sigma = \bar{y} - \hat{\bar{y}}$ is the difference between the performance output $\bar{y} = C_g\bar{x}_g$ of the true (convergent and nonlinear) system and the estimate $\hat{\bar{y}} = C_g\hat{\bar{x}}_g$, and where

$$\Lambda(t) = \hat{\bar{x}}_h(t) - \hat{u}(t) \tag{6.55}$$

with $\hat{u}$ defined in (6.52). It is important to realize that since $\hat{\bar{y}}$ is known, $\hat{\bar{x}}_h$ is completely known, and the expression in (6.55) can always be computed, either numerically or analytically. Moreover, since $\hat{\bar{y}}(t) = \hat{\bar{y}}(t + T)$, it follows that $\hat{\bar{x}}_h(t) = \hat{\bar{x}}_h(t + T)$, and since $\hat{u}(t) = \hat{u}(t + T)$ it holds that $\Lambda(t) = \Lambda(t + T)$ for all $t \in \mathbb{R}_{\geq 0}$. As such, one can consider (6.54) as the incremental system (6.17) (with performance output $q = y = C_gx_g$) subject to a periodic input $\Lambda(t)$. The following theorem presents a method for quantifying the mismatch between the linear approximation in (6.53) and the true nonlinear response of (6.7) in terms of an upper-bound on the root-mean-square (RMS) value of $\sigma = \bar{y} - \hat{\bar{y}}$.

Theorem 6.8 *Consider the closed-loop system in (6.7) with $q = y$, and subject to a bounded T-periodic input w. Suppose that this system satisfies Assumptions 6.1 and 6.2 and is input-to-state convergent. Suppose there exist constants $k > -1/2$, $\tau_1, \tau_2 \geq 0$, $\mu > 0$ and a matrix $M \in \mathbb{S}^{m\times m}$ that satisfy the conditions*

$$M + kk_h^2C_g^\top C_g \succ 0, \tag{6.56a}$$

and

$$\begin{bmatrix} A_g^\top M + M A_g + C_g^\top C_g & M B_g & M B_g \\ B_g^\top M & 0 & 0 \\ B_g^\top M & 0 & -\mu^2 \end{bmatrix} + kR + \tau_1 S_1 \prec 0, \tag{6.56b}$$

$$\begin{bmatrix} A_g^\top M + M A_g + C_g^\top C_g & M B_g & M B_g \\ B_g^\top M & 0 & 0 \\ B_g^\top M & 0 & -\mu^2 \end{bmatrix} + kR + \tau_2 S_2 + T \prec 0, \tag{6.56c}$$

$$\begin{bmatrix} A_g^\top M + M A_g + C_g^\top C_g & M B_g & M B_g \\ B_g^\top M & 0 & 0 \\ B_g^\top M & 0 & -\mu^2 \end{bmatrix} + k \begin{bmatrix} 0 & -c_2 C_g^\top & 0 \\ -c_2 C_g & 2c_1 & 0 \\ 0 & 0 & 0 \end{bmatrix} + T \prec 0, \tag{6.56d}$$

where $R = diag\left(k_h^2(A_g^\top C_g^\top C_g + C_g^\top C_g A_g), 0, 0\right)$,

$$T = \begin{bmatrix} -k_h c_2 C_g^\top C_g & \left(\frac{k_h c_1}{2} - c_2\right) C_g^\top + \frac{k_h}{2} A_g^\top C_g^\top & 0 \\ C_g \left(\frac{k_h c_1}{2} - c_2\right) + \frac{k_h}{2} C_g A_g & 2c_1 & 0 \\ 0 & 0 & 0 \end{bmatrix}, \tag{6.57a}$$

and with

$$S_1 = \begin{bmatrix} 0 & -k_h C_g^\top & 0 \\ -k_h C_g & -2 & 0 \\ 0 & 0 & 0 \end{bmatrix}, \quad S_2 = \begin{bmatrix} 0 & k_h C_g^\top & 0 \\ k_h C_g & -2 & 0 \\ 0 & 0 & 0 \end{bmatrix}. \tag{6.57b}$$

Then, the approximation error $\sigma = \bar{y} - \hat{\bar{y}}$ *satisfies*

$$\sqrt{\frac{1}{T}\int_0^T |\sigma(t)|^2 dt} \leq \mu \nu(T), \quad \text{with} \quad \nu(T) = \sqrt{\frac{1}{T}\int_0^T \Lambda^2(t) dt}. \tag{6.58}$$

Proof Consider the piecewise quadratic function

$$V(\xi, \zeta) = \xi^\top P \xi + k \max\left\{\zeta^2, k_h e^2\right\} + \max\left\{0, v(\zeta, e)\right\}, \tag{6.59}$$

with $e = -\sigma = -\bar{y} + \hat{\bar{y}}$, and $v(\zeta, e) = \zeta(\zeta - k_h e)$. Similar to the proof of Theorem 6.5 it can be shown that, under the hypothesis of the theorem, V is continuous, positive definite, and its time-derivative satisfies for almost all time

$$\frac{d}{dt} V(\xi(t), \zeta(t)) \leq -\sigma^2(t) + \mu^2 \Lambda^2(t). \tag{6.60}$$

Since by assumption the closed-loop system (6.7) is input-to-state convergent, the steady-state solution $\bar{x}$ is T-periodic, i.e., $\bar{x}(t) = \bar{x}(t + T)$. Moreover, as the steady-state solutions of the LTI system in (6.51) are T-periodic, the differences ξ and ζ are also T-periodic. For that reason, V is periodic and satisfies $V(t) = V(t + T)$. Integrating (6.60) from t to $t + T$, and using periodicity of V yields the inequality in (6.58). □

By computing the approximation error, one can find an upper- and lower-bound for the root-mean-square (RMS) value of the performance variable $\bar{y}$ as

$$\|\hat{\bar{y}}\|_{2,T} - \|\bar{y} - \hat{\bar{y}}\|_{2,T} \leq \|\bar{y}\|_{2,T} \leq \|\hat{\bar{y}}\|_{2,T} + \|\bar{y} - \hat{\bar{y}}\|_{2,T}, \tag{6.61}$$

where $\|v\|_{2,T} = \sqrt{\frac{1}{T}\int_0^T |v(t)|^2 dt}$, and use is made of the (reverse) triangle inequality. From $\|\bar{y} - \hat{\bar{y}}\|_{2,T} = \|\sigma\|_{2,T}$, the result in (6.61) allows, for instance, for computing upper- and lower-bounds on the RMS ratio of an input $w(t) = \sin(\omega t)$ with $\omega \in \mathbb{R}_{>0}$ and the corresponding steady-state response $\bar{y}$ as

$$\hat{\mathcal{M}}(\omega) - d(\omega) \leq \mathcal{M}(\omega) \leq \hat{\mathcal{M}}(\omega) + d(\omega), \tag{6.62}$$

in which

$$\mathcal{M}(\omega) := \frac{\|\bar{y}\|_{2,T}}{\|w\|_{2,T}}, \quad \text{and} \quad \hat{\mathcal{M}}(\omega) := \frac{\|\hat{\bar{y}}\|_{2,T}}{\|w\|_{2,T}} = \left\| \frac{G_{qw}(j\omega)}{1 + \Psi(j\omega)G_{qu}(j\omega)} \right\|, \tag{6.63}$$

and where $d(\omega) := \mu\nu(\omega)/\|w\|_{2,T}$. Note that, in this context, $\mathcal{M}(\omega)$ can be seen as a generalization of a traditional transfer function for LTI systems [27], and $\hat{\mathcal{M}}(\omega)$ is a harmonic approximation with $\Psi(j\omega)$ being the sinusoidal input describing function. The term $d(\omega)$ estimates the mismatch between $\mathcal{M}(j\omega)$ and $\hat{\mathcal{M}}(j\omega)$ due to, for instance, the presence of higher-order harmonics in the true nonlinear response. Note that for LTI systems one has $d(\omega) = 0$ for all $\omega \in \mathbb{R}_{\geq 0}$. Application of the above results to an example will be discussed in the next section.

6.7 A Numerical Case-Study[4]

To demonstrate the applicability of the tools that are presented in this chapter, consider the feedback interconnection as depicted in Fig. 6.4.

In this figure, the plant P is a second-order LTI system that is described by the transfer function

[4] The introductory text in Sect. 6.7 is partly reproduced with permission from IEEE Control Systems Letters, Vol 4, van den Eijnden S, Heertjes M, Nijmeijer H, Heemels M, A small-gain approach to incremental input-to-state stability of hybrid integrator-gain systems, 2443–2448. ©2023 IEEE.

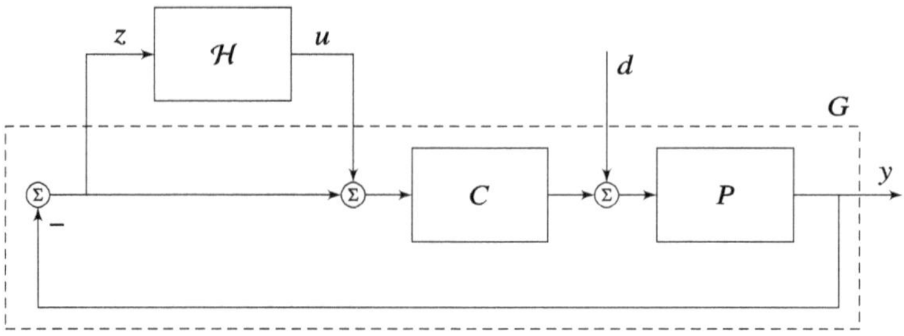

Fig. 6.4 Feedback interconnection with nonlinearity $\mathcal{H}$

$$P(s) = \frac{1}{s^2 + 2\beta_0\omega_0 s + \omega_0^2}, \tag{6.64}$$

with natural frequency $\omega_0 = 54 \cdot 2\pi$ rad/s and dimensionless damping coefficient $\beta_0 = 0.009$. The nominal LTI feedback controller C is given by

$$C(s) = k_p \left(\frac{s + \omega_i}{s}\right) \left(\frac{\omega_{lp}^2}{s^2 + 2\beta\omega_{lp} s + \omega_{lp}^2}\right), \tag{6.65}$$

with $k_p = 7 \cdot 10^4$ N/m, $\omega_i = 4.75 \cdot 2\pi$ rad/s, $\omega_{lp} = 6.5 \cdot 2\pi$ rad/s, and $\beta_{lp} = 0.8$.

The nonlinearity $\mathcal{H}$ is given in (6.1) with $f_h(x_h, z) = -\alpha x_h + \omega_h z$, and $\alpha, \omega_h \in \mathbb{R}_{>0}$, and, furthermore, $k_h = 0.6$. Note that f_h satisfies Assumptions 6.1 and 6.2, the latter with $c_1 = -\alpha$ and $c_2 = \omega_h$. The purpose of adding $\mathcal{H}$ to the existing LTI controller C is to balance between additional suppression of low-frequency disturbances and transient time-domain response in a more desirable manner (see also [9, 15, 41] for comparable strategies). A derivation of the sinusoidal input describing function for the nonlinearity $\mathcal{H}$ with the specific choice for f_h as above is given in Sect. B.3.1 and is used later to estimate steady-state performance properties of the closed-loop system in Fig. 6.4.

The feedback configuration in Fig. 6.4 can be rearranged into an equivalent Lur'e form as depicted earlier in Fig. 6.1. In the context of this figure one finds $q = y$, $w = d$ and $G(s) = [G_{yu}(s), G_{yw}(s)]$, with

$$G_{yu}(s) = \frac{P(s)C(s)}{1 + P(s)C(s)}, \quad \text{and} \quad G_{yw}(s) = \frac{P(s)}{1 + P(s)C(s)}. \tag{6.66}$$

By design, these transfer functions are stable and satisfy the standing assumption on the relative degree (Assumption 2.5).

6.7.1 Convergence

Input-to-state convergence of the closed-loop system is studied for different values of $\omega_h \in (0, 10 \cdot 2\pi]$ rad/s and different ratios for $\alpha/\omega_h \in [0, 2]$. The ratio α/ω_h determines the effect of an additional pole in the integrator dynamics. In principle, the smaller the ratio α/ω_h, the more the characteristics of $\mathcal{H}$ tend to that of HIGS as considered in Chaps. 2–5. The corresponding results from evaluating the LMI conditions in Theorem 6.5 as well as the frequency-domain conditions in Corollary 6.1 are shown in Fig. 6.5.

As expected, the range for which convergence can be shown by means of Theorem 6.5 is larger than the one resulting from Corollary 6.1 as the latter assumes a specific solution to the LMIs in (6.40) (note in this regard that in Fig. 6.5 the region indicated by the black dots is a subregion of the region indicated by the grey dots). Furthermore, note that since $\text{Re}\left\{G_{yu}(j\omega)\right\} \geq -0.209$, by virtue of the circle criterion the closed-loop system in Fig. 6.4 is ISS, and, more specifically, admits an ISS-Lyapunov function for all $k_h < \frac{1}{0.209} = 3.37$. For the SISO LTI system described by G_{yu} it is easy to verify that the $\mathcal{L}_1$-norm of $G_{yu}(s)$, denoted by γ_1, satisfies $\gamma_1 = 1.419$. Since $\gamma_1 k_h = 0.85 < 1$, it then follows that the small-gain condition in Theorem 6.3 is satisfied for all $\alpha/\omega_h > 1.419$. This region is indicated by the region to the right of the red line in Fig. 6.5. It is interesting to remark that for small values of ω_h and large ratios of α/ω_h, convergence is guaranteed by the small-gain condition in Theorem 6.3, but not by the LMIs and the frequency-domain

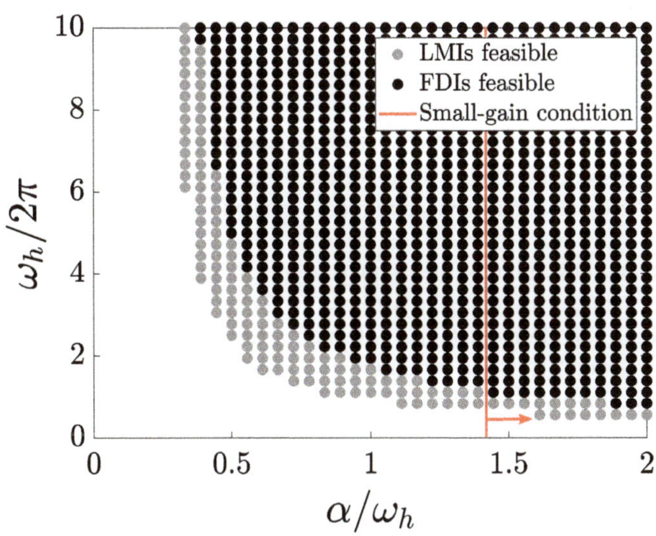

Fig. 6.5 A comparison between feasibility of the LMI conditions in Theorem 6.5 (grey dots), the frequency-domain conditions in Corollary 6.1 (black dots), and the small-gain condition in Theorem 6.3 (indicated by the region to the right of the red line) for different values of ω_h and ratios α/ω_h

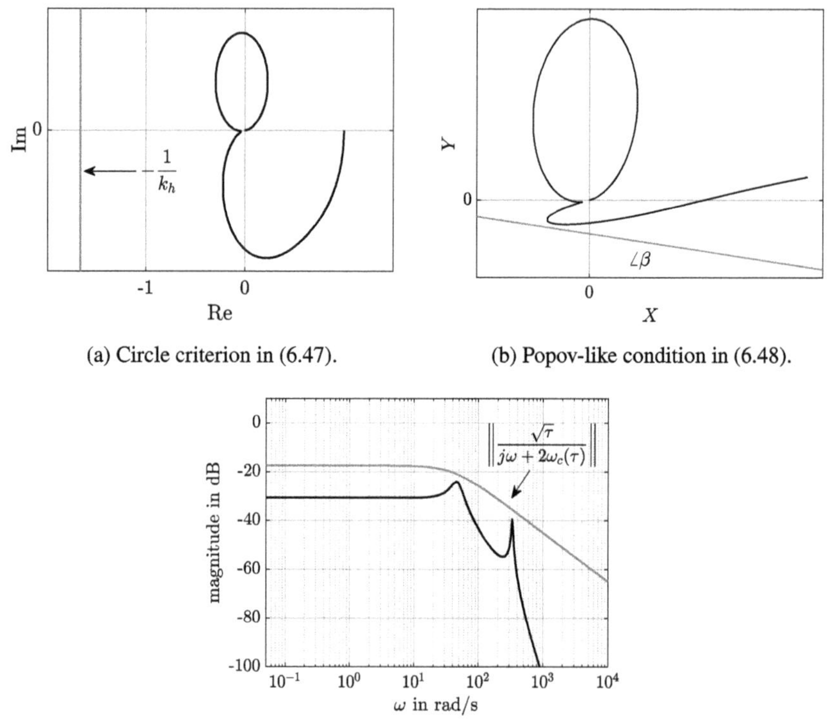

(a) Circle criterion in (6.47).

(b) Popov-like condition in (6.48).

(c) Small-gain-like condition in (6.50).

Fig. 6.6 Graphical interpretation of the three frequency-domain conditions in Corollary 6.1 for $\omega_h = 4 \cdot 2\pi$ rad/s, $\alpha/\omega_h = 0.6$, $\beta = -1/\alpha$, and $\tau = 67.2$

conditions in Theorem 6.5 and Corollary 6.1, respectively. This indicates that for the example discussed, the small-gain condition is not necessarily more restrictive than the Lyapunov-based conditions, and in that sense both conditions complement each other.

A graphical illustration of the frequency-domain conditions in Corollary 6.1 for $\omega_h = 4 \cdot 2\pi$ rad/s and $\alpha/\omega_h = 0.6$ with $\tau = 67.2$ is provided in Fig. 6.6. Note that the circle criterion condition is satisfied with significant margin, whereas the Popov-like condition and small-gain like condition appear more restrictive.

The closed-loop system with $\omega_h = 4 \cdot 2\pi$ rad/s and $\alpha/\omega_h = 0.6$ is simulated for different initial conditions, and with a sinusoidal input disturbance given by $d(t) = \sin(\omega t)$ with input frequency $\omega = 6 \cdot 2\pi$ rad/s. The corresponding steady-state output y of the plant and the output u of the nonlinearity are shown in Fig. 6.7. Clearly, the solutions converge to a unique steady-state periodic solution (shown in black) that has a fundamental period of 6 Hz, which is the same as that of the input w. Moreover, note that the output u is continuous, but it is not smooth.

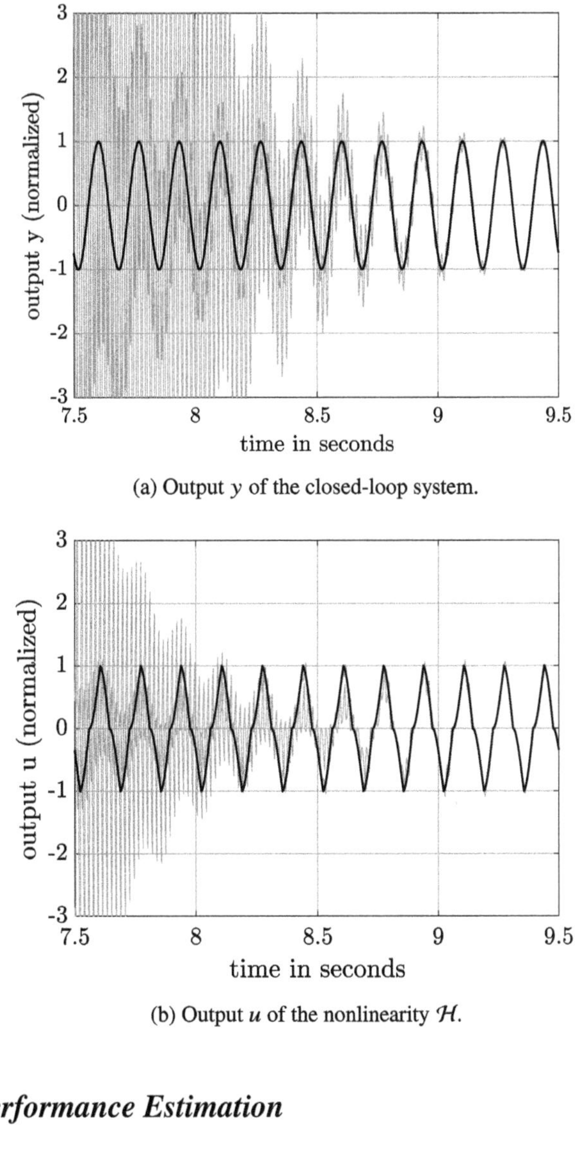

(a) Output y of the closed-loop system.

(b) Output u of the nonlinearity $\mathcal{H}$.

Fig. 6.7 Time-series simulation of the closed-loop system for different initial conditions and input $w(t) = \sin(\omega t)$ with $\omega = 6 \cdot 2\pi$ rad/s. The (transient) responses for different initial conditions are indicated in grey, and the steady-state solution is indicated in black

6.7.2 Steady-State Performance Estimation

Next it is shown how the tools from Sect. 6.6 can be applied for estimating performance of the closed-loop system from Fig. 6.4 on the basis of a describing function approximation with guaranteed error margins. For this purpose, suppose that the closed-loop system is subject to a sinusoidal input disturbance $d(t) = \sin(\omega t)$. Steady-state performance is quantified by the RMS ratio between the sinusoidal input d and the corresponding unique steady-state response $\bar{y}$ at various input frequencies ω. For LTI systems, this ratio corresponds to the process sensitivity which indicates how a harmonic input disturbance affects the steady-state output $\bar{y}$, i.e., how well disturbances are rejected by the closed-loop system. For the nonlinear system in Fig. 6.4 disturbance rejection properties can be estimated through a describing function-based approximation of the process sensitivity as given by

$$\hat{\mathcal{M}}(j\omega) = \frac{P(j\omega)}{1 + (1 + \Psi(j\omega))P(j\omega)C(j\omega)}, \tag{6.67}$$

in which $\Psi(j\omega) = \mathcal{D}(j\omega)$, with $\mathcal{D}(j\omega) \in \mathbb{C}$ the sinusoidal input describing function of $\mathcal{H}$ in (6.1) with $f_h(x_h, z) = -\alpha x_h + \omega_h z$ as provided in Sect. B.3.1.

Error bounds for the approximation in (6.67) can be established via the inequalities in (6.62). Note that in case of a sinusoidal input and for $\Psi(j\omega) = \mathcal{D}(j\omega)$, the term $\Lambda(t)$ that is used for constructing the bounds is specifically computed as

$$\Lambda(t) = \|\hat{\mathcal{M}}(j\omega)\| \left(\hat{\bar{u}}(t) - \|\mathcal{D}(j\omega)\| \sin(\omega t + \angle\mathcal{D}(j\omega))\right), \tag{6.68}$$

where $\hat{\bar{u}}$ is the response of $\mathcal{H}$ to a sinusoidal input, and is given by

$$\hat{\bar{u}}(t) = \begin{cases} \frac{\omega_h}{\sqrt{\omega^2+\alpha^2}} \sin(\omega t + \varphi) + \frac{\omega_h}{\omega^2+\alpha^2}\omega e^{-\alpha t}, & \text{if } \ 0 \le t < \frac{\gamma(\omega)}{\omega}, \\ k_h \sin(\omega t), & \text{if } \ \frac{\gamma(\omega)}{\omega} \le t < \frac{\pi}{\omega}, \\ \frac{\omega_h}{\sqrt{\omega^2+\alpha^2}} \sin(\omega t + \varphi) - \frac{\omega_h}{\omega^2+\alpha^2}\omega e^{-\alpha(t-\frac{\pi}{\omega})}, & \text{if } \ \frac{\pi}{\omega} \le t < \frac{\pi+\gamma(\omega)}{\omega}, \\ k_h \sin(\omega t), & \text{if } \ \frac{\pi+\gamma(\omega)}{\omega} \le t < \frac{2\pi}{\omega} \end{cases}$$

with $\gamma(\omega)$ the (numerically obtained) switching instance and $\varphi = \arctan(-\frac{\omega}{\alpha})$. The smallest value for μ appearing in (6.62) that leads to a solution to the matrix inequalities in Theorem 6.8 can be obtained numerically.

A steady-state performance estimate for the closed-loop system with $\omega_h = 4 \cdot 2\pi$ rad/s and $\alpha = 0.6\omega_h$ at various input frequencies is illustrated in Fig. 6.8. In this figure, an estimate for the RMS ratio $\|\bar{y}\|_{2,T}/\|d\|_{2,T}$ by means of the approximation in (6.67) is depicted by the red line, and the error bounds with $\mu = 6.101$ are indicated by the dotted black lines. The true RMS ratio obtained by means of time-series simulations is shown by the red dots.

It can be seen that the error bounds provide a useful uncertainty interval for steady-state performance and are computationally far less demanding to obtain than the exact RMS ratios via time-series simulations. Furthermore, for this example the describing function approximation seems to provide an accurate prediction of the closed-loop system performance when considering sinusoidal inputs. For each input frequency ω, the error $\|\bar{y} - \hat{\bar{y}}\|_{2,T}$ is shown in Fig. 6.9. The largest value for $\|\bar{y} - \hat{\bar{y}}\|_{2,T}$, i.e., the largest discrepancy (in terms of RMS value) between the unique steady-state response and its linear approximation is found to be $2.37 \cdot 10^{-7}$ and occurs at an input frequency of $\omega = 18 \cdot 2\pi$ rad/s. This frequency coincides with 1/3-rd of the resonance frequency at 54 Hz in the plant P, such that the third harmonic in the output of $\mathcal{H}$ is amplified in this case. Note that peaks are also visible at 1/5-th and 1/7-th of this resonance frequency. Despite these amplifications, the contribution

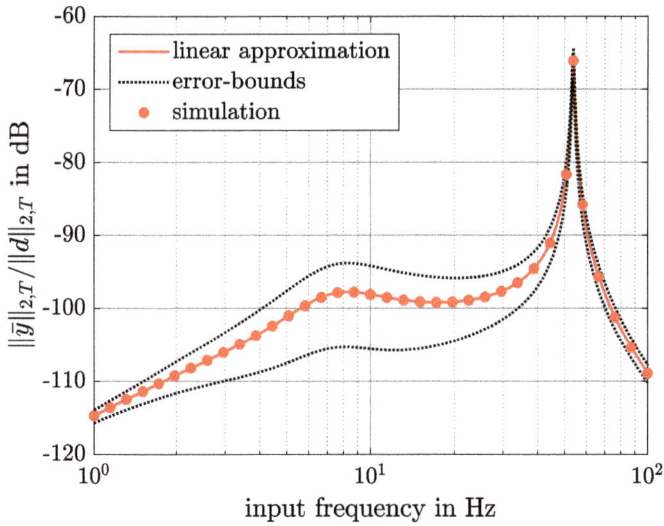

Fig. 6.8 Describing function approximation of the RMS ratio between a sinusoidal input $d(t) = \sin(\omega t)$ at various input frequencies and the steady-state response $\bar{y}$ (red line), error bounds (dotted black), and time-series simulations (red dots)

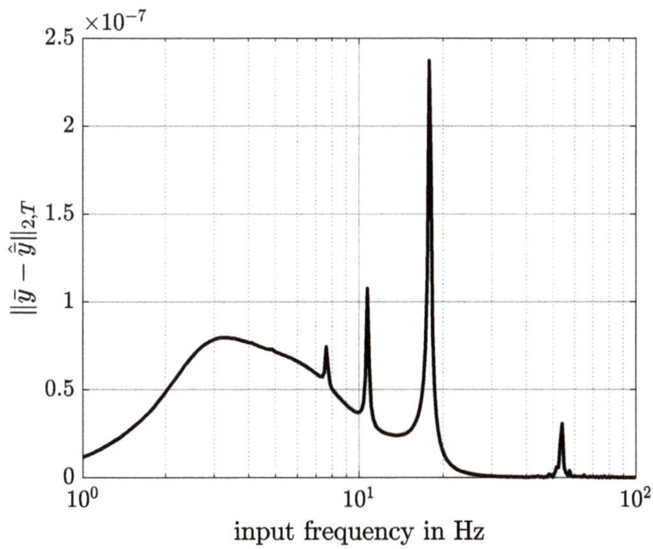

Fig. 6.9 RMS value of the steady-state approximation error $\|\bar{y} - \hat{\bar{y}}\|_{2,T}$ at various input frequencies

of higher harmonics in the output remains small, which may be attributed to the fact that the closed-loop transfer function $G_{yu}(s)$ from u to $z = -y$ has significant low-pass filtering characteristics, thereby largely filtering the higher-order harmonics generated by $\mathcal{H}$.

6.8 Summary

In this chapter, time- and frequency-domain conditions for (input-to-state) convergence of systems consisting of the feedback interconnection of an LTI plant and a hybrid integrator are presented. The hybrid integrator considered in this chapter can be seen as a generalization of HIGS discussed in earlier chapters. By exploiting incremental properties of the hybrid integrator that hold in a subset of the state space, two time-domain conditions for (input-to-state) convergence are formulated: an input–output small-gain condition, and a Lyapunov-based condition. The latter is formulated as a set of numerically tractable LMI conditions that, when feasible, guarantee the existence of a specific piecewise quadratic incremental Lyapunov function for the system. A frequency-domain interpretation of the LMIs is given, which allows for a graphical verification of the conditions and may provide further insights in robustness and parameter tuning. The convergence property is exploited for estimating steady-state performance on the basis of the describing function with guaranteed error margins.

The tools that have been developed so far are particularly useful for analysis of a given system containing HIGS (or a generalization thereof). Some of these tools also provide a useful direction for parameter tuning, but do not immediately support intuitive design of a closed-loop controller in its totality for the best interplay between linear and nonlinear parts. To deal with this, in the next chapter two approaches toward the design of HIGS-based controllers are pursued. The first approach exploits rigorous LMIs for controller synthesis, whereas the second approach is based on heuristics and exploits the describing function within a loop-shaping framework.

References

1. Alcorta Garcia, E., Rodriguez-Alfaro, L.H., Diaz-Romero, D., Posadas-Castillo, C.: Characterization of frequency response functions for nonlinear convergent systems. In: Conference on Control and Fault-Tolerant Systems, pp. 630–635 (2010)
2. Angeli, D.: A Lyapunov approach to incremental stability properties. IEEE Trans. Autom. Control **47**, 410–421 (2002)
3. Ascher, U., Mattheij, R., Russell, R.: Numerical Solution of Boundary Value Problems for Ordinary Differential Equations, 1st edn. Prentice-Hall Series in Computational Mathematics, Prentice-Hall (1988)
4. Bertollo, R., van den Eijnden, S., Heemels, W.M.: Incremental stability of discrete-time first-order projection elements. IFAC-PapersOnLine **58**(11), 123–128 (2024). (8th IFAC Conference on Analysis and Design of Hybrid Systems ADHS 2024)

5. Brogliato, B., Tanwani, A.: Dynamical systems coupled with monotone set-valued operators: formalisms, applications, well-posedness, and stability. SIAM Rev. **62** (2020)
6. Chaffey, T., Forni, F., Sepulchre, R.: Graphical nonlinear system analysis. IEEE Trans. Autom. Control **68**(10), 6067–6081 (2023)
7. Chait, Y., Hollot, C.V.: On Horowitz's contributions to reset control. Int. J. Robust Nonlinear Control **12**(4), 335–355 (2002)
8. Coogan, S.D., Margaliot, M.: Approximating the steady-state periodic solutions of contractive systems. IEEE Trans. Autom. Control **64**, 847–853 (2019)
9. Deenen, D.A., Heertjes, M.F., Heemels, W.P.M.H., Nijmeijer, H.: Hybrid integrator design for enhanced tracking in motion control. In: Proceedings of the American Control Conference (ACC'17), pp. 2863–2868 (2017)
10. Demidovich, B.P.: Lectures on Stability Theory. Moscow (1967)
11. Forni, F., Sepulchre, R.: A differential Lyapunov framework for contraction analysis. IEEE Trans. Autom. Control **59**(3), 614–628 (2012)
12. Fromion, V., Monaco, S., Normand-Cyrot, D.: Asymptotic properties of incrementally stable systems. IEEE Trans. Autom. Control **41**(5), 721–723 (1996)
13. Gelb, A., Vander Velde, W.: Multiple-Input Describing Functions and Nonlinear Systems Design. McGraw Hill, New York (1968)
14. Ghodrat, M., Marquez, H.: L 2-incremental gain stability of linear systems with nonlinear Lipschitz actuator. Syst. Control Lett. **147**, 104846 (2021)
15. Gruntjens,K,. Heertjes, M., Van Loon, S., Van De Wouw, N., Heemels, W.: Hybrid integral reset control with application to a lens motion system. In: 2019 American Control Conference (ACC), pp. 2408–2413 (2019)
16. Heemels, W.P.M.H., Tanwani, A.: Existence and completeness of solutions to extended projected dynamical systems and sector-bounded projection-based controllers. IEEE Control Syst. Lett. **7**, 1590–1595 (2023)
17. Heemels, W.P.M.H., Camlibel, M.K., Heertjes, M.F.: Oblique projected dynamical systems and incremental stability under state constraints. IEEE Control Syst. Lett. **4**(4), 1060–1065 (2020)
18. Horowitz, I., Rosenbaum, P.: Non-linear design for cost of feedback reduction in systems with large parameter uncertainty. Int. J. Control **21**(6), 977–1001 (1975)
19. Hu, G.D.: Observers for one-sided Lipschitz non-linear systems. IMA J. Math. Control. Inf. **23**(4), 395–401 (2006)
20. Khalil, H.K.: Nonlinear Systems, 3rd edn. Prentice-Hall, Upper Saddle River, New Jersey (2002)
21. Li, Y., Phillips, S., Sanfelice, R.: Results on incremental stability for a class of hybrid systems. In: 53rd IEEE Conference on Decision and Control, pp. 3089–3094 (2014)
22. Lohmiller, W., Slotine, J.J.E.: On contraction analysis for non-linear systems. Automatica **34**(6), 683–696 (1998)
23. Morinaga, E., Hirata, K., Ohta, Y.: Performance analysis of control systems with input constraints via piecewise quadratic storage functions. Asian J. Control **6**, 34–45 (2004)
24. Pavlov, A., Pogromsky, A., van de Wouw, N., Nijmeijer, H., Rooda, K.: Convergent piecewise affine systems: analysis and design part ii: discontinuous case. In: Proceedings of the 44th IEEE Conference on Decision and Control, pp. 5397–5402 (2005)
25. Pavlov, A., van de Wouw, N., Nijmeijer, H.: Uniform output regulation of nonlinear systems: a convergent dynamics approach. In: Systems and Control: Foundations and Applications. Birkhäuser Verlag, Switzerland (2006)
26. Pavlov, A., van de Wouw, N., Nijmeijer, H.: Frequency response functions for nonlinear convergent systems. IEEE Trans. Autom. Control **52**(6), 1159–1165 (2007)
27. Pavlov, A., Wouw, N., Pogromsky, A., Heertjes, M., Nijmeijer, H.: Frequency domain performance analysis of nonlinearly controlled motion systems. In: Proceedings of the IEEE Conference on Decision and Control, pp. 1621–1627 (2008)
28. Pavlov, A., Hunnekens, B., Wouw, N., Nijmeijer, H.: Steady-state performance optimization for nonlinear control systems of lur'e type. Automatica **49**(7), 2087–2097 (2013)

29. Pogromsky, A., van den Berg, R.: Frequency domain performance analysis of lur'e systems. IEEE Trans. Control Syst. Technol. **22**(5), 1949–1955 (2014)
30. Rantzer, A., Johansson, M.: Piecewise linear quadratic optimal control. IEEE Trans. Autom. Control **45**(4), 629–637 (2000)
31. Romanchuk, B.G., Smith, M.C.: Incremental gain analysis of piecewise linear systems and application to the antiwindup problem. Automatica **35**(7), 1275–1283 (1999)
32. Ruffer, B.S., van de Wouw, N., Mueller, M.: Convergent systems vs. incremental stability. Syst. Control Lett. **62**(3), 277–285 (2013)
33. Russo, G., Di Bernardo, M., Sontag, E.: Global entrainment of transcriptional systems to periodic inputs. PLoS Comput. Biol. **6**, 1000739 (2010)
34. Rutland, N.K., Lane, P.G.: Computing the 1-norm of the impulse response of linear time-invariant systems. Syst. Control Lett. **26**(3), 211–221 (1995)
35. Sharif, B., Heertjes, M., Nijmeijer, H., Heemels, W.: On the equivalence of extended and oblique projected dynamics with applications to hybrid integrator-gain systems. In: American Control Conference (ACC) 2021, New Orleans, USA (2021)
36. Skogestad, S., Postlethwaite, I.: Multivariable Feedback Control: Analysis and Design, 2nd ed. Wiley and Sons (2010)
37. Thenozhi, S., Tang, Y.: Frequency response function modeling of nonlinear convergent systems. In: 2016 IEEE 13th International Conference on Networking, Sensing, and Control (ICNSC), pp. 1–6 (2016)
38. van den Berg, R., Pogromsky, A., Rooda, K.: Well-posedness and accuracy of harmonic linearization for lur'e systems. In: In Proceedings of the 2008 IEEE Conference of Decision and Control (CDC'08), pp. 3841–3846 (2008)
39. van den Eijnden, S., Heertjes, M., Nijmeijer, H., Heemels, W.M.: On convergence of systems with sector-bounded hybrid integrators. In: 2022 IEEE 61st Conference on Decision and Control (CDC), pp. 7636–7641 (2022)
40. van den Eijnden, S.J.A.M., Heertjes, M.F., Nijmeijer, H., Heemels, W.P.M.H.: A small-gain approach to incremental input-to-state stability analysis of hybrid integrator-gain systems. IEEE Control Syst. Lett. **7**, 2443–2448 (2023)
41. Van Loon, S., Gruntjens, K., Heertjes, M., Wouw, N., Heemels, W.M.: Frequency-domain tools for stability analysis of reset control systems. Automatica **82**, 101–108 (2017)
42. Waitman, S., Massioni, P., Bako, L,. Scorletti, G., Fromion, V.: Incremental l_2-gain analysis of piecewise-affine systems using piecewise quadratic storage functions. In: 2016 IEEE 55th Conference on Decision and Control (CDC), pp. 1334–1339 (2016)
43. Waitman, S., Massioni, P., Bako, L., Scorletti, G.: Incremental l2-gain stability of piecewise-affine systems with piecewise-polynomial storage functions. Automatica **107**, 224–230 (2019)
44. Yakubovich, V.: Solution of certain matrix inequalities in the stability theory of nonlinear control systems (in Russian). Dokl. Akad. Nauk SSSR **143**, 1304–1307 (1962)
45. Yakubovich, V.A.: Method of matrix inequalities in stability theory for nonlinear control systems: I. absolute stability of forced vibrations. Autom. Remote. Control. **25**, 1017–1029 (1964)

Part III
Design and Applications

Chapter 7
HIGS-Based Controller Design

7.1 Introduction

The tools that have been developed in previous chapters are particularly useful for the purpose of analysis, but provide limited direction toward synthesis of HIGS-based controllers that can achieve improved system performance. As performance improvements do not result from blind application of HIGS, but rather from a careful interplay between its switching mechanism and the remaining linear time-invariant (LTI) parts of the system, there is a need for comprehensive design procedures.

This chapter considers two approaches for HIGS-based controller design. The first approach presents a method for robust controller synthesis by exploiting matrix inequalities that are derived from the frequency-domain conditions in Chap. 4, and combines these with the $\mathcal{L}_2$-gain discussed in Chap. 5. The approach links to the work in [3] where circle and Popov criteria are used as tools for nonlinear feedback design.

A possible drawback of an approach based on matrix inequalities, however, is the lack of a clear physical interpretation regarding parameter tuning, and limited guidance for redesign when the set of inequalities turns out to be infeasible. In an industrial setting, these aspects may be experienced as less desirable. Therefore, as an alternative, a second approach that exploits a describing function characterization of HIGS within a robust loop-shaping framework is presented. Key in this approach is to express the modelling error induced by the describing function approximation as an uncertainty in the controller [11, 21, 22]. Controller design with the describing function of HIGS as a frequency-domain approximation of its nonlinear steady-state input–output behaviour then proceeds in a manner similar to that for LTI systems, and entails robustness against both modelling errors in the plant as well as in the controller. One may argue that design based on describing functions is not relevant (if at all possible) in the given nonlinear control context, for example, due to the lack of mathematical rigour in the statements on stability and performance. But, as frequency-domain tools are widely adopted by control engineers in industry, such an

S. van den Eijnden et al., *Hybrid Integrator-Gain Systems*, Lecture Notes in Control and Information Sciences 499, https://doi.org/10.1007/978-3-032-18101-5_7

approach may be of great practical value and the benefits associated with it cannot be ignored. It is advocated that the approach is not only useful for HIGS-based control design, but applies to the design of other nonlinear strategies such as reset control [8, 24] as well. Also notice in this regard that the need for systematic design procedures in general is acknowledged as being one of the open problems in reset and hybrid control [41].

The remainder of this chapter is organized as follows. In Sect. 7.2, the control design problem formulation is specified. In Sect. 7.3, robust synthesis of HIGS-based controllers using matrix inequalities is discussed. Section 7.4 presents a design approach that exploits describing functions within a robust loop-shaping framework. The main findings of this chapter are summarized in Sect. 7.5.

7.2 System Description and Problem Formulation

In this section, the problem of controller synthesis in the presence of plant uncertainties is discussed. First, the generic system set-up is considered, after which the setting is tailored toward HIGS-based controller design.

7.2.1 General System Description

Consider the generalized feedback configuration as shown in Fig. 7.1. Here, G is the generalized plant, which is represented in state space as the linear time-invariant (LTI) multi-input multi-output (MIMO) system

$$G: \begin{cases} \dot{x}_g(t) &= A_g x_g(t) + B_g v(t) + B_w w(t) + B_{w_\Delta} w_\Delta(t), \\ y(t) &= C_g x_g(t) + D_g v(t) + D_w w(t) + D_{w_\Delta} w_\Delta(t), \\ q(t) &= \bar{C}_q x_g(t) + \bar{D}_q v(t) + \bar{D}_w w(t) + \bar{D}_{w_\Delta} w_\Delta(t), \\ q_\Delta(t) &= \hat{C}_q x_g(t) + \hat{D}_q v(t) + \hat{D}_w w(t) + \hat{D}_{w_\Delta} w_\Delta(t), \end{cases} \tag{7.1}$$

where $x_g(t) \in \mathbb{R}^m$ denotes the states of the generalized plant, $q(t) \in \mathbb{R}^r$ contains the performance variables, such as tracking errors and control signals, $w(t) \in \mathbb{R}^l$ are the exogenous input signals, such as disturbances, measurement noise, and reference commands. The matrices (A_g, B_g, C_g, D_g) are assumed to be minimal, and the remaining matrices in (7.1) are of appropriate dimension. The uncertainty channels $w_\Delta(t) \in \mathbb{R}^p, q_\Delta(t) \in \mathbb{R}^q$ are related by $w_\Delta = \Delta_G q_\Delta$, where $\Delta_G \in \bar{\Delta}$ is an unstructured LTI plant uncertainty that belongs to the set

$$\bar{\Delta} := \left\{ \Delta_G \in \mathcal{RH}_\infty^{p\times q} \mid \|\Delta_G\|_\infty < 1 \right\}. \tag{7.2}$$

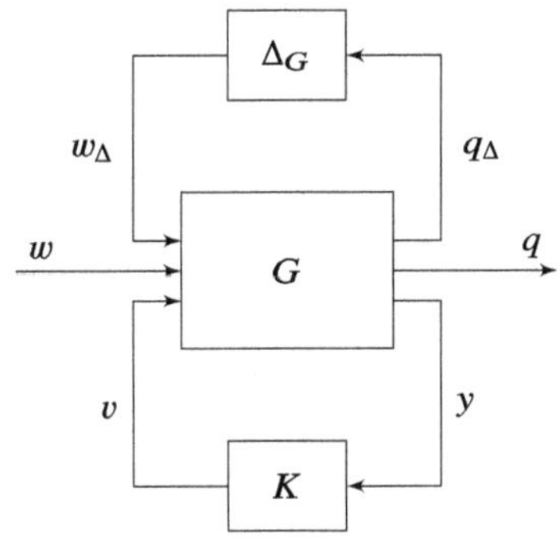

Fig. 7.1 Generalized feedback configuration with plant uncertainty

The measured output $y(t) \in \mathbb{R}$ is fed back to a (non)linear single-input single-output (SISO) feedback controller K that generates the control signal $v(t) \in \mathbb{R}$. The generalized plant G in Fig. 7.1 contains a nominal model of the plant to be controlled, scaling filters that normalize the uncertainty, and performance weighting filters [14, 32]. The latter filters are added for the purpose of including problem-specific input knowledge into the system description, and trade-off performance objectives between different input–output channels. In an LTI context, these filters can be used for specifying magnitude bounds on closed-loop transfer functions [26, 36].

7.2.2 Problem Formulation

The feedback controller K aims at stabilizing the closed-loop system, and achieving a certain level of system performance by minimizing (in some sense) the closed-loop mapping from exogenous inputs w to performance variables q. All of this must be achieved in the face of plant uncertainties, i.e., a level of robust closed-loop stability and performance must be achieved. Formally, the design of a (non)linear SISO feedback controller K can be formulated as the following constraint optimization problem:

$$\begin{aligned} \underset{K}{\text{minimize}} \quad & J(K) \\ \text{subject to} \quad & \left\| T_{qw}(K, \Delta_G) \right\|_\infty \leq \gamma, \\ & K \text{ stabilizes } G_\Delta, \\ & K \in \mathcal{K}, \end{aligned} \tag{7.3}$$

in which $T_{qw}(K, \Delta_G) : \mathbb{R}^n \to \mathbb{R}^m$ is the closed-loop mapping from inputs w to the performance channel q, for given K and $\Delta_G \in \bar{\Delta}$, which is desired to have an H_∞-norm of at most γ, with $\gamma > 0$. The H_∞-norm of a system is defined in terms of the $\mathcal{L}_2$-gain (see also Chap. 5 for a formal definition):

$$\|T_{qw}\|_\infty = \sup_{w \in \mathcal{L}_2 \setminus \{0\}} \frac{\|q\|_2}{\|w\|_2}. \tag{7.4}$$

In (7.3), G_Δ represents the interconnection of the generalized plant G in (7.1) and the uncertainty Δ_G, $J(K)$ is an objective function that resembles certain control system requirements (e.g., defined in terms of γ), and $K \in \mathcal{K}$ represents a structural constraint on the controller. In particular, the set $\mathcal{K}$ specifies the structure of the controller, as well as constraints on the associated tuning parameters. Structural constraints are added for accommodating practical situations, where it may be desired to find low-order controllers with a given structure, such as PID-based controllers, see, e.g., [1].

7.2.3 Toward HIGS-Based Controller Design

In an LTI context, two main directions can be found in the literature for solving the controller design problem (7.3): state space methods and frequency-domain methods. In a state space approach, it is customary to solve (7.3) using linear matrix inequalities (LMIs) [1, 2, 10, 12, 27, 42]. The main advantage of these methods is that the structure of the controller is completely left to the optimization routine, and the resulting controller is guaranteed to be the optimal one (with respect to the performance objective). However, these approaches all require a state space realization of the plant and controller. In a frequency-domain approach, (7.3) is solved via frequency-domain loop-shaping, done either manually [14, 33], or guided by an autotuner [4, 7]. Although a loop-shaping approach not necessarily leads to an optimal controller, and the structure of the controller should be determined beforehand by the control engineer, the main advantages over numerical state space methods are the possibility to directly use non-parametric models of the plant in the procedure, and the possibility for an intuitive/direct assessment of the effects from adjusting the controller parameters on the performance requirements. For these reasons, frequency-domain loop-shaping to this day remains the most popular method for LTI controller design in practice [29, 30].

The objective in this chapter is to formulate a procedure for effectively solving (7.3), in which $K = K(C_i, \mathcal{H})$ is a HIGS-based controller that is constructed by interconnecting HIGS $\mathcal{H}$ with several LTI filter elements $C_i(s)$, $i = 1, \ldots, m$. Note that the set $\mathcal{K}$ in (7.3) includes all controllers of this form. Inspired by the various directions for LTI controller design available in the literature, two approaches for solving (7.3) with K being a HIGS-based controller are presented. The first approach is based on matrix inequalities, and provides a synthesis procedure that leads to a closed-loop controller design with guaranteed robust stability and performance properties. This approach connects to state space methods for LTI design. The second approach exploits the describing function within a robust frequency-domain loop-shaping framework. This approach can be seen as a tuning procedure and connects to frequency-domain loop-shaping methods that are widely adopted in industry. Both approaches are discussed in detail in the next two sections.

7.3 Synthesis Using Matrix Inequalities

In this section, an approach toward synthesis of HIGS-based controllers for the purpose of input disturbance rejection is considered. The approach draws inspiration from the stability conditions in Chap. 4 and leads to (bilinear) matrix inequalities.

7.3.1 System Setting

Consider the setting in Fig. 7.2, where P is the LTI plant, C_1, C_2 are the LTI parts of the controller K, $\mathcal{H}$ represents HIGS, $d(t) \in \mathbb{R}$ is an input disturbance signal, $y(t) \in \mathbb{R}$ is the measured output signal used for feedback control, and $q(t) \in \mathbb{R}$ is the performance output signal. The (strictly) proper and stable SISO LTI weighting filters W and W_Δ are, respectively, used to include knowledge of the input disturbance and

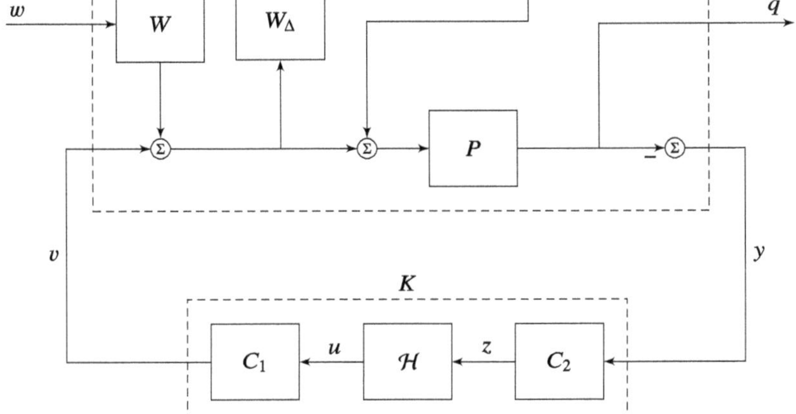

Fig. 7.2 Input disturbance rejection within the generalized plant set-up

plant uncertainty. Note that input multiplicative uncertainty is considered here, but, given the LTI characteristics of the SISO plant and uncertainty, output multiplicative uncertainty can be considered as well. It is assumed that the plant P in Fig. 7.2 has a relative degree of at least two and that the SISO LTI filters C_1 and C_2 are proper (causal) filters, such that, conform Assumption 2.5, the external inputs v, w, w_Δ as well as their time-derivatives do not appear directly in the input to HIGS.

The LTI filters in K are initially fixed to $C_2(s) = 1$ and $C_1(s) = C(s) = C_c(sI - A_c)^{-1}B_c + D_c$, leading to the equivalent state space description

$$C : \begin{cases} \dot{x}_c(t) &= A_c x_c(t) + B_c u(t), \\ v(t) &= C_c x_c(t) + D_c u(t), \end{cases} \tag{7.5}$$

with $x_c(t) \in \mathbb{R}^s$, $u(t) \in \mathbb{R}$ and $v(t) \in \mathbb{R}$ the filter state, input, and output, respectively, at time $t \in \mathbb{R}_{\geq 0}$, and where A_c, B_c, C_c, D_c are matrices of consistent dimensions. These specific choices for $C_1(s)$ and $C_2(s)$ are useful for deriving controller synthesis conditions, and will be reconsidered later.

Recall from earlier chapters (specifically Chap. 2) that HIGS is formulated as

$$\mathcal{H} : \begin{cases} \dot{x}_h(t) &= \omega_h z(t), \quad \text{if } (z(t), u(t), \dot{z}(t)) \in \mathcal{F}_1, \\ x_h(t) &= k_h z(t), \quad \text{if } (z(t), u(t), \dot{z}(t)) \in \mathcal{F}_2, \\ u(t) &= x_h(t), \end{cases} \tag{7.6}$$

where $x_h(t) \in \mathbb{R}$ is the integrator state, $z(t) \in \mathbb{R}$ is the input, $\dot{z}(t)$ is the corresponding time-derivative, $u(t) \in \mathbb{R}$ is the output at time $t \in \mathbb{R}_{\geq 0}$, and $\mathcal{F}_1$ and $\mathcal{F}_2$ are given by

$$\mathcal{F}_1 := \left\{(z, u, \dot{z}) \in \mathbb{R}^3 \mid (z, u, \dot{z}) \in \mathcal{F} \setminus \mathcal{F}_2\right\}, \tag{7.7a}$$

$$\mathcal{F}_2 := \left\{(z, u, \dot{z}) \in \mathbb{R}^3 \mid (z, u, \dot{z}) \in \mathcal{F} \vee u = k_h z \wedge \omega_h z^2 > k_h \dot{z} z\right\}, \tag{7.7b}$$

with the $[0, k_h]$-sector specified as

$$\mathcal{F} = \left\{(z, u, \dot{z}) \in \mathbb{R}^2 \mid zu \geq \frac{1}{k_h} u^2\right\}. \tag{7.8}$$

The design parameters of HIGS are given by the integrator frequency $\omega_h \in \mathbb{R}_{\geq 0}$ and the gain $k_h \in \mathbb{R}_{>0}$. Close inspection reveals that within the specific configuration of Fig. 7.2 the gain k_h is redundant when used in conjunction with additional filters, as its effect can always be obtained by appropriate scaling of the output u, and integrator parameter ω_h. Its effect can thus be encapsulated in the proportional gain of $C_1 = C$. In the remainder of this section, this parameter is therefore set at $k_h = 1$.

Interconnecting (7.1)–(7.6) in the manner as shown in Fig. 7.2, and treating q_Δ, w_Δ as additional input–output channels results in the standard closed-loop system description

$$\Sigma : \begin{cases} \dot{x}(t) &= A_i x(t) + Bw(t) + B_\Delta w_\Delta(t), \quad \text{if } x(t) \in \mathcal{X}_i,\ i \in \{1, 2\} \\ q(t) &= Cx(t), \\ q_\Delta(t) &= C_\Delta x(t) + D_\Delta w(t), \end{cases} \tag{7.9}$$

with states $x(t) = \left[x_g(t)^\top,\ x_c(t)^\top,\ x_h(t)\right]^\top \in \mathbb{R}^n$, $n = m + s + 1$, and $w_\Delta = \Delta_G q_\Delta$, with $\Delta_G \in \bar{\Delta}$, and where the set $\bar{\Delta}$ is specified in (7.2). The sets $\mathcal{X}_i, i = \{1, 2\}$, are given by

$$\mathcal{X}_i = \left\{x \in \mathbb{R}^n \mid Ex \in \mathcal{F}_i\right\}, \tag{7.10}$$

where the matrix E is given by

$$E = \begin{bmatrix} -C_g^\top & 0 & -(C_g A_g)^\top \\ 0 & 0 & 0 \\ 0 & 1 & 0 \end{bmatrix}. \tag{7.11}$$

The system matrices are written as

$$A_1 = \begin{bmatrix} A_g & B_g C_c & B_g D_c \\ 0 & A_c & B_c \\ -\omega_h C_g & 0 & 0 \end{bmatrix}, \quad A_2 = \begin{bmatrix} A_g & B_g C_c & B_g D_c \\ 0 & A_c & B_c \\ -k_h C_g A_g & 0 & 0 \end{bmatrix}, \tag{7.12a}$$

whereas the input matrices are given by

$$B = \left[B_w^\top\ 0\ 0\right]^\top, \quad B_\Delta = \left[B_{w_\Delta}^\top\ 0\ 0\right]^\top, \tag{7.12b}$$

and the output matrices are written as

$$C = \left[\bar{C}_q\ 0\ 0\right], \quad C_\Delta = \left[\hat{C}_q\ \hat{D}_q C_c\ \hat{D}_q D_c\right], \quad D_\Delta = \hat{D}_w. \tag{7.12c}$$

It is furthermore useful to define the matrices $C_z := \left[-C_g\ 0\ 0\right]$, and $C_u := \left[0\ 0\ 1\right]$.

7.3.2 From Analysis to Synthesis

Input-to-state stability (ISS) of the *nominal* closed-loop system in Fig. 7.2 (with $\Delta_G = 0$) can be assessed via Theorem 4.4, where the existence of a quadratic ISS-Lyapunov function is guaranteed via the frequency-domain inequality

$$1 + \mathrm{Re}\left\{\left(\lambda k C_u + F + \left(\lambda + k\frac{\omega_h}{k_h}\right) H\right)(j\omega I - \mathcal{A}_2)^{-1} b\right\} > 0, \tag{7.13}$$

with $\lambda \geq 0$, $k \geq 1$, and where the corresponding system matrices are given by $b = C_u^\top$, $H = k_h C_z - C_u$, $F = C_z(k_h A_1 - \omega_h I)$, and $\mathcal{A}_2 = A_2 + k\frac{\omega_h}{k_h} bH$. From the

Kalman–Yakubovich–Popov (KYP) lemma (see also Chap. 4), it readily follows that (7.13) is equivalent to the existence of a matrix $M = M^\top$ and numbers $\lambda \geq 0$, $k \geq 1$ that satisfy

$$M = M^\top \succ 0, \tag{7.14a}$$

$$\begin{bmatrix} \mathcal{A}_2^\top M + M\mathcal{A}_2 & Mb - c^\top \\ b^\top M - c & -2 \end{bmatrix} \prec 0, \tag{7.14b}$$

in which $c = \lambda k C_u + F + (\lambda + k\frac{\omega_h}{k_h})H$. Recall from Theorem 4.4 that (7.14) provides sufficient conditions for guaranteeing feasibility of the set of inequalities in (4.10) with $\tau_1, \tau_2 \geq 0$ and $\tau_3 = 0$, i.e., the existence of a common quadratic Lyapunov function (including S-procedure terms) for the closed-loop system in (7.9). Different from the set of "direct" inequalities in (4.10), however, the set of inequalities in (7.14) can also be used in an LTI context to show, e.g., passivity of the system $(\mathcal{A}_2, b, c, 2)$. In this regard, typical machinery used for synthesizing LTI controllers via matrix inequalities comparable with (7.14) can conveniently be exploited for synthesizing HIGS-based controllers. Before discussing this in more detail, the inequalities in (7.14) are extended toward conditions for robust performance of the closed-loop system in (7.9). Hereto, a small-gain theorem from the input channel w_Δ to the output q_Δ is exploited.

Proposition 7.1 *Consider the closed-loop system in (7.9) with $w_\Delta = \Delta_G q_\Delta$. Suppose there exist a symmetric matrix $M = M^\top \succ 0$ and numbers $\lambda \geq 0$, $k \geq 1$, and $\gamma > 0$ that satisfy the matrix inequality*

$$\begin{bmatrix} \mathcal{A}_2^\top M + M\mathcal{A}_2 & Mb - c^\top & M\bar{B} & \bar{C}^\top \\ b^\top M - c & -2 & 0 & 0 \\ \bar{B}^\top M & 0 & -\Gamma & \bar{D}^\top \\ \bar{C} & 0 & \bar{D} & -\Gamma \end{bmatrix} \prec 0, \tag{7.15}$$

in which $\bar{B} = [B \; B_\Delta]$, $\bar{C} = [C^\top \; C_\Delta^\top]^\top$, $\bar{D} = diag(0, D_\Delta)$, and $\Gamma = diag(\gamma, 1)$. Then for all $\Delta_G \in \bar{\Delta}$ with $\bar{\Delta}$ in (7.2), the closed-loop system in (7.9) is input-to-state stable and has a finite $\mathcal{L}_2$-gain from w to q as in (7.4) smaller than or equal to γ.

Proof By virtue of the Schur complement, (7.15) is equivalent to

$$\begin{bmatrix} \mathcal{A}_2^\top M + M\mathcal{A}_2 + Q & M\bar{B} \\ \bar{B}^\top M & -\Gamma \end{bmatrix} + \begin{bmatrix} \bar{C}^\top \\ \bar{D}^\top \end{bmatrix} \Gamma^{-1} \begin{bmatrix} \bar{C} & \bar{D} \end{bmatrix} \prec 0, \tag{7.16}$$

with $\mathcal{A}_2 = \left(A_2 + k\frac{\omega_h}{k_h} bH\right)$ and $Q = \frac{1}{2}\left(Mb - c^\top\right)\left(Mb - c^\top\right)^\top \succeq 0$. Pre- and post-multiplying (7.16) with $(x^\top, \bar{w}^\top) = (x^\top, w^\top, w_\Delta^\top)$ yields

$$x^\top \left(\mathcal{A}_2^\top M + M\mathcal{A}_2 + Q\right) x + 2x^\top M\bar{B}\bar{w} + \frac{1}{\gamma} q^2 - \gamma w^2 + q_\Delta^2 - w_\Delta^2 < 0. \tag{7.17}$$

Define $V(x) = x^\top P x$ with

$$P = M + \lambda H^\top H + \lambda k(C_u^\top H + H^\top C_u), \tag{7.18}$$

and observe that due to the assumption on the relative degree, one finds $P\bar{B} = M\bar{B}$. Then, using the result in Theorem 4.4 it follows that $t \mapsto V(x(t))$ satisfies for almost all times and all $x \in \mathcal{X}$

$$\dot{V} \leq -\epsilon V + \gamma w^2 - \frac{1}{\gamma} q^2 + w_\Delta^2 - q_\Delta^2, \tag{7.19}$$

with $\epsilon > 0$ sufficiently small. By virtue of the bounded real lemma [23] it follows that for all $\Delta_G \in \bar{\Delta}$ there exists a $V_\Delta(x_\Delta) > 0$ with $x_\Delta \in \mathbb{R}^s$ the (internal) states pertaining Δ_G that satisfies $\dot{V}_\Delta \leq -\mu V_\Delta + q_\Delta^2 - w_\Delta^2$. Hence, for the interconnection of the nominal system with the uncertainty Δ_G one finds $U(x, x_\Delta) = V(x) + V_\Delta(x_\Delta) > 0$ for all $(x, x_\Delta) \in \mathcal{X} \times \mathbb{R}^s$, and

$$\dot{U} \leq -\tau U - \frac{1}{\gamma} q^2 + \gamma w^2, \tag{7.20}$$

for all $(x, x_\Delta) \in \mathcal{X} \times \mathbb{R}^s$, and where $\tau = \min\{\epsilon, \mu\}$. The proof is then completed using standard arguments. □

The matrix inequality in (7.15) is instrumental in synthesizing robust HIGS-based controllers with guaranteed performance properties. In reformulating (7.15) into appropriate synthesis conditions, note that the parameters associated with the LTI part of the controller only appear in the matrix inequality (7.15) via the matrices A_2 and $\bar{C}$. By gathering all LTI controller parameters into the single variable

$$K_c = \begin{bmatrix} A_c & B_c \\ C_c & D_c \end{bmatrix}, \tag{7.21}$$

and introducing the notation

$$A = \begin{bmatrix} A_g & 0\ 0 \\ 0 & 0\ 0 \\ -k_h C_g A_g & 0\ 0 \end{bmatrix}, \quad \bar{C}_0 = \begin{bmatrix} \bar{C}_q & 0\ 0 \\ \hat{C}_q & 0\ 0 \end{bmatrix}, \tag{7.22}$$

and

$$\mathcal{B} = \begin{bmatrix} 0 & B_g \\ I & 0 \\ 0 & 0 \end{bmatrix}, \quad \mathcal{C} = \begin{bmatrix} 0 & I & 0 \\ 0 & 0 & I \end{bmatrix}, \quad \mathcal{D} = \begin{bmatrix} 0 & 0 \\ 0 & \hat{D}_q \end{bmatrix}, \tag{7.23}$$

the matrix A_2 can be written more compactly as $A_2 = A + \mathcal{B}K_c\mathcal{C}$, whereas the matrix C is written as $C = \bar{C}_0 + \mathcal{D}K_c\mathcal{C}$. As such, one can write the corresponding inequality in (7.15) in the form

$$\begin{bmatrix} \mathcal{A}^\top M + M\mathcal{A} & Mb - c^\top & M\bar{B} & \bar{C}_0^\top \\ b^\top M - c & -2 & 0 & 0 \\ \bar{B}^\top M & 0 & -\Gamma & \bar{D}^\top \\ \bar{C}_0 & 0 & \bar{D} & -\Gamma \end{bmatrix} + \text{He}\left\{ \begin{bmatrix} M\mathcal{B} \\ 0 \\ \mathcal{D} \end{bmatrix} K_c \begin{bmatrix} C & 0 & 0 \end{bmatrix} \right\} \prec 0, \quad (7.24)$$

in which $\mathcal{A} = A + k\frac{\omega_h}{k_h}bH$, with A given in (7.22). The inequality in (7.24) is not an LMI and cannot be used directly for synthesis. However, it is in a standard form that allows for direct application of the elimination lemma [6]. By applying a convenient congruence transformation to (7.24) with the matrix $T := \text{diag}(1, J, I)$ with $J = \left[\begin{smallmatrix} 0 & 1 \\ 1 & 0 \end{smallmatrix}\right]$, and subsequently following the steps in [12, Theorems 4.2 and 4.3], it is possible to show that the conditions in Proposition 7.1 are equivalent to the following set of inequalities:

$$\begin{bmatrix} W_{\mathcal{B}_0} & 0 \\ 0 & I \\ I & 0 \end{bmatrix}^\top \begin{bmatrix} \mathcal{A}_0 R + R\mathcal{A}_0^\top & b_0 - Rc_0^\top & B_0 & RC_0^\top \\ b_0^\top - c_0^\top R & -2 & 0 & 0 \\ B_0^\top & 0 & -\Gamma & D_0^\top \\ C_0 R & 0 & D_0 & -\Gamma \end{bmatrix} \begin{bmatrix} W_{\mathcal{B}_0} & 0 \\ 0 & I \\ I & 0 \end{bmatrix} \prec 0, \quad (7.25a)$$

$$\begin{bmatrix} W_{C_0} & 0 \\ 0 & I \\ I & 0 \end{bmatrix}^\top \begin{bmatrix} S\mathcal{A}_0 + \mathcal{A}_0^\top S & Sb_0 - c_0^\top & SB_0 & C_0^\top \\ b_0^\top S - c_0 & -2 & 0 & 0 \\ B_0^\top S & 0 & -\Gamma & D_0^\top \\ C_0 & 0 & D_0 & -\Gamma \end{bmatrix} \begin{bmatrix} W_{C_0} & 0 \\ 0 & I \\ I & 0 \end{bmatrix} \prec 0, \quad (7.25b)$$

$$\begin{bmatrix} R & I \\ I & S \end{bmatrix} \succ 0, \quad (7.25c)$$

in which $\mathcal{A}_0 = A_0 + k\frac{\omega_h}{k_h}b_0 H_0$ with

$$A_0 = \begin{bmatrix} A_g & 0 \\ -\omega_h C_g & 0 \end{bmatrix}, \quad B_0 = \begin{bmatrix} B_w & B_{w_\Delta} \\ 0 & 0 \end{bmatrix}, \quad C_0 = \begin{bmatrix} \bar{C}_q & 0 \\ \hat{C}_q & 0 \end{bmatrix}, \quad D_0 = \bar{D}, \quad (7.26)$$

and $b_0 = [0, 1]^\top$, $H_0 = [-k_h C_g, -1]$, $c_0 = \lambda k b_0^\top + F_0 + (\lambda + k\frac{\omega_h}{k_h})H_0$ with $F_0 = [C_g(\omega_h I - k_h A_g), 0]$. Furthermore, the matrices $W_{\mathcal{B}_0}$ and W_{C_0} satisfy

$$\text{im } W_{\mathcal{B}_0} = \ker\,[B_g^\top, \hat{D}_q], \quad \text{and} \quad \text{im } W_{C_0} = \ker\,[0, I]. \quad (7.27)$$

Note that the above inequalities result from assuming the following structure on the matrix $\hat{M} = T^\top M T$:

$$\hat{M} = \begin{bmatrix} R & X \\ X^\top & \star \end{bmatrix}, \quad \text{and} \quad \hat{M}^{-1} = \begin{bmatrix} S & Y \\ Y^\top & \star \end{bmatrix}, \quad (7.28)$$

where $R, S \in \mathbb{R}^{(m+1)\times(m+1)}$ and $X, Y \in \mathbb{R}^{s\times s}$. For details regarding the structure in (7.28) and the explicit steps taken in reformulating the inequalities, the reader is referred to the seminal results in [12].

7.3.3 Controller Synthesis Algorithm

Based on the conditions derived above and under the assumption $k_h = 1$, the controller synthesis problem can be formulated as follows:

$$\begin{aligned} \underset{R,S,k,\lambda,\omega_h}{\text{minimize}} \quad & \gamma \\ \text{subject to} \quad & (7.25) \\ & \lambda \geq 0,\ k \geq 1,\ \omega_h > 0. \end{aligned} \tag{7.29}$$

The inequalities in (7.25) are not LMIs due to products of the variables R, S with λ, k, ω_h, such that (7.29) is a non-convex optimization problem, potentially with multiple local minima. Clearly, one could perform an exhaustive search over the parameters λ, k, and ω_h to find a (local) minimum to (7.29). Such an approach, however, may be computationally intensive. To potentially improve the computational efficiency, one could exploit additional structure in (7.25). Namely, from the observation that $b_0^\top W_{C_0} = 0$ the inequality in (7.25b) appears to be quasi-convex, in the sense that the inequality is convex (in the variables S, λ, ω_h, and γ) for fixed $\tau := k\omega_h$. This allows for a simple and intuitive two-step approach for finding a solution to the inequalities in (7.25), which is summarized as follows:

- **Step 1:** Fix τ, and minimize γ subject to the LMIs (7.25b), $S \succ 0$, $\lambda \geq 0$, and $\tau \geq \omega_h > 0$ in which the free variables are S, λ, and ω_h.
- **Step 2:** With the obtained values for S, λ, ω_h, and $k = \tau/\omega_h$, minimize γ subject to the LMIs (7.25a), and (7.25c).

If feasible, the resulting controller has guaranteed robust performance indicated by the maximum γ obtained in the above two steps and with respect to all uncertainties $\Delta_G \in \bar{\Delta}$. Although numerically efficient, the solution of this approach is not guaranteed to be a (local) minimum to (7.29). To refine the result, one can perform a bisection over the variable γ, or resort to more sophisticated (iterative) schemes [15, 16, 35].

When a feasible solution is found, the LTI controller parameters K_c in (7.21) can be retrieved from the obtained matrices R and S by reconstructing $\hat{M}$ in (7.28). Hereto, the equality $XY^\top = I - RS$ should be solved first, e.g., by using a singular value decomposition of $I - RS$, to obtain X and Y. Then, $\hat{M}$ is recovered as (see [12, Sect. 7])

$$\hat{M} = \begin{bmatrix} S & I \\ Y^\top & 0 \end{bmatrix} \begin{bmatrix} I & R \\ 0 & X^\top \end{bmatrix}^{-1}. \tag{7.30}$$

The matrix $M = T^{-\top}\hat{M}T$, and parameters λ, k, ω_h and γ can then be used in (7.24) to compute K_c. Note that a solution to (7.24) is guaranteed to exist when (7.25) is feasible.

The resulting LTI part of the controller can be written as $C(s) = C_c(sI - A_c)^{-1}B_c + D_c$. Referring to Fig. 7.2, the a priori choice $C_1(s) = C(s)$ and $C_2(s) = 1$ will yield a robustly stable closed-loop system with guaranteed performance indicated by γ. However, note that any (strictly) proper and stable $C_1(s)$ and $C_2(s)$ satisfying $C_1(s)C_2(s) = C(s)$ will yield a robust input-to-state stable closed-loop system, since the transfer function from u to z (i.e., the transfer function that the nonlinearity "sees") does not change. However, an $\mathcal{L}_2$-gain smaller than γ may not be guaranteed. In principle, one can set $C_1(s) = C(s)F^{-1}(s)$ and $C_2(s) = F(s)$ with $F(s)$ a (strictly) proper and stable filter. Clearly, the choice for $F(s)$ has an effect on switching properties of HIGS, and may benefit performance (recall the results in Chap. 3). At this point, it is not clear what an appropriate choice for $F(s)$ would be in general, but this is expected to be highly dependent on the performance objective, as well as the nature of the external input signals. One approach for tuning the filter F could be through online data-based optimization techniques [18].

7.3.4 Example

To demonstrate the merit of the above synthesis procedure, consider the plant

$$P(s) = \frac{1}{s^2 + 0.18\,s + 1}, \tag{7.31}$$

for which a robustly stabilizing HIGS-based controller must be designed. For taking into account unmodelled dynamics at high frequencies, the uncertainty is characterized through the weighting filter

$$W_\Delta(s) = \frac{0.045\,s + 0.2}{0.02\,s + 1}. \tag{7.32}$$

To enforce integral action in the controller and obtain a gain of at least 20 dB at lower frequencies, the performance filter W is set to $W(s) = 10^{-3}/(s + \epsilon)$ with $\epsilon = 10^{-5}$. Interconnecting the plant and weighting filters results in the generalized plant description

$$G : \begin{cases} \dot{x}_g(t) &= A_g x_g(t) + B_g v(t) + B_w w(t) + B_{w_\Delta} w_\Delta(t), \\ y(t) &= C_g x_g(t), \\ q(t) &= C_q x_g(t), \\ q_\Delta(t) &= \hat{C}_q x_g(t) + \hat{D}_q v(t), \end{cases} \tag{7.33}$$

with plant and filter states $x_g^\top = [x_p^\top, x_w, x_{w_\Delta}]^\top \in \mathbb{R}^4$, system matrices

$$\left[\begin{array}{c|c} A_g & B_g \\ \hline C_g & D_g \end{array}\right] = \left[\begin{array}{cccc|c} 0 & 1 & 0 & 0 & 0 \\ -1 & -0.18 & 1 & 0 & 1 \\ 0 & 0 & -10^{-5} & 0 & 0 \\ 0 & 0 & 1 & -50 & 1 \\ \hline -1 & 0 & 0 & 0 & 0 \end{array}\right],$$

and input–output matrices $B_w = [0\ 0\ 10^{-3}\ 0]^\top$, $B_{w_\Delta} = [0\ 1\ 0\ 0]^\top$, $C_q = -C_g$, $\hat{C}_q = [0\ 0\ 2.25\ \ -102.5]$, and $\hat{D}_q = 2.25$. A controller is synthesized by solving the conditions in (7.25) through the two-step approach proposed in Sect. 7.3.3. The obtained controller is given by

$$A_c = \begin{bmatrix} -73.03 & 8.59 & -5.91 & -21.57 & -2.27 \cdot 10^3 \\ 6.69 \cdot 10^{-2} & 110.58 & 6.88 & 600.47 & -3.41 \cdot 10^4 \\ -2.71 \cdot 10^{-2} & -68.02 & -10.33 & -3.06 \cdot 10^2 & 2.06 \cdot 10^4 \\ 4.22 \cdot 10^{-4} & 1.10 & 9.75 \cdot 10^{-2} & 4.65 & -3.37 \cdot 10^2 \\ 3.45 \cdot 10^{-2} & 0.95 & 4.70 \cdot 10^{-2} & 5.25 & -2.93 \cdot 10^2 \end{bmatrix},$$

$$\begin{aligned} B_c &= \left[-4.48 \cdot 10^5\ -1.54 \cdot 10^5\ 7.72 \cdot 10^4\ -724.28\ 1.35 \cdot 10^3\right]^\top, \\ C_c &= \left[-1.99 \cdot 10^{-2}\ -6.15 \cdot 10^{-2}\ 6.10 \cdot 10^{-3}\ -0.48\ 19.86\right], \\ D_c &= 9.51, \end{aligned}$$

with $k_h = 1$, $\omega_h = 0.25$ rad/s, and is guaranteed to achieve an $\mathcal{L}_2$-gain smaller than $\gamma = 4.18$. Note that this can be a conservative estimate as the conditions in (7.25) are only sufficient. The frequency-domain characteristics of the LTI part of the HIGS-based controller are shown in Fig. 7.3 in solid black.

Interestingly, the LTI part by itself demonstrates PID-like characteristics having high (constant) gain at high frequencies. However, from a describing function perspective, HIGS acts as low-pass filter and provides desired roll-off at high frequencies while inducing less phase lag as compared to a linear low-pass filter. The frequency-domain condition in (7.13) with $\lambda = 0.03$ and $k = 2.6$ is shown in Fig. 7.4, and is satisfied with a certain margin, indeed indicating some room for uncertainty in the plant.

7.4 Robust Loop-Shaping Design

In this section, an approach toward the design of HIGS-based controllers is presented that exploits a frequency-domain approximation of HIGS within a robust loop-shaping design framework. Key in the approach is to regard the modelling error induced by the controller approximation as an additional uncertainty in the system, next to possible uncertainties in the plant. Before presenting the approach in detail, it is stressed that the aim is not to provide a rigorous procedure as in the previous section, but rather a practically helpful design procedure.

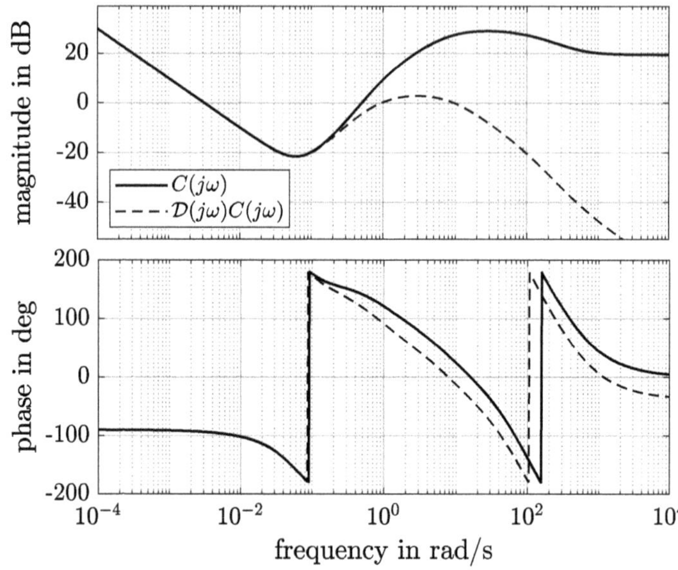

Fig. 7.3 Frequency-domain characteristics of $C(j\omega) = C_c(j\omega I - A_c)^{-1}B_c + D_c$ (solid) and describing function of the HIGS-based controller $\mathcal{H}C$ (dashed)

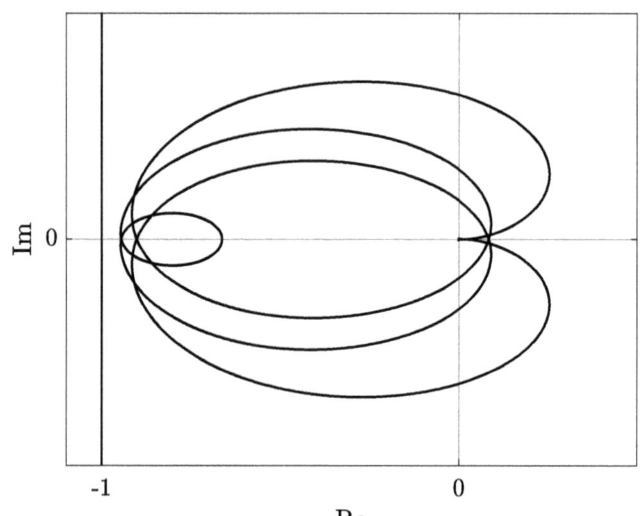

Fig. 7.4 Frequency-domain check in (7.13) with $\lambda = 0.03$ and $k = 2.6$

7.4.1 A Conceptual Loop-Shaping Idea

Consider the generalized plant set-up as depicted in Fig. 7.1 in which $K = K(C_i, \mathcal{H})$ represents a HIGS-based controller, being the interconnection of HIGS $\mathcal{H}$ with additional LTI filters C_i, $i = 1, \ldots, m$. By separating HIGS from the LTI filters and embedding the latter into the generalized plant G, which results in the LTI interconnection $L = L(C_i, G), i = 1, \ldots, m$ the original plant description can be rearranged into an equivalent Lur'e form as depicted in Fig. 7.5a.

A classical way for design and analysis of a nonlinear system as in Fig. 7.5a is to replace the corresponding nonlinearity by a describing function approximation [5, 13]. In general, a describing function can be defined for each harmonic input component of the nonlinearity to the component of the same frequency at the output of the nonlinearity. For example, for an input to HIGS of the form

$$z(t) = \sum_{i=1}^{n} A_i \sin(\omega_i t + \varphi_i) \tag{7.34}$$

with $n \in \mathbb{N}_{>0}$, frequencies $\omega_i \in \mathbb{R}_{\geq 0}$, amplitudes $A_i \in \mathbb{R}$, and phase $\varphi \in [0, 2\pi)$ the output u of HIGS will contain, among components at many frequencies, components at frequencies ω_i, $i = 1, \ldots, n$. Let the complex Fourier coefficient of these components be denoted by $U_i(j\omega_i) \in \mathbb{C}$. Then, the general describing function for each component is defined as [13]

$$\Psi(j\omega_i, z) = \frac{U_i(j\omega_i)}{A_i e^{j\varphi_i}}. \tag{7.35}$$

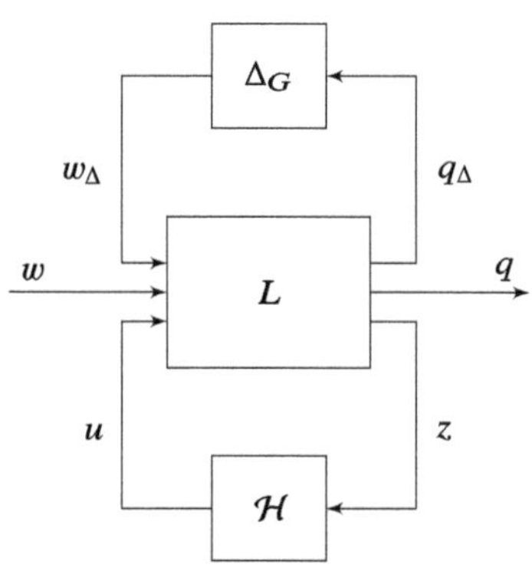

(a) Lur'e representation.

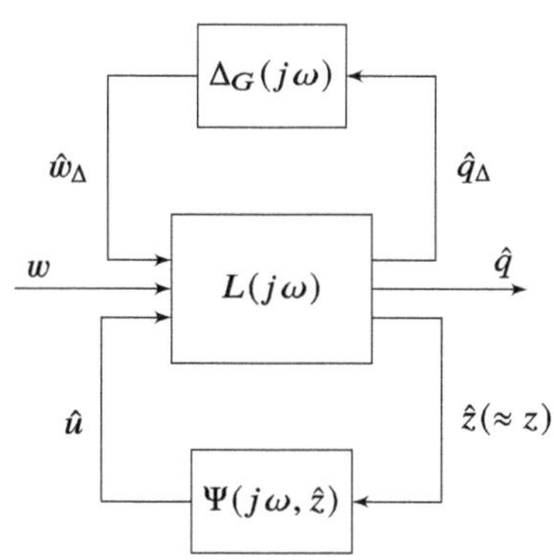

(b) Quasi-linear approximation.

Fig. 7.5 Lur'e representation (left), and quasi-linear approximation (right)

For $n = 1$, (7.35) will coincide with the sinusoidal input describing function, which for HIGS has been explicitly derived in Chap. 2, and is given by

$$\mathcal{D}(j\omega) = \frac{\omega_h}{j\omega}\left(\frac{\gamma}{\pi} + j\frac{e^{-2j\gamma} - 4e^{-j\gamma} + 3}{2\pi}\right) + k_h\left(\frac{\pi - \gamma}{\pi} + j\frac{e^{-2j\gamma} - 1}{2\pi}\right), \tag{7.36}$$

with $\gamma(\omega) = 2\arctan(k_h\omega/\omega_h)$. When an explicit expression for (7.35) is available, it can aid the design and analysis of the nonlinear system using (approximate) frequency-domain loop-shaping techniques. This setting is depicted in Fig. 7.5b, where $\hat{q}, \hat{u}, \hat{z}, \hat{w}_\Delta$, and $\hat{q}_\Delta$ are approximations of the corresponding nonlinear closed-loop signals in Fig. 7.5a in steady-state. For a single sinusoidal input to the nonlinearity, loop-shaping based on sinusoidal input describing functions has been considered in the literature for the design of nonlinear systems, see, e.g., [9, 13, 28]. For such an approach to be meaningful, the input to the nonlinearity must be nearly sinusoidal. In practice, this is often not the case, for example, when the external input w in Fig. 7.5 contains multiple-harmonic components. In case of multi-harmonic input signals to the nonlinearity, however, an exact analytical expression for $\Psi(j\omega, \hat{z})$ in (7.35) may be difficult (or even impossible) to obtain, because in general it not only depends on the input frequencies, but also on the amplitude distribution of harmonic components in the input, as well as the phase distribution. For that reason, direct use of multiple-harmonic describing functions may be of limited value in practice.

In order to still be able to exploit frequency-domain reasoning for HIGS-based controller design, it is proposed to approximate (7.35) by the sinusoidal input describing function in (7.36), and lump the modelling error induced by this approximation in an additive uncertainty in the controller. This uncertainty is characterized as

$$\Delta_\Psi(j\omega, \hat{z}) = \Psi(j\omega, \hat{z}) - \mathcal{D}(j\omega), \tag{7.37}$$

where $\hat{z}$ provides an approximation of the closed-loop signal z to HIGS in Fig. 7.7a. This idea is depicted in Fig. 7.6, in which $\mathcal{N}$ represents the nominal model resulting from the interconnection of $L(C_i, G)$, $i = 1, \ldots, m$ and the sinusoidal input describing function $\mathcal{D}(j\omega)$ in (7.36). Controller design now entails loop-shaping with the describing function to obtain robustness of the *nominal* closed-loop system $\mathcal{N}$ against both uncertainty in the plant as well as uncertainty in the controller. One may argue that the difficulty in finding an appropriate expression for Ψ is now only shifted toward the characterization of Δ_Ψ. But, within a robust control design framework there is no need to find Δ_Ψ exactly, as one can exploit an estimate for the characteristics of Δ_Ψ (e.g., in terms of an upper-bound on its magnitude), which may be easier to find. Interesting work related to this approach can be found in [34].

Approximating the multi-harmonic describing function Ψ in (7.35) by the sinusoidal input describing function $\mathcal{D}(j\omega)$ in (7.36) seems rather naive given the absence of superposition in HIGS. However, there are good reasons for doing this. First of all, $\mathcal{D}(j\omega)$ in (7.36) closely relates to standard frequency-response functions for LTI systems. It can be easily embedded into existing and intuitive frequency-domain loop-shaping tools which are well understood by practicing control engineers. Second, as

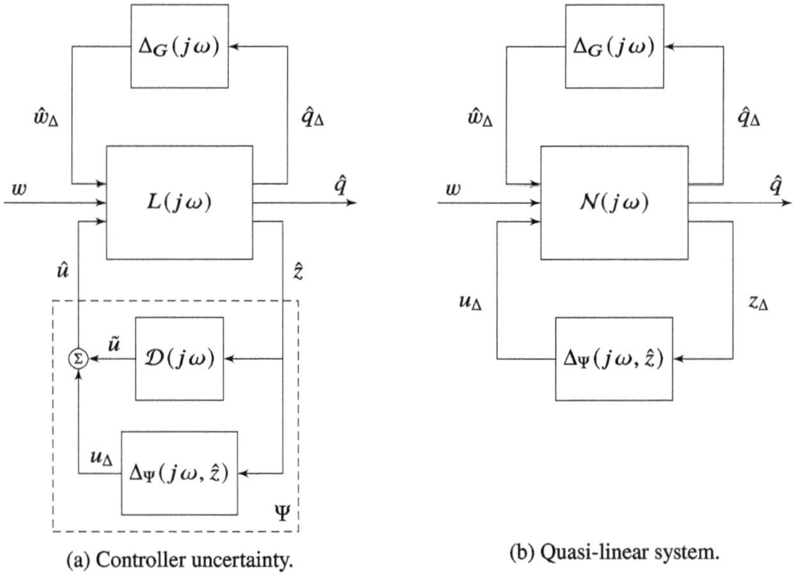

Fig. 7.6 Approximation with controller uncertainty (left), and quasi-linear system representation with uncertainties (right)

observed in many previous examples (see also [37, 38]), the describing function in (7.36) provides a surprisingly accurate prediction for stability of a motion system in feedback with a HIGS-based controller. This heuristic observation suggests the possibility for separating design aspects for robust stability from design aspects for robust performance. Namely, design for robust stability involves only $\mathcal{N}$ and Δ_G, whereas design for performance involves $\mathcal{N}$, Δ_G, and Δ_Ψ. As such, Δ_Ψ is considered relevant only for performance-based controller design.

7.4.2 Characterizing Uncertainty

For developing insights into properties of the uncertainty Δ_Ψ, it is instructive to consider an input signal to HIGS of the form

$$\hat{z}(t) = A_1 \sin(\omega_1 t) + A_2 \sin(\omega_2 t + \varphi), \tag{7.38}$$

with frequencies $\omega_1, \omega_2 \in \mathbb{R}_{\geq 0}$, amplitudes $A_1, A_2 \in \mathbb{R}$, and phase $\varphi \in [0, 2\pi)$. For such inputs, $\Psi(j\omega, \hat{z})$ in (7.35) corresponds to the two-sinusoidal input describing function [13, Chap. 5]. Although an explicit expression for this describing function cannot be derived, assuming $\omega_1 < \omega_2$ some practically useful insights can be obtained from considering two cases: i) the low-frequency component at ω_1 rad/s dominates

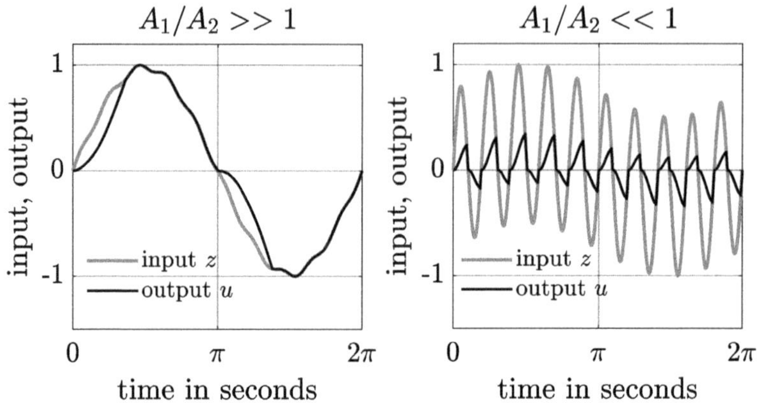

Fig. 7.7 Input–output characteristics of HIGS for an input of the form $\hat{z}(t) = A_1 \sin(2\pi t) + A_2 \sin(20\pi t)$ with $A_1/A_2 = \{0.33, 33\}$, $k_h = 1$, and $\omega_h = 3\pi$ rad/s. In case the low-frequency content dominates in the input to HIGS ($A_1/A_2 = 33 \gg 1$, in the left part), switching is mainly triggered by the low-frequency component. The presence of a dominating high-frequency term ($A_1/A_2 = 0.33 \ll 1$, in the right part) triggers high-frequency switching

in the input to HIGS, i.e., $A_1/A_2 \gg 1$, and ii) the high-frequency component at ω_2 rad/s dominates in the input to HIGS, i.e., $A_1/A_2 \ll 1$. Input–output characteristics typical for these two situations are depicted in Fig. 7.7.

From Fig. 7.7, it can be seen that the dominating harmonic in the input to HIGS triggers the switching mechanism of HIGS, thereby largely determining its response. This is a direct consequence of the loss of superposition. Regarding the modelling error in (7.37), two worst-case situations can be recognized. First, for $A_1/A_2 \gg 1$ and $\omega_1 \to 0$ one finds $u(t) \approx k_h \hat{z}(t)$ in which case HIGS essentially behaves as a linear gain, and $\Psi(j\omega_1, \hat{z}) \approx \Psi(j\omega_2, \hat{z}) \approx k_h$, leading to

$$\Delta_\Psi(j\omega_i, \hat{z}) \approx k_h - \mathcal{D}(j\omega_i), \quad i = \{1, 2\}. \tag{7.39}$$

Second, for $A_1/A_2 \ll 1$ and $\omega_2 \to \infty$ one finds $u(t) \approx 0$ such that $\Psi(j\omega_1, \hat{z}) \approx \Psi(j\omega_2, \hat{z}) \approx 0$, and

$$\Delta_\Psi(j\omega_i, \hat{z}) \approx 0 - \mathcal{D}(j\omega_i), \quad i = \{1, 2\}. \tag{7.40}$$

Although discussed for a two-sinusoidal input, the above rationale holds true for an input with an arbitrary number of harmonics, where the dominating harmonic in the input to HIGS largely determines its response. Note that (7.39) is typical for HIGS and results from the presence of the static gain mode, whereas (7.40) results from the dynamic integrator mode and can also be found to occur in other type of hybrid/switching nonlinearities such as reset integrators [8, 20, 40] or split-path integrators [31, 39]. From the above rationale, it is possible to estimate an upper-bound on the magnitude of Δ_Ψ in (7.37) for any input $\hat{z}$ consisting of multiple harmonics, namely

$$\|\Delta_\Psi(j\omega,\hat{z})\|_\infty \leq \max\{\|k_h - \mathcal{D}(j\omega)\|, \|0 - \mathcal{D}(j\omega)\|\} \tag{7.41}$$

for all $\omega \in \mathbb{R}_{\geq 0}$. However, the upper-bound in (7.41) may be restrictive since it assumes the presence of harmonics at $\omega = 0$ and $\omega = \infty$. In practice, the frequencies of the present harmonics are restricted to a finite range. Also, the upper-bound on Δ_Ψ to some extent should reflect the nature of the input $\hat{z}$. For example, for $\hat{z}(t) = \sin(\omega t)$ one finds $\Delta_\Psi(j\omega,\hat{z}) = 0$. Noting that $\lim_{\omega\to 0}\mathcal{D}(j\omega) = k_h$ and $\lim_{\omega\to\infty}\mathcal{D}(j\omega) = 0$, and from the observation that the dominant harmonic largely dictates properties of the response of HIGS, an alternative, potentially less conservative estimate for Δ_Ψ is conjectured as follows.

Conjecture 7.1 For an input signal of the form $\hat{z}(t) = \sum_{i=1}^{n} A_i \sin(\omega_i t + \varphi_i)$ with amplitude $A_i \in \mathbb{R}$, phase $\varphi_i \in [0, 2\pi)$, and non-harmonically related input frequencies $\omega_i \in [\underline{\omega}, \overline{\omega}]$, where $\underline{\omega}, \overline{\omega} \in \mathbb{R}_{\geq 0}$ denote the smallest and largest frequencies present in $\hat{z}$, the approximation error Δ_Ψ in (7.37) satisfies for all $\omega = \omega_i$, $i = 1, \ldots, n$

$$\|\Delta_\Psi(j\omega,\hat{z})\|_\infty \leq \max\left\{\|\mathcal{D}(j\underline{\omega}) - \mathcal{D}(j\omega)\|, \|\mathcal{D}(j\overline{\omega}) - \mathcal{D}(j\omega)\|\right\}. \tag{7.42}$$

For a sinusoidal input $\hat{z}(t) = A\sin(\omega t)$, one has $\underline{\omega} = \omega = \overline{\omega}$ such that the upper-bound in (7.42) reduces to zero, whereas for a two-sinusoidal input as in (7.38) one has $\underline{\omega} = \omega_1$ and $\overline{\omega} = \omega_2$, leading to $\|\mathcal{D}(j\omega_2) - \mathcal{D}(j\omega_1)\|$ when considering $\omega = \omega_1$, and $\omega = \omega_2$. Though the estimation in (7.42) is based on heuristics, and a formal justification is lacking, the practical value of this bound is supported by the numerical results presented in Fig. 7.8.

Figure 7.8 depicts the modelling error $\Delta_\Psi(j\omega,\hat{z})$ for the component at a frequency $\omega = \omega_1$ of a two-sinusoidal input $\hat{z}(t) = A_1\sin(\omega_1 t) + A_2\sin(\omega_2 t + \varphi)$ to HIGS, with a fixed frequency $\omega_2 = \{0.2\pi, 2\pi, 20\pi, 200\pi\}$ rad/s, for different amplitude ratios $A_1/A_2 = \{0.1, 1, 10\}$, and with $\varphi = 0$. The upper-bound given in (7.42) is depicted in black in Fig. 7.8. It can be seen that in all cases considered, the inequality in (7.42) is satisfied, and the largest values are indeed obtained for smaller values of the ratio A_1/A_2, i.e., when the harmonic at ω_2 dominates the input. The contribution of a non-zero phase φ is not depicted in Fig. 7.8 as this parameter is observed to have little effect on the results. These observations are in line with [13, Chap. 5] where it is argued that for non-harmonically related input sinusoids the phase φ can be seen as a random variable, making the describing function independent of φ.

Remark 7.1 For harmonically related input frequencies, Conjecture 7.1 may be false. Namely, given $\Psi(j\omega_i,\hat{z}) = U_i(j\omega_i)/A_i e^{j\varphi_i}$, for $A_i \to 0$, U_i may be non-zero when related to higher harmonics generated by input components at frequencies lower than ω_i. In that case $\Psi(j\omega_i,\hat{z}) \to \infty$, such that (7.42) is violated. Note that Fig. 7.8 depicts numerical results for both harmonically and non-harmonically related input

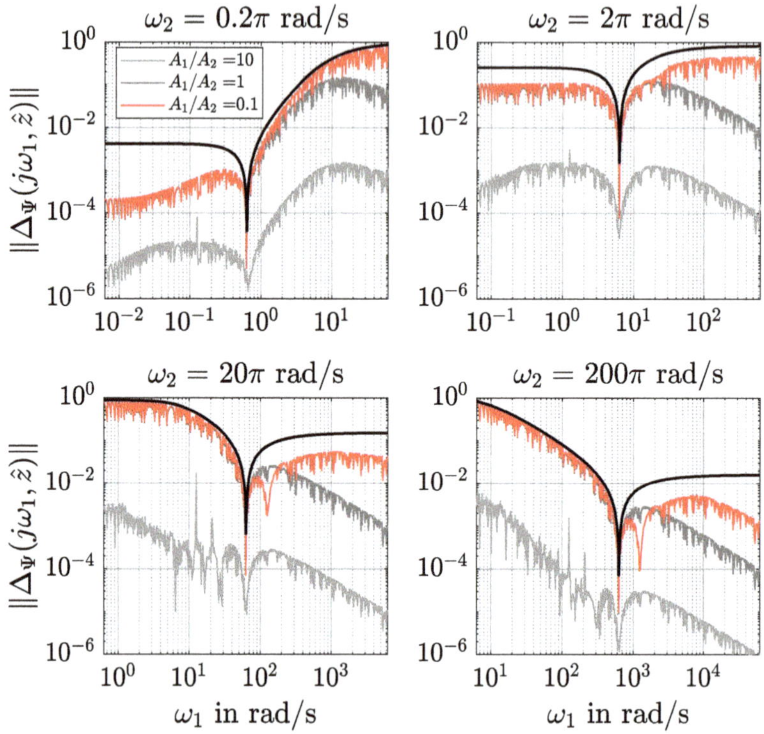

Fig. 7.8 Numerically obtained modelling error $\|\Delta_\Psi(j\omega_1, \hat{z})\| = \|\Psi(j\omega_1, \hat{z}) - \mathcal{D}(j\omega_1)\|$ as a function of ω_1 for an input of the form $\hat{z}(t) = A_1 \sin(\omega_1 t) + A_2 \sin(\omega_2 t + \varphi)$, with a fixed frequency $\omega_2 = \{0.2\pi, 2\pi, 20\pi, 200\pi\}$ rad/s, various amplitude ratios $A_1/A_2 = \{0.1, 1, 10\}$, and phase $\varphi = 0$. The estimated upper-bound $\|\Delta_\Psi(j\omega_1, \hat{z})\|_\infty \leq \|\mathcal{D}(j\omega_1) - \mathcal{D}(j\omega_2)\|$ is shown in solid black. The parameters of HIGS are set to $k_h = 1$ and $\omega_h = 2\pi$ rad/s

components. As in all cases considered the upper-bound is satisfied, these results indicate that, at least to some extent, (7.42) can be of use for harmonically related input components as well.

The following example illustrates several phenomena that can occur in a closed-loop system with HIGS, and which are captured effectively through the modelling error Δ_Ψ.

Example 7.1 Consider the system in Fig. 7.9, where the LTI plant P and LTI part of the controller C are given by

$$P(s) = \frac{1}{s^2 + 0.18\,s + 1}, \quad \text{and} \quad C(s) = \frac{\omega_i}{s} \cdot \frac{\omega_{lp}}{(s + \omega_{lp})}. \tag{7.43}$$

The closed-loop system is subject to an input of the form $r(t) = r_1 \sin(\omega_1 t) + r_2 \sin(\omega_2 t)$, with amplitudes $r_1, r_2 \in \mathbb{R}$, and frequencies $\omega_1, \omega_2 \in [0.01, 2]$ rad/s.

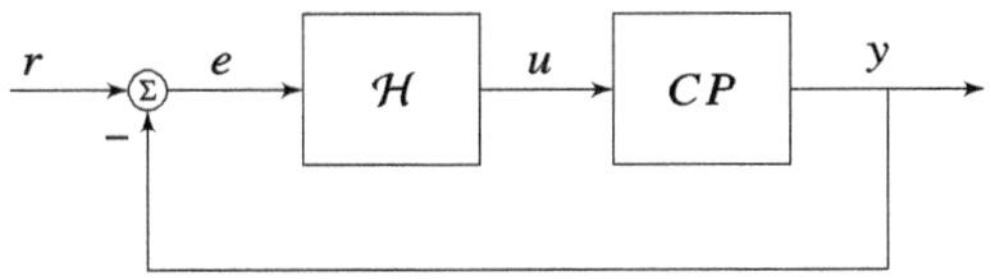

Fig. 7.9 Nonlinear system configuration with multi-sine input r

A nominal controller is designed using the describing function within a standard loop-shaping setting, resulting in $\omega_i = 0.24$ rad/s, $\omega_{lp} = 1.25$ rad/s, $k_h = 1$, and $\omega_h = 0.2$ rad/s. The nominal quasi-linear sensitivity characteristics

$$\mathcal{S}(j\omega) = \frac{1}{1 + \mathcal{D}(j\omega)C(j\omega)P(j\omega)} \tag{7.44}$$

are shown in Fig. 7.10 by the bold black curve. From the nature of the external input r, it is reasonable to approximate the input to HIGS by $e(t) \approx \hat{z}(t) = A_1 \sin(\omega_1 t + \varphi_1) + A_2 \sin(\omega_2 t + \varphi_2)$. Based on this approximation, one can characterize the uncertainty in the controller by $\Delta_\Psi(j\omega, \hat{z}) = W_\Delta(j\omega, \hat{z})\Delta_{\Psi_0}(j\omega, \hat{z})$, with $\|\Delta_{\Psi_0}(j\omega, \hat{z})\| \leq 1$ accounting for possible deviations in phase, and where

$$W_\Delta(j\omega, \hat{z}) = \max\left\{\|\mathcal{D}(0.01j) - \mathcal{D}(j\omega)\|, \|\mathcal{D}(j\omega) - \mathcal{D}(2j)\|\right\} \tag{7.45}$$

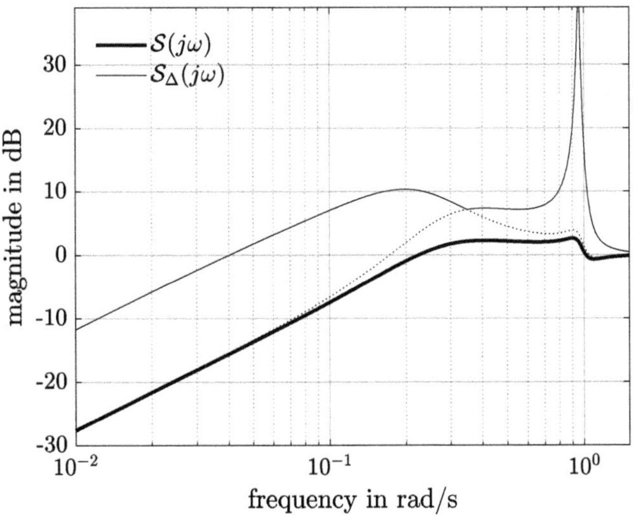

Fig. 7.10 Nominal sensitivity characteristics (bold black) and predicted worst-case characteristics (thin black)

accounts for magnitude deviations. The worst-case sensitivity magnitude characteristics are then given by

$$\|\mathcal{S}_{\Delta}(j\omega, \hat{z})\| = \frac{1}{\|1 + \mathcal{D}(j\omega)C(j\omega)P(j\omega)\| - \|W_{\Delta}(j\omega, \hat{z})C(j\omega)P(j\omega)\|}, \tag{7.46}$$

and are depicted in Fig. 7.10 by the thin black curve.

Figure 7.10 predicts a significant variation in both low-frequency disturbance attenuation as well as sensitivity to high frequencies. The steady-state error $e = r - y$ resulting from a two-sinusoidal input r can be written as

$$e(t) = e_1 \sin(\omega_1 t + \varphi_1) + e_2 \sin(\omega_2 t + \varphi_2) + e^*(t),$$

where e^* contains the harmonics at frequencies different from those in the input. Note that Fig. 7.10 predicts $\|\mathcal{S}(j\omega_i)\| \leq |e_i/r_i| \leq \|\mathcal{S}_{\Delta}(j\omega_i)\|$, $i = \{1, 2\}$. The numerically obtained amplification ratios $|e_1/r_1|$ and $|e_2/r_2|$ with their predicted minimum and maximum, as well as the relative contribution of higher harmonics to the total energy in the error response for an input with $\omega_1 = 0.03$ rad/s, $\omega_2 = 0.94$ rad/s, and various amplitude ratios r_1/r_2 are reported in Table 7.1. The numerical values in Table 7.1 are in accordance with the predictions based on the sensitivity characteristics in Fig. 7.10. Note that the predicted worst-case amplifications appear rather conservative.

The numerical results in Table 7.1 demonstrate a large dependency on the input amplitude ratio r_1/r_2. For decreasing values of r_1/r_2, the high-frequency content in the input r is becoming more dominant, which results in the input to HIGS to have an increasingly dominating component at 0.94 rad/s. As a consequence, switching of HIGS is mainly triggered by this high-frequency component, leading to a decrease of low-frequency suppression properties at 0.03 rad/s (recall Fig. 7.7). On the other hand, for increasing values of r_1/r_2, the low-frequency component at 0.03 rad/s is dominantly present in the input to HIGS, resulting in significant amplification of high-frequency content at 0.94 rad/s. In all cases considered, the relative contribution of higher harmonics in the error remains contained.

Table 7.1 Predicted and numerically obtained error characteristics of the nominal design with $\omega_1 = 0.03$ rad/s, $\omega_2 = 0.94$ rad/s, and $r_1/r_2 = \{10^4, 10^2, 1\}$

		$\|e_1/r_1\|$ in dB	$\|e_2/r_2\|$ in dB	$\|e^*\|_2/\|e\|_2$ in %
Prediction	Nominal	−18.10	2.41	–
	Worst-case	−2.39	33.89	–
Simulation	$r_1/r_2 = 10^4$	−18.03	20.61	2.51
	$r_1/r_2 = 10^2$	−17.66	10.21	10.27
	$r_1/r_2 = 1$	−10.13	2.43	5.22

7.4.3 Design Procedure

From the results in Example 7.1, it is clear that performance of a closed-loop system with HIGS largely depends on the distribution of input amplitudes. This amplitude dependency is an immediate consequence of the loss of superposition in HIGS, and may compromise low-frequency tracking properties or high-frequency sensitivity. Two complementary design solutions for effectively dealing with these potential issues are discussed next. In particular, a solution for the reduced low-frequency disturbance suppression is explicitly considered in time-domain, whereas a solution for the increased sensitivity at high frequencies is proposed in terms of a design constraint in the frequency-domain.

Pre- and Post-filtering Strategy

Deteriorated suppression properties of the closed-loop system at low frequencies stem from the fact that for $A_1/A_2 \ll 1$ the high-frequency component at ω_2 rad/s in the input to HIGS acts as a carrier wave for the low-frequency component at ω_1 rad/s. As a consequence of high-frequency switching, the output of HIGS is significantly smaller than what is required for achieving the anticipated low-frequency suppression properties, recall the diagram on the right-hand side of Fig. 7.7. This is comparable to low-gain feedback control.

A solution to this problem of gain–loss at low frequencies may be found in a pre- and post-filtering strategy as depicted in Fig. 7.11. Such a strategy has been applied effectively within the context of audio applications and variable-gain control [17, 25].

The LTI weighting filter $W(s)$, $s \in \mathbb{C}$ in Fig. 7.11 should be chosen such that it lifts the relevant frequency content in the input to HIGS, and switching is triggered in an appropriate manner. With such an approach, one effectively enforces the situation $A_1/A_2 \gg 1$. Possible design choices for W include (skew) notch filters, low-pass filters, and bandpass filters. In any case, the use of weighting filters that sustain a constant (DC) output, like an integrator, should generally be avoided. The reason for this being that input signals having an offset can cause undesired (asymmetric) switching of HIGS, or in the absence of any zero-crossings may prevent switching at all [19].

The inverse operation W^{-1} in Fig. 7.11 is included to compensate for the additional gain and phase introduced by W, so that the overall describing function characteristics are preserved. Furthermore, in a closed-loop setting it assures that the loop transfer connecting the output u to the input z (see also Fig. 7.7a) remains unaffected, so that closed-loop stability properties remain unaffected as well. The justification for applying the inverse operation W^{-1} on the output u, which could lead to the

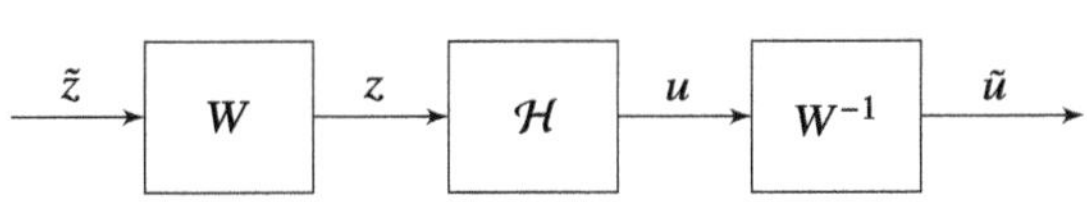

Fig. 7.11 Pre- and post-filtering strategy

amplification of higher harmonics, should be in accordance with the low-pass filter hypothesis in [13, Chap. 3]. This renders a well-balanced choice for pre- and post-filters dependent on the interconnected system, and exogenous inputs. It is important to realize that the weighting filter W will not play a role in the frequency-domain loop-shaping procedure discussed in the previous section due to its cancellation by the exact inverse W^{-1} in the chain of filters. This makes these filters susceptible to time-domain tuning only. It is argued that this weighting filter strategy is not only effective for HIGS, but is sufficiently generic to allow for extension to various other types of nonlinear integrators, including reset and split-path integrators.

Sensitivity Constraints

As mentioned before, applying a pre- and post-filter strategy enforces the situation $A_1/A_2 \gg 1$, in which case the only relevant component in (7.42) to consider is

$$\|\Delta_\Psi(j\omega, \hat{z})\|_\infty \leq \|\mathcal{D}(j\underline{\omega}) - \mathcal{D}(j\omega)\|, \tag{7.47}$$

which implies small modelling error at low frequencies, and larger error at high frequencies. In limiting the sensitivity peaking associated with such modelling error (see Fig. 7.10), one may consider $\Delta_\Psi(j\omega, \hat{z}) = W_\Delta(j\omega, \hat{z})\Delta_{\Psi_0}(j\omega, \hat{z})$, with normalized uncertainty $\|\Delta_{\Psi_0}(j\omega, \hat{z})\|_\infty \leq 1$ for all $\omega \in \mathbb{R}_{\geq 0} \cup \{\infty\}$, and

$$W_\Delta(j\omega, \hat{z}) = \mathcal{D}(j\underline{\omega}) - \mathcal{D}(j\omega), \tag{7.48}$$

where the latter filter can be used as a robust performance weighting filter to be included in the plant $\mathcal{N}$ in Fig. 7.6b. Design of a HIGS-based controller can then be formulated as a quasi-linear loop-shaping procedure as:

$$\begin{aligned} \underset{K(C_i,\mathcal{D})}{\text{minimize}} \quad & J(K(C_i, \mathcal{D})) \\ \text{subject to} \quad & \|\mathcal{M}(j\omega, \hat{z})\|_\infty \leq 1, \quad \text{for all } \omega \in \mathbb{R} \cup \{\infty\}, \\ & K(C_i, \mathcal{D}) \text{ stabilizes } G, \\ & K(C_i, \mathcal{D}) \in \mathcal{K}, \end{aligned} \tag{7.49}$$

in which $K(C_i, \mathcal{D})$ denotes the quasi-linear controller description (where HIGS is replaced by $\mathcal{D}$), and $\mathcal{M}$ represents the nominal system $\mathcal{N}$ augmented with the weighting filter W_Δ in (7.48). As in the LTI case, (7.49) can be solved by a standard manual/automated loop-shaping approach.

Remark 7.2 By the approach in (7.49) no formal guarantees are given for the resulting nonlinear system in terms of robust closed-loop stability and performance. To deal with this shortcoming, the results from Proposition 7.1 can be applied. In fact, by the KYP-lemma, the matrix inequality in (7.15) admits an equivalent frequency-domain interpretation that can be embedded in the quasi-linear optimization procedure (7.49). Specifically, the quasi-linear constraint "$K(C_i, \mathcal{D})$ stabilizes G" can be replaced with this sufficient frequency-domain condition. The latter can be integrated effectively in the design procedure (7.49) without the need for a parametric model. In this way, robust stability with respect to plant uncertainties $\Delta_G \in \bar{\Delta}$ is guaranteed for the true nonlinear system, and the worst-case $\mathcal{L}_2$-gain is given by γ. Additional performance criteria are enforced through describing function characteristics. Hence, such an approach combines the rigorous design procedure in Sect. 7.3 with the description of describing function-based reasoning. The main difference with the approach in Sect. 7.3, however, is that the structure of the controller should be selected a priori. Also, including the rigorous frequency-domain constraints in (7.49) may lead to a conservative design that limits the potential of HIGS. An interesting direction to explore further in this regard is to translate the matrix inequalities in Chap. 5 into frequency-domain conditions, potentially reducing this conservatism.

Summarizing, design of a HIGS-based (motion) controller using describing function heuristics may proceed according to the following steps:

- **Step 1:** Robust frequency-domain design based on the quasi-linear loop-shaping procedure outlined in (7.49), in which high-frequency modelling error is accounted for by the weighting filter in (7.48).
- **Step 2:** Augment the controller with pre- and post-weighting filters that are designed to enforce switching of HIGS within the relevant frequency range (for instance, a range below the bandwidth).

Example 7.2 (*Example* 7.1 *revisited*) Returning to the previous example in Fig. 7.9 with P given in (7.43), the controller is redesigned on the basis of the above design procedure with $W_\Delta(j\omega, \hat{z}) = \mathcal{D}(0.01j) - \mathcal{D}(j\omega)$, resulting in

$$C(s) = \frac{0.22}{s} \cdot \frac{0.53}{(s + 0.53)}, \tag{7.50}$$

with $k_h = 1$, and $\omega_h = 0.25$ rad/s. The resulting sensitivity characteristics are shown in Fig. 7.12 and demonstrate a significant reduction of high-frequency sensitivity as compared to the initial design in Fig. 7.10. Key in this example for achieving reduced sensitivity is the sufficient rotation of the resonance of the plant in the open-loop toward the complex right-half plane in the Nyquist plot by reducing the cut-off frequency of the low-pass filter in (7.50).

The numerical results presented in Table 7.2 for $\omega_1 = 0.03$ rad/s, $\omega_2 = 0.94$ rad/s, and various r_1/r_2 support the effectiveness of the robust loop-shaping design approach, as for all r_1/r_2 the amplification factor $|e_2/r_2|$ at 0.94 rad/s remains significantly more contained as compared to the initial design in Table 7.1. However, low-frequency suppression properties are still compromised for an increasingly dominant high-frequency component in the input, i.e., for decreasing values of r_1/r_2. To account for this, a pre-filter is chosen as

$$W(s) = \frac{s^2 + 2\beta_z\omega_z s + \omega_z}{s^2 + 2\beta_p\omega_p s + \omega_p^2}, \tag{7.51}$$

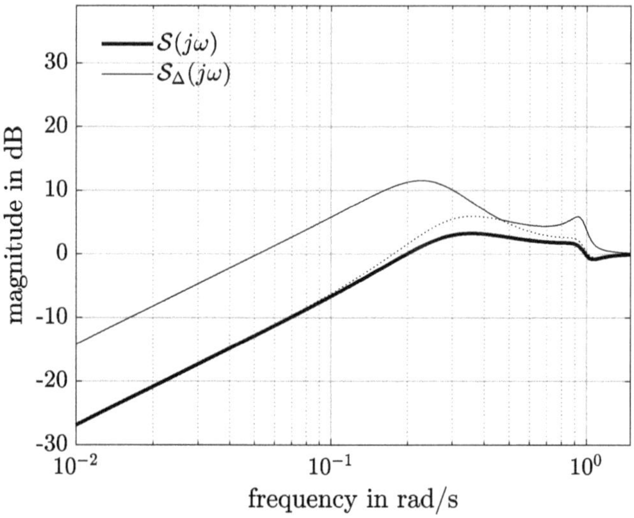

Fig. 7.12 Nominal sensitivity characteristics (bold black) for the system with redesigned controller, and predicted worst-case characteristics (thin black)

Table 7.2 Predicted and numerically obtained error characteristics of the robust redesign with $\omega_1 = 0.03$ rad/s, $\omega_2 = 0.94$ rad/s, and $r_1/r_2 = \{10^4, 10^2, 1\}$

		$\lvert e_1/r_1 \rvert$ in dB	$\lvert e_2/r_2 \rvert$ in dB	$\lVert e^* \rVert_2 / \lVert e \rVert_2$ in %
Prediction	Nominal	−17.20	1.10	–
	Worst-case	−3.25	6.01	–
Without pre- and post-filtering strategy				
Simulation	$r_1/r_2 = 10^4$	−17.19	4.72	2.21
	$r_1/r_2 = 10^2$	−17.18	4.47	2.49
	$r_1/r_2 = 1$	−10.85	1.14	2.01
With pre- and post-filtering strategy				
Simulation	$r_1/r_2 = 10^4$	−17.19	4.40	2.63
	$r_1/r_2 = 10^2$	−17.19	4.22	3.13
	$r_1/r_2 = 1$	−16.10	3.39	2.39

with $\omega_z = \omega_p = \omega_2 = 0.94$ rad/s, $\beta_z = 0.005$, and $\beta_p = 0.1$. This filter suppresses content at 0.94 rad/s in the input to HIGS, thereby enforcing the switching mechanism to be primarily triggered by the component at 0.03 rad/s. Effectiveness of the filtering strategy for improving low-frequency rejection properties is demonstrated by the numerical results in Table 7.2. Compared to the previous designs, dependency on the amplitude distribution of the input signal is reduced, and low-frequency disturbance rejection is achieved without significant high-frequency amplification. Compared to the results in Table 7.1, the robust redesign provides some additional robustness against the effects of higher harmonics. An industrial example illustrating the effectiveness of the design approach for a broader spectrum of input frequencies is discussed in Chap. 8.

7.5 Summary

In this chapter, two methods for HIGS-based controller design are presented. The first method provides a rigorous way for robust controller synthesis. In particular, the frequency-domain stability conditions as derived in Chap. 4 are reformulated into matrix inequalities which allow for a direct translation toward appropriate synthesis conditions. The latter conditions show a strong resemblance with synthesis conditions typically used in an LTI setting. The presented approach returns a robustly stabilizing HIGS-based controller with guaranteed $\mathcal{L}_2$-performance properties. The approach, however, requires the use of a parametric model and provides limited insight in parameter tuning and robust redesign, which in an industrial setting is sometimes considered to be a drawback.

As an alternative, a practical tuning approach is presented that hinges on the use of the describing function within a robust loop-shaping framework. Key in the approach is to approximate the closed-loop system by a quasi-linear system, and capture the modelling error induced by the describing function approximation as an uncertainty in the controller, for which a heuristic bound on the magnitude characteristics is derived. With this bound, relevant performance issues are revealed for which effective solutions are proposed in the form of both time- and frequency-domain design guidelines. The approach connects to the industrial practice where frequency-domain loop-shaping design with non-parametric models are common practice.

Given the practical applicability of the loop-shaping design approach guided by describing functions, in the next chapter this approach is used for the design of a HIGS-based PID controller for a scanning stage system of an industrial wafer scanner.

References

1. Apkarian, P., Noll, D.: Nonsmooth h_∞ synthesis. IEEE Trans. Autom. Control **51**(1), 71–86 (2006)
2. Apkarian, P., Dao, M.N., Noll, D.: Parametric robust structured control design. IEEE Trans. Autom. Control **60**(7), 1857–1869 (2015)
3. Arcak, M., Larsen, M., Kokotović, P.: Circle and Popov criteria as tools for nonlinear feedback design. Automatica **39**(4), 643–650 (2003)
4. Åström, K., Hägglund, T., Hang, C., Ho, W.: Automatic tuning and adaptation for PID controllers - a survey. IFAC Proc. Vol. **25**(14), 371–376, 4th IFAC Symposium on Adaptive Systems in Control and Signal Processing 1992, Grenoble, France, 1–3 July (1992)
5. Atherton, D.: Nonlinear Control Engineering. Van Nostrand Reinhold (1982)
6. Boyd, S., El Ghaoui, L., Feron, E., Balakrishnan, V.: Linear Matrix Inequalities in System and Control Theory, Studies in Applied Mathematics, vol. 15. SIAM, Philadelphia, PA (1994)

7. Bruijnen, D., van de Molengraft, R., Steinbuch, M.: Optimization aided loop shaping for motion systems. In: 2006 IEEE Conference on Computer Aided Control System Design, 2006 IEEE International Conference on Control Applications, 2006 IEEE International Symposium on Intelligent Control, pp. 255–260 (2006)
8. Clegg, J.C.: A nonlinear integrator for servomechanisms. Trans. Am. Inst. Electr. Eng., Part II: Appl. Ind. **77**(1), 41–42 (1958)
9. Colgren, R., Jonckheere, E.: h_∞ control of a class of nonlinear systems using describing functions and simplicial algorithms. IEEE Trans. Autom. Control **42**(5), 707–712 (1997)
10. Doyle, J., Glover, K., Khargonekar, P., Francis, B.: State space solution to standard h2 and h_∞ control problem. IEEE Trans. Autom. Control **34**, 831–847 (1989)
11. Ferreres, G., Fromion, V.: Nonlinear analysis in the presence of parametric uncertainties. Int. J. Control **69**, 695–716 (1998)
12. Gahinet, P., Apkarian, P.: A linear matrix inequality approach to h_∞ control. Int. J. Robust Nonlinear Control **4**(4), 421–448 (1994)
13. Gelb, A., Vander Velde, W.: Multiple-input Describing Functions and Nonlinear Systems Design. McGraw Hill, New York (1968)
14. Glover, K., McFarlane, D.: Robust stabilization of normalized coprime factor plant descriptions with h_∞-bounded uncertainty. IEEE Trans. Autom. Control **34**(8), 821–830 (1989)
15. Goh, K., Turan, L., Safonov, M., Papavassilopoulos, G., Ly, J.: Biaffine matrix inequality properties and computational methods. In: Proceedings of 1994 American Control Conference, vol. 1, pp. 850–855 (1994)
16. Hassibi, A., How, J., Boyd, S.: A path-following method for solving BMI problems in control. In: Proceedings of the 1999 American Control Conference, vol. 2, pp. 1385–1389 (1999)
17. Heertjes, M., Schuurbiers, X., Nijmeijer, H.: Performance-improved design of N-PID controlled motion systems with applications to wafer stages. IEEE Trans. Industr. Electron. **56**, 1347–1355 (2009)
18. Heertjes, M., Irigoyen Perdiguero, N., Deenen, D.: Robust control and data-driven tuning of a hybrid integrator-gain system with applications to wafer scanners. Int. J. Adapt. Control Signal Process. **33**(2), 371–387 (2019)
19. Heertjes, M., van den Eijnden, S., Heemels, W., Nijmeijer, H.: A solution to gain loss in hybrid integrator-gain systems. In: 2021 IEEE Conference on Control Technology and Applications (CCTA), pp. 1179–1184 (2021)
20. Horowitz, I., Rosenbaum, P.: Non-linear design for cost of feedback reduction in systems with large parameter uncertainty. Int. J. Control **21**(6), 977–1001 (1975)
21. Impram, S., Munro, N.: Describing functions in non-linear systems with structured and unstructured uncertainties. Int. J. Control **74**, 600–608 (2001)
22. Katebi, M., Zhang, Y.: h_∞ control analysis and design for nonlinear systems. Int. J. Control **61**(2), 459–474 (1995)
23. Khalil, H.K.: Nonlinear Systems, 3rd edn. Prentice-Hall, Upper Saddle River, New Jersey (2002)
24. Nešić, D., Zaccarian, L., Teel, A.: Stability properties of reset systems. Automatica **44**(8), 2019–2026 (2008)
25. Oppenheim, A., Schafer, R., Stockham, T.: Nonlinear filtering of multiplied and convolved signals. Proc. IEEE **56**(8), 1264–1291 (1968)
26. Ortega, M., Vargas, M., Castaño, F., Rubio, F.: Improved design of the weighting matrices for the S/KS/T mixed sensitivity problem - application to a multivariable thermodynamic system. IEEE Trans. Control Syst. Technol. **14**, 82–90 (2006)
27. Packard, A., Doyle, J.: The complex structured singular value. Automatica **29**(1), 71–109 (1993)
28. Saikumar, N., Sinha, R.K., HosseinNia, S.H.: “Constant in gain lead in phase” element – application in precision motion control. IEEE/ASME Trans. Mechatron. **24**(3), 1176–1185 (2019)
29. Samad, T.: A survey on industry impact and challenges thereof. IEEE Control Syst. Mag. **37**(1), 17–18 (2017)

30. Samad, T., Bauer, M., Bortoff, S., Di Cairano, S., Fagiano, L., Odgaard, P.F., Rhinehart, R.R., Sánchez-Peña, R., Serbezov, A., Ankersen, F., Goupil, P., Grosman, B., Heertjes, M., Mareels, I., Sosseh, R.: Industry engagement with control research: Perspective and messages. Annu. Rev. Control **49**, 1–14 (2020)
31. Sharif, B., van der Maas, A., van de Wouw, N., Heemels, W.P.M.H.: Filtered split-path nonlinear integrator: a hybrid controller for transient performance improvement. IEEE Trans. Control. Syst. Technol. 1–13 (2021)
32. Skogestad, S., Postlethwaite, I.: Multivariable Feedback Control: Analysis and Design, 2nd edn. Wiley and Sons (2010)
33. Steinbuch, M., Norg, M.: Advanced motion control: An industrial perspective. Eur. J. Control **4**(4), 278–293 (1998)
34. Taylor, L.O.: Error bounds for the dual input describing function technique. PhD thesis, Purdue university (1970)
35. Tran Dinh, Q., Gumussoy, S., Michiels, W., Diehl, M.: Combining convex–concave decompositions and linearization approaches for solving BMIs, with application to static output feedback. IEEE Trans. Autom. Control **57**(6), 1377–1390 (2012)
36. van de Wal, M., Baars, G., Sperling, F., Bosgra, O.: Multivariable h_∞/μ feedback control design for high-precision wafer stage motion. Control Eng. Pract. **10**, 739–755 (2002)
37. van den Eijnden, S., Heertjes, M., Nijmeijer, H.: Robust stability and nonlinear loop-shaping design for hybrid integrator-gain-based control systems. In: 2019 American Control Conference (ACC), pp. 3063–3068 (2019)
38. van den Eijnden, S., Heemels, W., Nijmeijer, H., Heertjes, M.: Stability and performance analysis of hybrid integrator-gain systems: a linear matrix inequality approach. Nonlinear Anal. Hybrid Syst. **45**, 101192 (2022)
39. van Loon, S., Hunnekens, B., Heemels, W., van de Wouw, N., Nijmeijer, H.: Split-path nonlinear integral control for transient performance improvement. Automatica **66**, 262–270 (2016)
40. Zaccarian, L., Nešić, D., Teel, A.: First order reset elements and the Clegg integrator revisited. In: Proceedings of the 2005, American Control Conference, 2005, vol. 1, pp. 563–568 (2005)
41. Zhao, G., Nešić, D., Tan, Y., Wang, J.: Open problems in reset control. In: 52nd IEEE Conference on Decision and Control, pp. 3326–3331 (2013)
42. Zhou, K., Doyle, J.C., Glover, K.: Robust and Optimal Control. Prentice-Hall Inc., USA (1996)

Chapter 8
Case-Study: An Industrial Wafer Scanner

8.1 Introduction

Wafer scanners employ a crucial step during the manufacturing process of integrated circuits (ICs), namely, transferring the circuit topology obtained from a reticle onto the photosensitive layers of a silicon wafer. Both reticle and wafer are part of separate high-precision positioning systems: the reticle stage and the wafer stage. The dynamical behaviour of these stages can be accurately modelled according to the linear paradigm, allowing for the application of well-developed linear feedforward and feedback control methods [7, 21]. Most of the feedback control loops in wafer scanners originate from a decoupling design in combination with proportional–integral–derivative (PID) control [16, 23]. Despite the clear advantages of PID control in terms of simplicity and predictability, servo performance is usually compromised as a result of inherent design trade-offs. The fact that modern linear techniques such as H_∞ control cannot overcome these trade-offs poses a serious challenge for meeting the ever growing demands on scanning accuracy and throughput in wafer scanners.

As a potential means to alleviate the problem, this chapter considers the use of multiple HIGS-based filters within an otherwise linear PID configuration that is commonly used for feedback control of the reticle and wafer stages. From a describing function perspective, the HIGS-based filters demonstrate similar magnitude characteristics as their linear counterparts, but with preferable phase properties that can be tuned independently of the magnitude. This extra design freedom facilitates increased controller bandwidths, giving rise to improved disturbance rejection and transient properties of the closed-loop system. Performance is evaluated through measurement results obtained from a wafer-stage system of an industrial wafer scanner. In order to show the relative improvements in both time- and frequency-domain, a comparison is made with a baseline linear PID design. Besides demonstrating the potential benefits of HIGS-based control in practice, the experiments are valuable for identifying practically relevant factors that may affect controller performance within an industrial setting, and, therefore, require specific attention in the early phases of the design

S. van den Eijnden et al., *Hybrid Integrator-Gain Systems*, Lecture Notes in Control and Information Sciences 499, https://doi.org/10.1007/978-3-032-18101-5_8

process. Next to wafer scanners, HIGS-based control has become the emerging control technology in metrology inspection machines [2] and has been applied to wire bonders [11], vibration isolation systems [15], micro-electro-mechanical positioning systems [27], and medical devices [4].

This chapter is organized as follows. To develop some intuition for the wafer scanning application at hand, Sect. 8.2 introduces relevant scanning principles and performance measures. The wafer-stage dynamics and current feedforward and feedback control methods are discussed in Sect. 8.3, whereas the HIGS-based feedback control strategy is presented in Sect. 8.4. Practically relevant design aspects are considered in Sect. 8.5, and measurement results are presented in Sect. 8.6. Sections 8.7 and 8.8 summarize the main takeaways and conclusions of this chapter.

8.2 Wafer Scanning Principles

In this section, wafer scanning principles are discussed. This includes a brief explanation of the scanning procedure, generation of the setpoint used to transfer the reticle image onto the wafer, and the definition of performance measures.

8.2.1 Scanning Procedure and Setpoint Definition

The case-study considered in this chapter is an industrial wafer scanner, of which an artist impression and schematics are provided in Fig. 8.1.

Many wafer scanners exploit the lithographic principle in which deep or extreme ultraviolet (DUV/EUV) light emitted by a source falls onto a reticle (or mask), being a quartz plate which contains a blueprint of the integrated circuit to be processed. After passing through the reticle, light travels through an optical column which scales down the image of the circuit topology by a factor of four, and projects it onto the photosensitive layers of a silicon wafer. A schematic illustration of the exposure process is shown in Fig. 8.1. Both reticle and wafer are part of separate motion systems: the reticle stage and the wafer stage. Each of these stages employs a dual-stroke strategy. A long-stroke stage is used for coarse positioning with accuracy at the micrometer level, whereas a short-stroke stage is used for fine positioning with accuracy at the nanometer level. For realizing accurate and fast projection of the circuit topology onto the wafer, both reticle and wafer stage perform a synchronized meandering motion in which exposure by the light beam takes place during repeated intervals of zero and constant velocity in x- and y-direction, respectively. A typical meandering profile for the wafer stage is shown in Fig. 8.2. The scanning phases, indicated in the figure by the solid arrows, are alternated with preparation phases: intervals with non-zero acceleration in both directions, allowing the system to reach the next exposure area, the latter also referred to as field or die. In order for setpoint-induced transients to sufficiently die out before starting a new scan, an additional

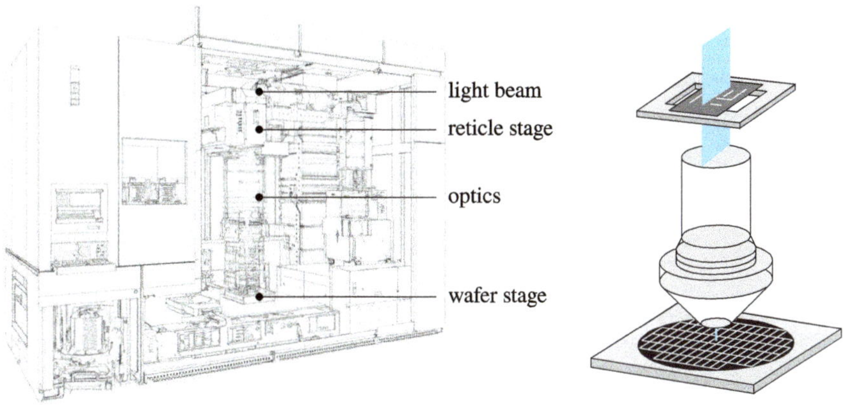

Fig. 8.1 Left: artist impression of a DUV wafer scanner. Right: simplified visualization of an exposure process showing from top to bottom a reticle atop a reticle stage, an optics column, and a wafer atop a wafer stage. Right-hand panel reprinted with permission from IEEE Transactions on Control Systems Technology, Vol 22, Oomen T, van Herpen R, Quist S, van de Wal M, Bosgra O, Steinbuch M, Connecting system identification and robust control for next-generation motion control of a wafer stage, 102–118. ©2014 IEEE

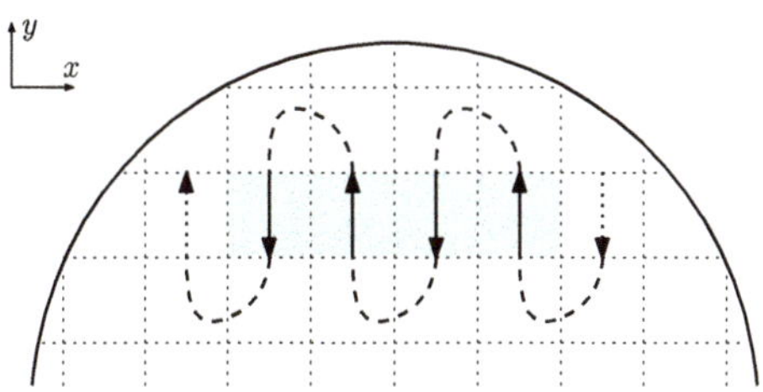

Fig. 8.2 Top view of a meandering profile for the wafer-stage system. The scanning phases, indicated by the solid arrows, are alternated with preparation phases, indicated by the dashed lines. Reprinted with permission from IEEE Control Systems, Vol 31, Butler H, Position control in lithographic equipment, 28–47. ©2011 IEEE

(zero-acceleration) settling phase is often used. In the z-direction, the wafer stage generally tracks a non-zero reference, representing the measured wafer topology.

For the wafer-stage application considered in this chapter, a third-order motion profile is considered in x- and y-direction. Third order refers to the fact that the position, velocity, and acceleration profiles are continuous, and only the jerk profile (third derivative) contains discontinuities. That is, the jerk profile consists of a sequence of step functions. This results in the generation of less high-frequency input content compared to a second-order setpoint. Since the setpoint/reference profile can be seen as the key deterministic input that determines closed-loop performance, this results in a tracking error that is generally less affected by high-frequency signals and therefore leads to improved scanning performance. Some representative parameters used for generating the third-order setpoint profile in scanning y-direction are listed in

Table 8.1 Third-order setpoint parameters

Limit	Value
Maximum scanning velocity	0.8 m/s
Maximum acceleration	30 m/s^2
Maximum jerk	6600 m/s^3
Acceleration/velocity ratio	37.5 Hz
Jerk/acceleration ratio	220 Hz
Settling time	0 s

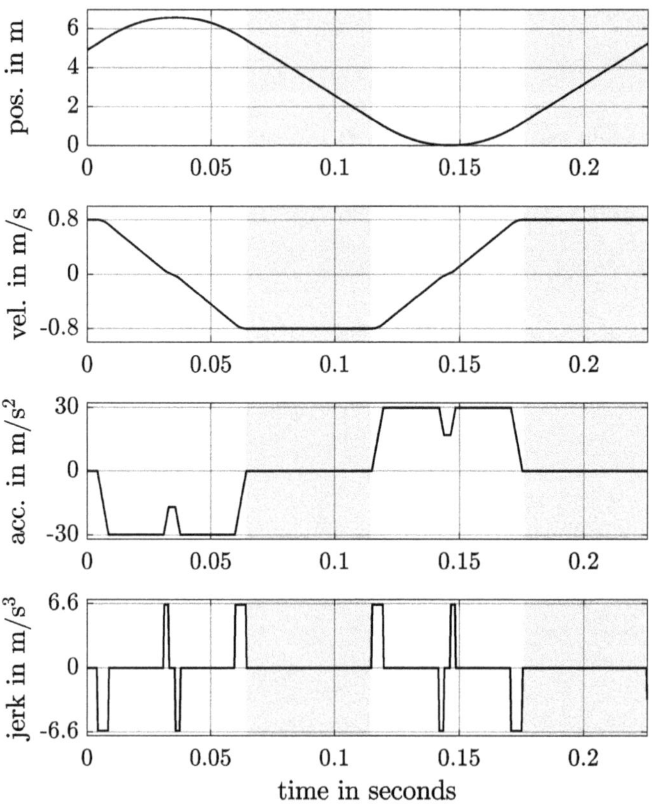

Fig. 8.3 Time-domain characteristics of the third-order setpoint in scanning y-direction. From top to bottom: position, velocity, acceleration, and jerk. The scanning intervals are indicated in grey

Table 8.1, and the corresponding time-domain profiles are depicted in Fig. 8.3. For this setpoint profile the settling time is set to zero seconds such that exposure starts directly after the acceleration phase. It should be mentioned that for the application considered, this setpoint profile is considered rather aggressive (high jerk), but allows for a significant increase in wafer throughput.

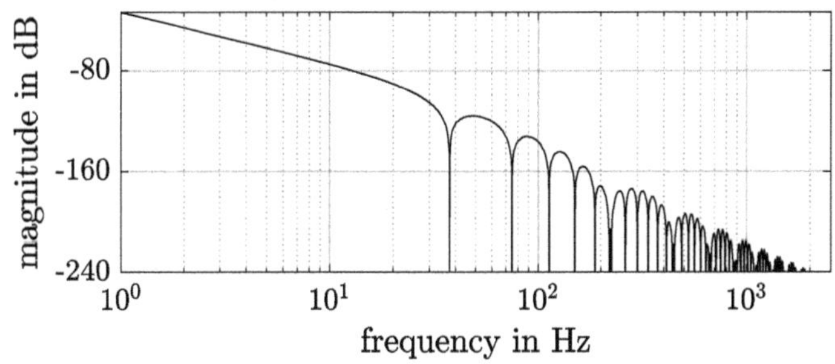

Fig. 8.4 Amplitude spectrum of the third-order position setpoint profile in scanning y-direction, as provided in (8.1)

The acceleration/speed and jerk/acceleration ratios listed in Table 8.1 are of special interest, as they determine the location of zeros in the frequency spectrum of the setpoint profile. This becomes clear from studying the Fourier transform of the reference position $r(t)$, which, given the typical block-shaped profile of the jerk signal as depicted in the lower plot in Fig. 8.3, can be derived analytically as (see [7] for further details)

$$\hat{r}(\omega) = \int_{-\infty}^{\infty} r(t)e^{-j\omega t}dt = \frac{v_{\max}}{\omega^2} \cdot e^{-j\frac{\omega\tau_v}{2}} \cdot \text{sinc}\left(\frac{\omega\tau_a}{2}\right) \cdot \text{sinc}\left(\frac{\omega\tau_j}{2}\right), \quad (8.1)$$

with time constants $\tau_a = v_{\max}/a_{\max}$, $\tau_j = a_{\max}/j_{\max}$, $\tau_v = \tau_a + \tau_j$, and $\text{sinc}(x) = \sin(x)/x$. By appropriate tuning of τ_a and τ_j, the zeros in the input spectrum of (8.1) can be set to match the frequencies where in the physical plant structural modes like a saddle and/or torsion mode occur. This gives the ability to minimize excitation of such modes. Moreover, in the HIGS-based feedback control context as considered in this chapter, different setpoint characteristics may affect switching of the HIGS elements, and thus effectiveness of the overall controller. The frequency spectrum of the considered setpoint profile in scanning y-direction is depicted in Fig. 8.4. Note the 80 dB amplitude decrease per decade at higher frequencies, which is a typical feature of the third-order setpoint profile that readily follows from (8.1).

8.2.2 *Scanning Performance Measures*

Performance of wafer scanners is driven by three measures: overlay, imaging, and throughput. Overlay refers to the ability to expose two layers on top of each other, imaging refers to image quality in terms of uniformity or line-edge roughness, and throughput refers to the number of processed wafers per hour. Throughput is mainly related to the settling phase prior to scanning, whereas overlay and imaging relate to the scanning interval in which exposure takes place and translate into servo specifications.

The actual position of the stage during exposure determines the location to which the circuit topology is transferred onto the wafer, and thus is critical for dedicated chuck overlay, i.e., layer-to-layer accuracy. This low-frequency effect directly translates to the moving average (MA) of the positioning error $e(t) = r(t) - y(t)$, being the difference between the reference r and the actual position y, which is defined as

$$\mathrm{MA}(t) := \frac{1}{T_e} \int_{t-T_e/2}^{t+T_e/2} e(\tau) d\tau. \tag{8.2}$$

Here, T_e is the exposure time, i.e., the time it takes for a single point on the wafer to receive a full dose of the DUV/EUV light emitted by the source. The quality of a single layer, on the other hand, is affected by stage-positioning accuracy, since positioning errors reduce contrast [7]. Even when the average position is correct, high-frequency vibrations of the stage result in the projection of a blurry image onto the wafer and deteriorate the imaging quality. This effect is quantified through the moving standard deviation (MSD) of the positioning error $e(t)$, which is computed by

$$\mathrm{MSD}(t) := \sqrt{\frac{1}{T_e} \int_{t-T_e/2}^{t+T_e/2} \left(e(\tau) - \mathrm{MA}(t)\right)^2 d\tau}. \tag{8.3}$$

Note that the MA and MSD values in (8.2) and (8.3) are computed over an exposure window of length T_e seconds. Hence, the computation of MA and MSD values requires the positioning error in the exposure window to be known upfront, which makes (8.2) and (8.3) non-causal operations that can only be performed off-line (after exposure has taken place).

8.3 Wafer-Stage Dynamics and Control

The process of wafer scanning requires accurate and fast positioning of the wafer-stage system in six degrees-of-freedom (DOF): three planar DOFs (x, y, z), and three rotational DOFs (r_x, r_y, r_z). Through an appropriate (position-dependent) decoupling strategy, the to-be-controlled system becomes diagonally dominant so that multivariable control design aspects can be handled through a multiloop single-input single-output (SISO) control design. This section reviews SISO strategies that are currently used for wafer-stage control.

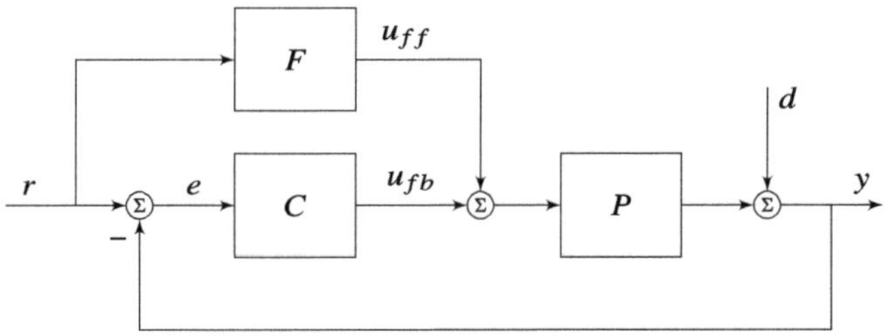

Fig. 8.5 Simplified wafer-stage control architecture

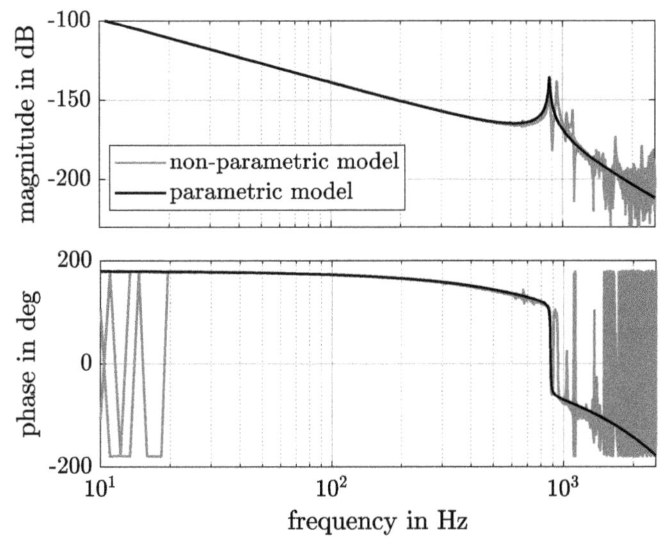

Fig. 8.6 Measured FRF characteristics of the short-stroke wafer-stage system in scanning y-direction (grey) and parametric model (black)

8.3.1 Control Problem Formulation

For stage control, one basically distinguishes between two problems: (i) the servo (tracking) problem in which the output y of the stage should track the setpoint profile r, and (ii) the disturbance rejection (regulator) problem in which the tracking error $e = r - y$ should be brought close to zero in the presence of plant uncertainty and external disturbances d, the latter which are assumed to be uncorrelated with r [30]. The servo problem can largely be solved by an LTI feedforward controller $F(s)$, whereas the remaining disturbance rejection problem is dealt with by an LTI feedback controller $C(s)$ as illustrated in Fig. 8.5. In this figure, P represents the wafer-stage dynamics in one DOF. Representative frequency-response function (FRF) characteristics of the stage in scanning y-direction, measured at different positions on the wafer are shown in Fig. 8.6.

Figure 8.6 indicates that up to a frequency of 500 Hz, the stage behaves mainly as a floating mass system, i.e., a double integrator characterized by a 40 dB/decade amplitude decrease combined with a phase lag of 180°. Beyond 500 Hz, the rigid body assumption is no longer valid as several resonances appear that associate with the non-rigid body dynamics of the stage system. Also note the additional phase lag due to the sampling rate of 5 kHz which is observed at higher frequencies. To provide some further intuition for the system dynamics at hand, a parametric model of the plant with delay is given by

$$P(s) = \frac{bs + k}{m_1 m_2 s^4 + (m_1 + m_2)bs^3 + (m_1 + m_2)ks^2} \cdot e^{-T_s s}. \tag{8.4}$$

Here, $m_1 = 5$ kg represents the mass of the motor yoke driving the stage, $m_2 = 17.7$ kg mainly represents the mass of the mirror block carrying the wafer and used for interferometry position measurement, and $m = m_1 + m_2$ is the total mass. The stiffness coefficient $k = 1.186 \cdot 10^8$ N/m and damping coefficient $b = 193.6$ Ns/m represent properties of the coupling interface between motor and stage, whereas $T_s = 2 \cdot 10^{-4}$ s is the sampling time. The model characteristics are depicted in Fig. 8.6 in black.

Given the LTI context of Fig. 8.5, the servo error signal $e := r - (y + d)$ satisfies (in the Laplace domain)

$$e(s) = S(s)(1 - P(s)F(s))r(s) - S(s)d(s), \tag{8.5}$$

where $S(s) = (1 + P(s)C(s))^{-1}$ is the closed-loop sensitivity function. In view of minimizing the error contribution in (8.5), three main directions are taken: (i) input shaping in which the setpoint r is designed to minimize the excitation of higher order dynamics (recall Sect. 8.2.1, and see, e.g., [6, 10]), (ii) feedforward control design to ideally achieve $F = P^{-1}$, and (iii) feedback control design to minimize S. Wafer-stage feedforward and feedback control design is discussed in more detail next.

8.3.2 *Feedforward Control*

As mentioned before, for solving the servo problem, it easily follows from (8.5) that F should ideally match the inverse of the plant, i.e., $F = P^{-1}$. However, in practice there are numerous reasons for why an exact inverse cannot be realized, implying that P^{-1} may only be approximated. In this regard, the dominant double integrator characteristics appearing in Fig. 8.6 and (8.4) motivate the use of acceleration feedforward control [5, 7], resulting in

$$F(s) = ms^2. \tag{8.6}$$

Atop the feedforward structure in (8.6) typical advances found in wafer scanners include jerk derivative (snap) feedforward control [5], finite impulse response (FIR) feedforward control [17, 24], adaptive feedforward control [8], and compliance compensation techniques [19].

It is interesting to observe that during the acceleration phase, at least 99.9% of the performance is achieved through acceleration feedforward control. This is illustrated by the following example. For the performance in scanning y-direction, a mass of around $m = 20$ kg is accelerated with a maximum acceleration of 30 m/s^2, which accounts for an actuator force of 600 N. The closed-loop error is typically found in the order of $e \approx 10$ nm, and feedback control gains are in the order of $2 \cdot 10^7$ N/m, leading to a feedback force of 0.2 N. In comparison to the feedforward force of 600 N, this provides only 0.033% of the total control output. Note that while feedforward control is crucial during the acceleration phase for tracking the setpoint profiles, during the scanning phase feedback control becomes essential for suppressing unknown disturbances that cannot be compensated for by feedforward control, as well as uncertainty in the plant that through a mismatch in feedforward signals results in an error residue that is left to the feedback controller. In the remainder of this chapter, the availability of a well-designed (advanced) feedforward controller is assumed.

8.3.3 Feedback Control

Concerning (8.5), the LTI feedback controller C should be designed so that the magnitude of the sensitivity function S is as small as possible, at least in a frequency range of interest. For achieving this, the structure that is commonly used in wafer-stage control is a PID-based feedback controller of the form:

$$C(s) = C_{pid}(s) C_{lp}(s) \prod_{i=1}^{n} C_{n,i}(s), \tag{8.7}$$

in which

$$C_{pid}(s) = k_p \left(1 + \frac{\omega_i}{s} + \frac{s}{\omega_d}\right) \tag{8.8a}$$

is a PID filter with proportional gain $k_p \in \mathbb{R}_{>0}$, integrator frequency $\omega_i \in \mathbb{R}_{>0}$, and differentiator frequency $\omega_d \in \mathbb{R}_{>0}$,

$$C_{lp}(s) = \frac{\omega_{lp}^2}{s^2 + 2\beta_{lp}\omega_{lp} + \omega_{lp}^2} \tag{8.8b}$$

is a second-order low-pass filter with cut-off frequency $\omega_{lp} \in \mathbb{R}_{>0}$ and dimensionless damping coefficient $\beta_{lp} \in \mathbb{R}$, and where

$$C_{n,i}(s) = \frac{s^2 + 2\beta_{z,i}\omega_{z,i} + \omega_{z,i}^2}{s^2 + 2\beta_{p,i}\omega_{p,i} + \omega_{p,i}^2} \tag{8.8c}$$

is a notch filter with zero and pole frequencies $\omega_{z,i}, \omega_{p,i} \in \mathbb{R}_{>0}$, and dimensionless damping coefficients $\beta_{z,i}, \beta_{p,i} \in \mathbb{R}$. The PID filter C_{pid} aims at low-frequency/constant disturbance suppression coming from the integrator, and robust stability provided by the differentiator. The second-order low-pass filter C_{lp} is used to avoid high-frequency noise amplification. Several notch filters $C_{n,i}$ are added to counteract the effect of performance limiting resonances in the plant as well as to create loop enhancements.

In tuning the controller in (8.7), it is aimed to maximize the controller bandwidth, defined as the frequency where the open-loop system $L(s) = P(s)C(s)$ crosses one (zero dB) for the first time, while simultaneously satisfying a peaking constraint on the closed-loop sensitivity function $S(s)$. In this way, the region where low-frequency disturbance suppression is achieved is maximized, while the high-frequency noise response is kept within specifications. Recall from Sect. 8.2 that low-frequency error content directly affects MA values and thus overlay, hence the need for good low-frequency disturbance suppression. High-frequency content affects MSD values and the related imaging quality, thereby requiring limited noise amplification. Although crucial in the controller design, the use of an integrator as well as a low-pass filter in (8.7) leads to the introduction of phase lag in the feedback loop as an inevitable consequence of Bode's gain–phase relationship [26, 28]. This leads to trade-offs between, for example, (i) improved low-frequency disturbance rejection by increasing the bandwidth, (ii) robust stability properties and (iii) transient response in terms of overshoot and settling times.

8.4 HIGS in a PID Architecture

To allow for more design freedom and potentially balance between the aforementioned trade-offs in a more desirable manner, this section discusses the use of integrators and low-pass filters along with their HIGS-based counterparts in the PID structure of (8.7).

8.4.1 HIGS-Based Integrator and Low-Pass Filter Design

Consider the generic interconnection as depicted in Fig. 8.7 where HIGS is augmented with LTI filters $W(s)$, $V(s)$, $s \in \mathbb{C}$.

In Fig. 8.7, one may recognize the pre- and post-filtering strategy as discussed in Chap. 7 for modulating the switching characteristics of HIGS. The LTI filter V is

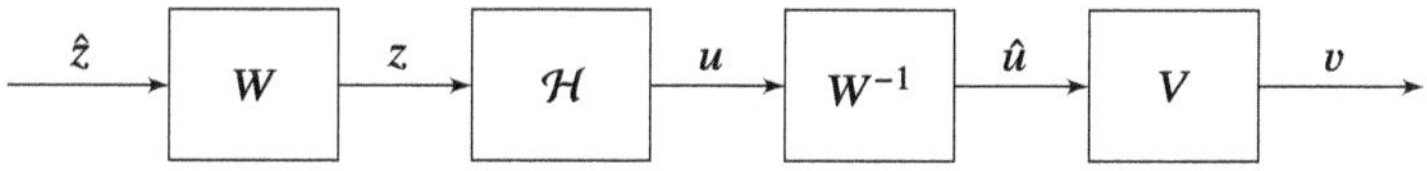

Fig. 8.7 Generic HIGS-based filter design with weighting filter $W(s)$ and loop-shaping filter $V(s)$

added for the specific purpose of shaping the overall describing function of the interconnection to achieve desirable magnitude and phase characteristics. In this context, this filter can be regarded as a "loop-shaping" filter. Through linear reasoning, the describing function characteristics of the nonlinear filter in Fig. 8.7 read

$$\mathcal{D}^*(j\omega) = \mathcal{D}(j\omega)V(j\omega), \tag{8.9}$$

where $\mathcal{D}(j\omega)$ is the describing function of HIGS, derived in Chap. 2 as

$$\mathcal{D}(j\omega) = \frac{\omega_h}{j\omega}\left(\frac{\gamma}{\pi} + j\frac{e^{-2j\gamma} - 4e^{-j\gamma} + 3}{2\pi}\right) + k_h\left(\frac{\pi - \gamma}{\pi} + j\frac{e^{-2j\gamma} - 1}{2\pi}\right), \tag{8.10}$$

with $\gamma(\omega) = 2\arctan\left(k_h\omega/\omega_h\right)$, and V is chosen as

$$V(s) = H^{-1}(s)C^*(s), \quad \text{with} \quad H(s) = k_h\left(\frac{\omega_c}{s + \omega_c}\right), \tag{8.11}$$

where $\omega_c = \frac{\omega_h}{k_h}|1 + \frac{4j}{\pi}|$ rad/s, and in which $C^*(s) \in \mathbb{C}$ is an arbitrary LTI filter. Recall that the describing function $\mathcal{D}(j\omega)$ in (8.10) resembles first-order low-pass characteristics with low-frequency (DC) gain k_h, cross-over frequency ω_c rad/s, and a phase lag that does not exceed 38.15°. The filter $H(s)$ in (8.11) then provides an approximation of $\mathcal{D}(j\omega)$ in terms of its magnitude. As such, the non-realizable filter $H^{-1}(s)$ "inverts" the magnitude characteristics of $\mathcal{D}(j\omega)$, and provides additional phase lead up to 90°. Hence, for all $\omega \in \mathbb{R}$ one achieves

$$\|\mathcal{D}^*(j\omega)\| \approx \|C^*(j\omega)\|, \quad \text{and} \quad \angle(\mathcal{D}^*(j\omega)) = \angle(C^*(j\omega)) + \varphi(\omega),$$

with $\varphi(\omega) = \angle(H^{-1}(j\omega)) + \angle(\mathcal{D}(j\omega)) \in (0, 58.85)°$. In other words, the describing function $\mathcal{D}^*(j\omega)$ in (8.9) possesses similar magnitude characteristics as the LTI filter $C^*(s)$, but with a phase lag reduction up to 58.85°. Comparable strategies that combine linear and nonlinear filters into quasi-linear elements with improved phase properties as observed from their describing function can be found in, e.g., [9, 13, 18, 20, 25]. By selecting $C^*(s) = C_i(s) = \omega_i/s$ or $C^*(s) = C_{lp}(s)$ with C_{lp} in (8.8b) one obtains, respectively, a HIGS-based integrator or HIGS-based low-pass filter for which the characteristics of the describing function are depicted in Fig. 8.8 in black. Note that the filter $C^*(s)H^{-1}(s)$ is realizable. The characteristics of the corresponding linear filters are shown in grey. It can be seen that the nonlinear filters demonstrate a significant phase advantage over their linear counterparts.

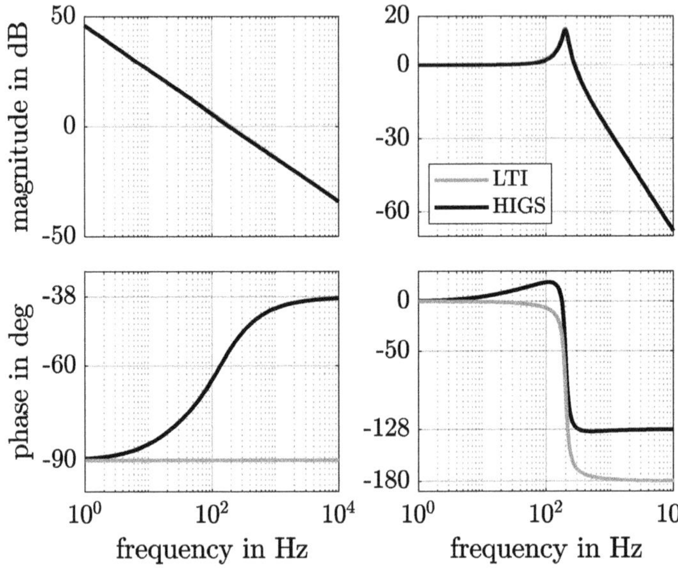

Fig. 8.8 Describing function of an LTI integrator and its HIGS-equivalent design (left), and an LTI low-pass filter and its HIGS-equivalent design (right)

Remark 8.1 Regarding the above design philosophy, it is important to point out that the freedom in designing the magnitude characteristics of $\mathcal{D}^*(j\omega)$ is completely determined by the parameters associated with $C^*(s)$, while the freedom to design the phase is determined by both the parameters of $C^*(s)$ and those related to $\varphi(\omega)$. As the only parameters that appear in $\varphi(\omega)$ are ω_h and k_h associated with HIGS, magnitude and phase are no longer uniquely related, but, to some extent, can be designed separately.

Remark 8.2 It should be mentioned that in practice both time- and frequency-domain properties of the linear equivalent filter C^* can be recovered by selecting $\omega_h \to \infty$. Indeed, in the frequency-domain this choice yields $H^{-1}(j\omega)\mathcal{D}(j\omega) \to 1$ such that $\mathcal{D}^*(j\omega) \to C^*(j\omega)$ for all $\omega \in \mathbb{R}$. In the time-domain, it can be argued that for $\omega_h \to \infty$ the integrator mode in HIGS is simply too fast, such that HIGS effectively reduces to a linear gain k_h. Together with the previous observations, this implies that the input–output signals of the filter in Fig. 8.7 satisfy $v \to C^*\hat{z}$, that is, the linear filter operation is closely approximated.

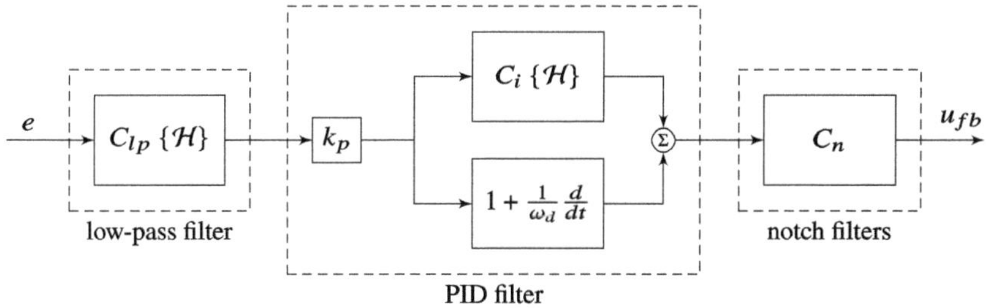

Fig. 8.9 Structure of a HIGS-based PID controller $C\{\mathcal{H}\}$ with HIGS-based integrator $C_i\{\mathcal{H}\}$ and HIGS-based low-pass filter $C_{lp}\{\mathcal{H}\}$

8.4.2 HIGS-Based PID Scheme

A configuration that adopts both a HIGS-based integrator, denoted by $C_i\{\mathcal{H}\}$, and a HIGS-based low-pass filter, denoted by $C_{lp}\{\mathcal{H}\}$ within the PID structure of (8.7) is depicted in Fig. 8.9.

Different from the LTI setting, the sequence in which linear and nonlinear filters appear with respect to each other affects the input–output characteristics of the controller, see also Chaps. 3 and 7. The specific choice for placing the integrator branch $C_i\{\mathcal{H}\}$ after the low-pass filter $C_{lp}\{\mathcal{H}\}$ in Fig. 8.9 is motivated by the consideration that filters sustaining a constant output should be placed after HIGS as to avoid asymmetric switching or no switching at all due to an input signal with an offset. It is furthermore required that the linear characteristics $W_{lp}^{-1}V_{lp}W_i$ separating each HIGS element have a relative degree of at least one for ensuring the input signal to HIGS associated with $C_i\{\mathcal{H}\}$ to be one time continuously differentiable. This can be realized by choosing $V_{lp}(s) = H_{lp}^{-1}(s)C_{lp}(s)$ and restricting W_{lp} and W_i to be proper filters.

Remark 8.3 Well-posedness of generic interconnections containing multiple HIGS elements remains an open issue. To guarantee the global existence of solutions for the interconnection in Fig. 8.5, in which $C = C\{\mathcal{H}\}$ is a HIGS-based PID controller as depicted in Fig. 8.9, however, the results in [14] can be extended provided the inputs belong to the class of piecewise continuous functions, and the HIGS elements are separated by an LTI filter having a relative degree of at least one.

The engineering heuristic underlying the use of two HIGS-based filters in the configuration of Fig. 8.9 is explained as follows. As argued in Chap. 7, HIGS relies on the spectral content of its input to generate effective outputs with anticipated gain and phase characteristics. Phase advantages for an integrator should be accessed in the low-frequency range, allowing for larger integrator gains and increased low-frequency disturbance rejection. The functionality of HIGS in the integrator is therefore purely performance-based. On the other hand, for the low-pass filter phase advantages are to be accessed around the system bandwidth to increase robust stability margins. A clear distinction can be made in frequency content on the basis of which HIGS should primarily switch. However, because of the lack of superposition, HIGS can only generate outputs that are effective in one region, thereby compromising properties in the other region. For example, choosing $\omega_{h,lp}$ associated with HIGS in $C_{lp}\{\mathcal{H}\}$ small yields broad-band phase advantages, but at the same time can result in low-frequency gain–loss (recall Chap. 7). Using both $C_i\{\mathcal{H}\}$ as well as $C_{lp}\{\mathcal{H}\}$ allows for separating the switching objectives such that HIGS associated with each filter can be made effective in its respective frequency region by selecting appropriate weighting filters.

8.4.3 Frequency-Domain Approximation

For tuning purposes, it is useful to have a describing function characterization of the controller in Fig. 8.9. The use of multiple nonlinearities, however, complicates an analytical derivation. In [12, Chap. 3, Sect. 3.6], it is argued that replacing the nonlinearities with their respective individual describing function is permitted if the LTI filters separating them have sufficient low-pass filter characteristics, thereby filtering higher-order harmonics generated by the preceding nonlinear element. In that case, one finds the approximation

$$C(j\omega) \approx \mathcal{D}_{pid}(j\omega)\mathcal{D}_{lp}(j\omega) = k_p\left(1 + \mathcal{D}_i(j\omega) + \frac{j\omega}{\omega_d}\right)\mathcal{D}_{lp}(j\omega), \tag{8.12}$$

in which $\mathcal{D}_i(j\omega)$ and $\mathcal{D}_{lp}(j\omega)$ denote the describing function of $C_i\{\mathcal{H}\}$ and $C_{lp}\{\mathcal{H}\}$, respectively. However, this filtering assumption may be limiting or partially invalid. An alternative route for justifying the approximation in (8.12) exploits the following frequency-domain separation principle [32]:

- In PID tuning, it approximately holds true that $\gamma\omega_i < \omega_b < \omega_{lp}/\gamma$, with ω_b the system bandwidth and $\gamma > 1$, see [29]. Phase benefits with $C_{lp}\{\mathcal{H}\}$ are to be obtained around the bandwidth, such that an appropriate tuning for the parameter $\omega_{h,lp}$ associated with HIGS in $C_{lp}\{\mathcal{H}\}$ should roughly satisfy $\omega_{h,lp} \geq \omega_b$.
- For sinusoidal inputs satisfying $\omega < \omega_b$, HIGS associated with $C_{lp}\{\mathcal{H}\}$ is dominantly in gain mode such that the input to $C_i\{\mathcal{H}\}$ is (nearly) sinusoidal, i.e., the nonlinear effect of $C_{lp}\{\mathcal{H}\}$ is sufficiently contained for low-frequency inputs to avoid interference with $C_i\{\mathcal{H}\}$. Hence, $\mathcal{D}_i(j\omega)$ plays the main role in the describing function of the controller.
- For inputs satisfying $\omega \geq \omega_b$, the proportional–derivative branch of the PID part in Fig. 8.9 largely dominates over the fundamental harmonic in the response of $C_i\{\mathcal{H}\}$, so that $\mathcal{D}_{lp}(j\omega)$ plays the main role in the describing function of the controller.

Typical frequency-domain characteristics for both linear and HIGS-based PID controllers are shown in Fig. 8.10. The parameters are chosen to match the magnitude characteristics of both designs. To indicate those regions where phase benefits are

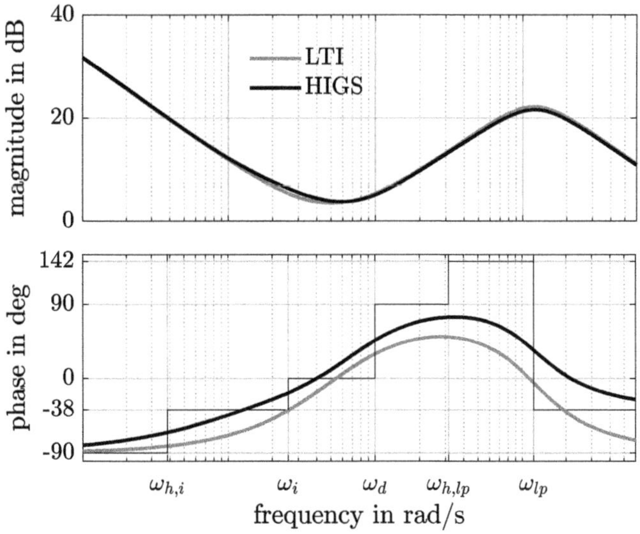

Fig. 8.10 Frequency-domain characteristics of the linear and HIGS-based PID controllers in (8.7) and (8.12), respectively. Asymptotic behaviour of the latter is indicated by thin grey lines. Phase benefits due to $C_i\,\{\mathcal{H}\}$ are mainly obtained for all $\omega \in [\omega_{h,i}, \omega_i]$, whereas benefits due to $C_{lp}\,\{\mathcal{H}\}$ are obtained for $\omega \geq \omega_{h,lp}$. Note that both regions are sufficiently separated

expected from $C_i\,\{\mathcal{H}\}$ and $C_{lp}\,\{\mathcal{H}\}$, the asymptotic phase behaviour of $C(j\omega)$ is shown by thin black lines. Note the clear separation of both regions for which phase advantages are obtained by the integrator and low-pass filter, which results from the fact that $\omega_{h,lp}$ is chosen sufficiently large.

8.5 Closed-Loop Design Aspects

In this section, practically relevant aspects for tuning a HIGS-based PID controller for an industrial wafer-stage system are discussed in detail. The procedure is based on the quasi-linear loop-shaping approach outlined in Chap. 7 and involves modelling uncertainty and tuning of the weighting filters.

8.5.1 *Modelling Uncertainty*

Characterizing the uncertainty of the modelling in the approximation of the describing function (8.12) requires approximate knowledge of the servo errors in terms of harmonic content. Representative characteristics of a servo error signal e in scanning

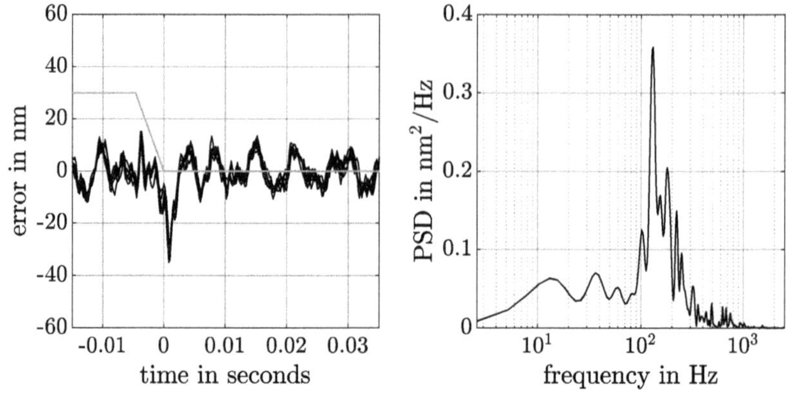

Fig. 8.11 Representative characteristics of a servo error signal e in both time- and frequency-domain when applying linear PID control

y-direction when applying acceleration feedforward and linear feedback control are shown in Fig. 8.11.

The main parts of the frequency content in the servo errors e are due to the setpoint profile, despite the use of an appropriate feedforward controller. It can be seen that the harmonic content in the 100–250 Hz region dominates the frequency spectrum, which can be linked to the 235 Hz bandwidth of the closed-loop system. In addition, in the 600–800 Hz region, harmonic contributions are present that result from the structural dynamics of the plant. Given the high reproducibility of the reference profiles, it is reasonable to expect that these frequencies dominantly persist in the servo error when applying a HIGS-based PID controller as well.

Remark 8.4 By virtue of the separation principle discussed in the previous section, the main source of modelling error at high frequencies may be identified to be the result of the HIGS-based low-pass filter $C_{lp}\{\mathcal{H}\}$. Namely, since for higher frequencies the output of the integrator is largely dominated by the proportional–derivative branch in Fig. 8.9, modelling uncertainty in the frequency-domain approximation of $C_i\{\mathcal{H}\}$ at high frequencies is found to become insignificant compared to modelling uncertainty in $C_{lp}\{\mathcal{H}\}$. Although both uncertainties can be taken into account, the latter will be of primary interest for subsequent robust controller design and is therefore discussed next.

Based on the above observations, a meaningful test signal to approximate the input to HIGS associated with $C_{lp}\{\mathcal{H}\}$ may be of the form

$$\hat{z}(t) = \sin(2\pi f_1 t) + \alpha \sin(2\pi f_2 t),$$

where $f_1 \in [100, 250]$ Hz, $f_2 \in [600, 800]$ Hz, and $\alpha \in (0, 1)$. Assuming f_1 and f_2 to be incommensurate, modelling uncertainty in the describing function approximation of the HIGS-based low-pass filter can be characterized by $\Delta_{\Psi}(j\omega, \hat{z}) =$

$\Delta_{\psi 0}(j\omega, \hat{z})W_\Delta(j\omega)$ with $\|\Delta_{\psi 0}(j\omega, \hat{z})\|_\infty \leq 1$, and $W_\Delta(j\omega) = \mathcal{D}(j\omega_{\min}) - \mathcal{D}(j\omega)$ for all $\omega \geq 600 \cdot 2\pi$ rad/s, and zero otherwise, and $\omega_{\min} = 100 \cdot 2\pi$ rad/s. Further details on this characterization can be found in Chap. 7.

To demonstrate the relevance of accounting for this modelling uncertainty in the wafer-stage control context, three HIGS-based PID controllers are designed by solving the following optimization problem:

$$\begin{aligned} &\underset{C}{\text{maximize}} && \omega_b, \\ &\text{subject to} && \|W_S(j\omega)\mathcal{S}(j\omega)\|_\infty + \lambda\|W_a(j\omega)\mathcal{S}(j\omega)P(j\omega)\|_\infty \leq 1, \\ &&& C \text{ stabilizes } P, \end{aligned} \tag{8.13}$$

where C is given in (8.12) and denotes the describing function approximation of the HIGS-based PID controller in Fig. 8.9, ω_b denotes the open-loop bandwidth, i.e., the frequency at which $\mathcal{L}(j\omega) = P(j\omega)C(j\omega)$ crosses the zero dB line for the first time, $\mathcal{S}(j\omega) = (1 + \mathcal{L}(j\omega))^{-1}$ is the quasi-linear sensitivity function, $W_a(j\omega) = W_\Delta(j\omega)\mathcal{D}_{pid}(j\omega)V_{lp}(j\omega)$ represents the additive modelling uncertainty in the controller due to the HIGS-based low-pass filter, and $W_S(j\omega) = 1/\sqrt{10}$ defines a 10 dB peaking constraint on the sensitivity. The additional factor $\lambda \in [0, 1]$ is used as a means to determine the degree to which modelling uncertainty is taken into account. In practice, it is found that the choice $\lambda = 1$ often leads to overly restrictive designs. A balanced choice is left to the control engineer and typically depends on the system and disturbances at hand. Remark that for $\lambda = 1$ the constraint in (8.13) implies that the magnitude of the worst-case sensitivity $|\mathcal{S}_\Delta| = |\mathcal{S}|/(1 - |W_a\mathcal{S}P|)$ does not exceed 10 dB. Note that $P(j\omega)$ in (8.13) is the *measured* frequency-response function (FRF) of the stage system, and the stability constraint can be verified on the basis of a Nyquist-like test on the open-loop characteristics $\mathcal{L}(j\omega)$. The resulting nominal and worst-case sensitivity characteristics for designs with $\lambda = \{0, 0.3, 0.6\}$ are shown in Fig. 8.12, along with a linear design that is considered for reference. The measurement results obtained from a wafer-stage system are presented in Fig. 8.13.

From the figures, a clear correlation can be seen between the worst-case predictions and the actual measurements in terms of increased error content around 700 Hz which translates to a deterioration of the MSD-values in Fig. 8.13, especially in the scanning interval. By partly taking into account the modelling uncertainty, this content can be significantly reduced, thereby demonstrating the merits of the proposed approach. In fact, the design with $\lambda = 0.6$ is already able to bring this content and the MSD values to about the same level as for the linear case, despite 12 dB peaking of the worst-case quasi-linear sensitivity. In addition to comparable MSD values, the cumulative error representation in Fig. 8.13 demonstrates improved low-frequency disturbance rejection properties (translating to improved MA values) as compared to the linear design. In the comparison, the setpoint signal is highly reproducible and constitutes the main part of the frequency content in e. This is different for the exogenous input that stems from multiple noise sources and that is generally non-reproducible. However, its contribution constitutes a minor part of the frequency

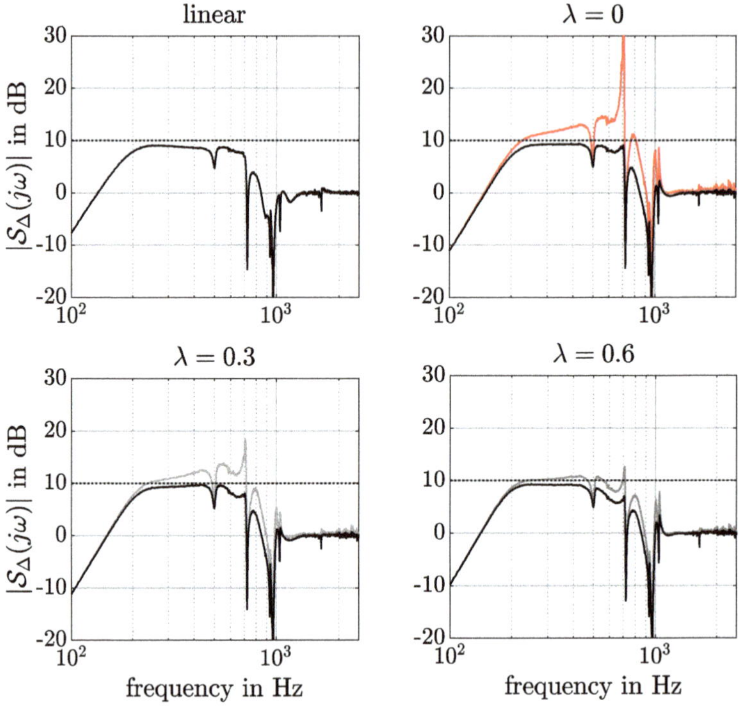

Fig. 8.12 Nominal (black) and worst-case (red, grey) quasi-linear sensitivities for three HIGS-PID designs with different levels of robustness ($\lambda = \{0, 0.3, 0.6\}$). A nominal linear design (top left) is considered for reference

content in e. In this regard, taking into account the uncertainty with $\lambda = 1$ may be too restrictive for the disturbances at hand.

8.5.2 Design and Tuning of the Pre- and Post-filters

As shown in Chap. 7, the content of high-frequency signals can compromise the anticipated gain and phase properties of the HIGS-based controller at low frequencies. Considering the content in Fig. 8.11, the use of pre- and post-filters (recall the structure in Fig. 8.7) becomes particularly relevant for the HIGS-based integrator $C_i\{\mathcal{H}\}$. That is, since $C_i\{\mathcal{H}\}$ is intended to be effective at low frequencies, typically below 100 Hz, dominating content beyond this frequency may compromise the switching of HIGS. Note that for the low-pass filter $C_{lp}\{\mathcal{H}\}$, HIGS should typically switch at frequencies around the bandwidth, such that the use of weighting filters in that case is of less significance.

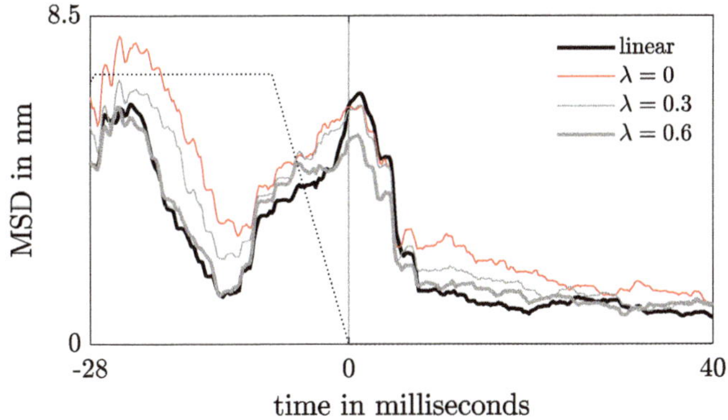

(a) Time-series measurements of the MSD-filtered servo errors.

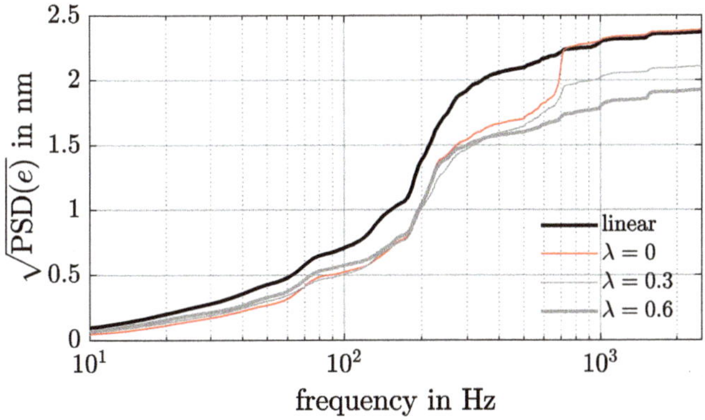

(b) Cumulative root-mean-square representation of the servo errors.

Fig. 8.13 Time- and frequency-domain representation of the measured servo errors for three HIGS-based PID designs with $\lambda = \{0, 0.3, 0.6\}$. Measurements with a linear design (black) are provided for reference

To illustrate the need for a weighting filter design in the context of wafer-stage feedback control, consider a HIGS-based PID controller that is tuned by solving the same optimization problem as in (8.13). The resulting describing function-based sensitivity characteristics are depicted in Fig. 8.14 in grey. For reference, a baseline linear design is provided in black.

When using no weighting filter in the integrator, i.e., $W_i(s) = 1$, the measurement results in Fig. 8.15 show inferior low-frequency disturbance rejection properties for the current HIGS-based design (red curve) compared to the baseline linear design (black curve). This is not in accordance with the predictions in Fig. 8.14. The discrepancy between the linear PID and the HIGS-based PID performance without filtering is particularly visible for low frequencies. Note, for example, the dominant 40 Hz component in the MA filtered error with HIGS-based control, which for the linear

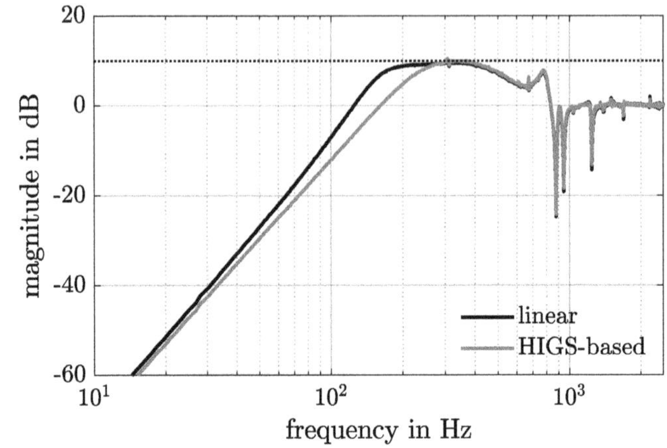

Fig. 8.14 Bode magnitude characteristics of the closed-loop sensitivity functions with linear PID control (black) and HIGS-based PID control (grey)

design appears to be largely suppressed. From the cumulative power spectrum significant gain–loss is found in the frequency range below 100 Hz, which is the result of frequency contributions above 100 Hz that induce too frequent switching of HIGS to gain mode.

As a solution to performance loss in terms of deteriorated low-frequency disturbance rejection, consider the HIGS-based integrator $C_i\{\mathcal{H}\}$ in which the weighting filter $W_i(s)$ is chosen as a skew notch filter of the form

$$W_i(s) = \left(\frac{\omega_{pi}^2}{\omega_{zi}^2}\right) \cdot \frac{s^2 + 2\beta_{zi}\omega_{zi}s + \omega_{zi}^2}{s^2 + 2\beta_{pi}\omega_{pi}s + \omega_{pi}^2}, \tag{8.14}$$

with $\omega_{pi} = 100 \cdot 2\pi$ rad/s, $\omega_{zi} = 1000 \cdot 2\pi$, $\beta_{zi} = \beta_{pi} = 1$. Note that (8.14) is proper, and its exact inverse is realizable. The choice for (8.14) along with the tuning of its parameters is motivated by the design argument to let $C_i\{\mathcal{H}\}$ operate (and thus switch) on frequency content below 200 Hz. Of course, mitigation of the problem with $W_i(s)$ is less effective if either the exogenous inputs have more high-frequency components or the existing high-frequency components become more dominant. Hence, the choice for $W_i(s)$ depends on the application at hand.

In Fig. 8.15, also the measurement result of the HIGS-based PID controller with weighting filter design as in (8.14) is shown through the grey curve. It can be seen that the result is well in line with the expectations raised from the sensitivity predictions in Fig. 8.14, thus demonstrating the effectiveness of the pre- and post-filtering approach. The describing function-based sensitivity function for the considered wafer-stage example appears to provide only a meaningful frequency-domain reflection of performance for the design with weighting filters.

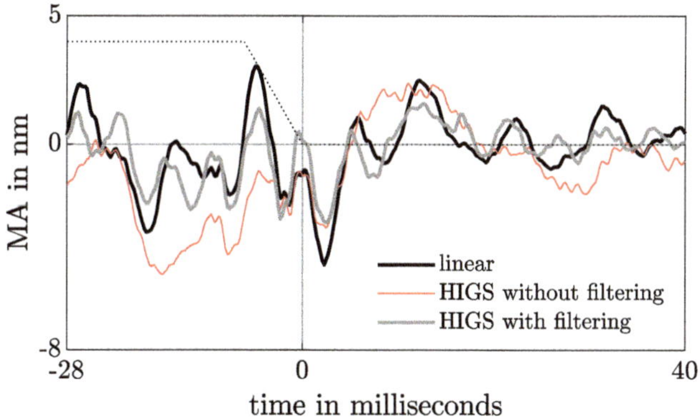

(a) Time-series measurements of the MA-filtered servo errors.

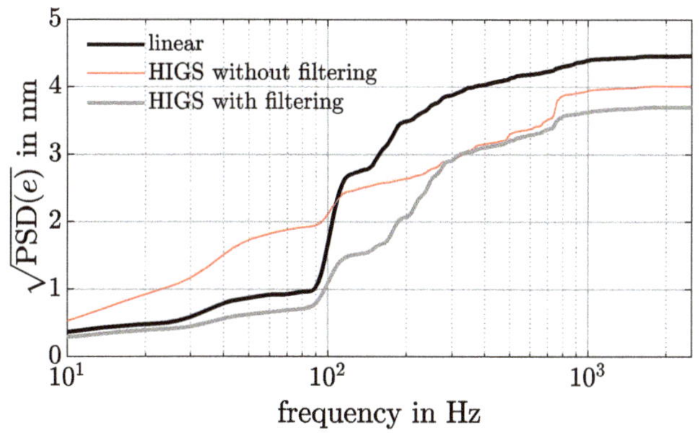

(b) Cumulative root-mean-square representation of the servo errors.

Fig. 8.15 Time- and frequency-domain representation of the measured servo errors with (grey) and without (red) weighting filter strategies. Measurements with a linear design (black) are provided for reference

For a final observation, remark that the case without filtering also induces increased sensitivity to frequencies around 800 Hz, which also affects the MA-filtered response. This increased sensitivity is a consequence of the fact that frequencies below the bandwidth (that is, frequencies around 40 Hz) are significantly present in the error, and thus also in the input to HIGS associated with $C_{lp}\,\{\mathcal{H}\}$, thus affecting switching in an undesirable manner.

Remark 8.5 In the examples above, an approximate Nyquist-like check on the quasi-linear open-loop characteristics $\mathcal{L}(j\omega) = P(j\omega)C(j\omega)$ is found to be particularly useful for guiding a stable closed-loop controller design. Still, there remains a need for a formal assessment of closed-loop stability. In principle, the tools presented in Chaps. 4 and 5 can be extended toward the interconnections of multiple HIGS elements. Preliminary results in this direction are presented in [3, 31]. However, note that the complexity of the conditions significantly increases with an increasing number of piecewise-linear subsystems. In this regard, a compositional approach such as presented in [1] may appear useful to derive more tractable conditions.

8.6 Comparative Design and Performance Evaluation

In this section, the obtained insights are combined into a HIGS-based PID design that aims at performance improvements of the wafer-stage system in scanning y-direction. Performance is evaluated on the basis of measurement results, and is compared to a baseline linear design that is tuned under identical constraints.

8.6.1 Design for Performance

Consider a HIGS-based PID controller as shown in Fig. 8.9 that contains three notch filters. Tuning for maximum bandwidth is done through an automated loop-shaping procedure formulated as the optimization problem in (8.13) with $\lambda = 0.4$, and where $W_S(j\omega)$ defines a piecewise-linear peaking constraint on the quasi-linear sensitivity function. Note that in defining this constraint, explicit knowledge regarding machine-to-machine variations is exploited. To account for position-dependent behaviour of the stage, five FRFs, measured at different (x, y)-locations on the wafer (see Fig. 8.6) are used in the tuning procedure. A linear PID controller with four notch filters is tuned under equivalent design constraints.

The resulting set of controller parameters along with the bandwidths obtained is listed in Table 8.2. A Nyquist plot of the (quasi-)linear open-loop characteristics and the corresponding closed-loop sensitivity characteristics are shown in Fig. 8.16. Table 8.2 shows a substantial increase in the proportional gain k_p for the quasi-linear design compared to the linear design. This increase in gain translates into a 25% increase in bandwidth, and a 4–6 dB improvement in low-frequency disturbance suppression as seen in the sensitivity characteristics in Fig. 8.16b.

8.6.2 Time-Domain Measurement Results

The linear and HIGS-based controllers are applied for feedback control of the wafer-stage system in scanning y-direction. In the actual implementation, discrete-time versions of the controllers are used on the basis of a sampling frequency of 5 kHz, see also Sect. 2.4. In both configurations, the wafer stage is subject to the setpoint signals depicted in Fig. 8.3, and an identical feedforward controller is applied.

Table 8.2 Parameter values for the controllers in scanning y-direction

		Linear design	HIGS design	
PID filter	k_p	$2.67 \cdot 10^7$	$3.38 \cdot 10^7$	N/m
	ω_i	$100 \cdot 2\pi$	$106 \cdot 2\pi$	rad/s
	ω_d	$127.5 \cdot 2\pi$	$169.7 \cdot 2\pi$	rad/s
HIGS (integrator)	k_h	–	1	–
	ω_h	–	$146 \cdot 2\pi$	rad/s
Weighting filter (integrator)	ω_z	–	$104 \cdot 2\pi$	rad/s
	β_z	–	1	–
	ω_p	–	$2000 \cdot 2\pi$	rad/s
	β_p	–	1	–
Low-pass filter	ω_{lp}	$850.2 \cdot 2\pi$	$907 \cdot 2\pi$	rad/s
	β	0.17	0.33	–
HIGS (low-pass filter)	k_h	–	1	–
	ω_h	–	$544.2 \cdot 2\pi$	rad/s
Notch filter	ω_z	$1034.63 \cdot 2\pi$	$717 \cdot 2\pi$	rad/s
	β_z	0.0092	0.119	–
	ω_p	$1027.83 \cdot 2\pi$	$766.6 \cdot 2\pi$	rad/s
	β_z	0.0089	0.184	–
Notch filter	ω_z	$1109.43 \cdot 2\pi$	$1020 \cdot 2\pi$	rad/s
	β_z	0.0061	0.0053	–
	ω_p	$1243.75 \cdot 2\pi$	$1606.4 \cdot 2\pi$	rad/s
	β_z	0.0028	0.1308	–
Notch filter	ω_z	$1005.6 \cdot 2\pi$	$1745.2 \cdot 2\pi$	rad/s
	β_z	0.068	0.1874	–
	ω_p	$979.57 \cdot 2\pi$	$1701 \cdot 2\pi$	rad/s
	β_z	0.467	0.3696	–
Notch filter	ω_z	$709.91 \cdot 2\pi$	–	rad/s
	β_z	0.131	–	–
	ω_p	$840 \cdot 2\pi$	–	rad/s
	β_z	0.333	–	–
Bandwidth	ω_b	$235 \cdot 2\pi$	$293 \cdot 2\pi$	rad/s

The MA- and MSD-filtered servo errors measured at the centre point of the wafer (die one) and the top right corner of the wafer (die three) during four consecutive scanning motions are shown in Fig. 8.17. The first point on a die is exposed in the interval from $t = 0$ to $t = T_e$ s, with T_e the exposure time, meaning that the computation of MA- and MSD-values relevant for scanning performance starts at $\frac{1}{2}T_e = 3.8$ milliseconds after the start of the scan.This is indicated in Fig. 8.17 by the vertical dotted line. The results in Fig. 8.17 show improvements in terms of MA-

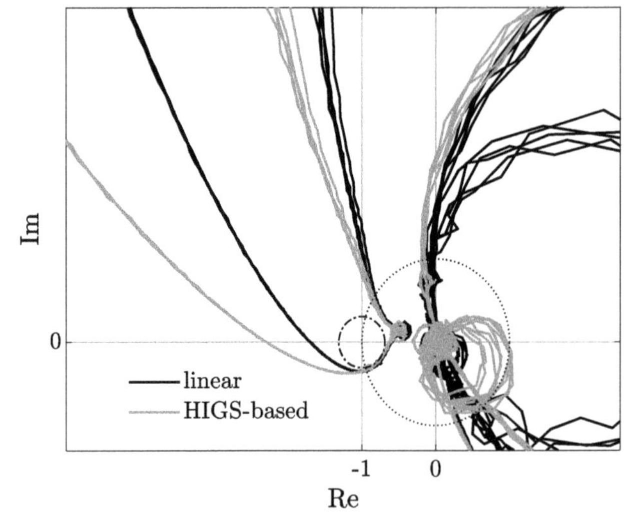

(a) Nyquist plots of the open-loop characteristics.

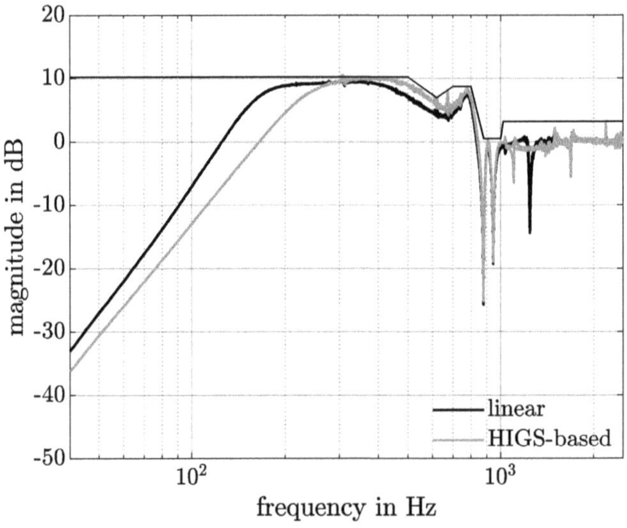

(b) Closed-loop sensitivity characteristics.

Fig. 8.16 Describing function-based Nyquist curves (top) and sensitivity characteristics (bottom) for the wafer-stage system with linear PID control (black) and HIGS-based PID control (grey). In the latter figure, the performance constraint defined by W_S is shown by the thin black curve

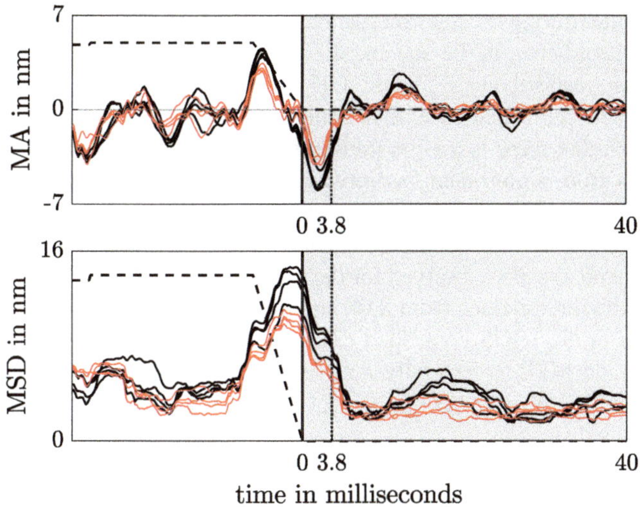

(a) Measurements at die one (center die).

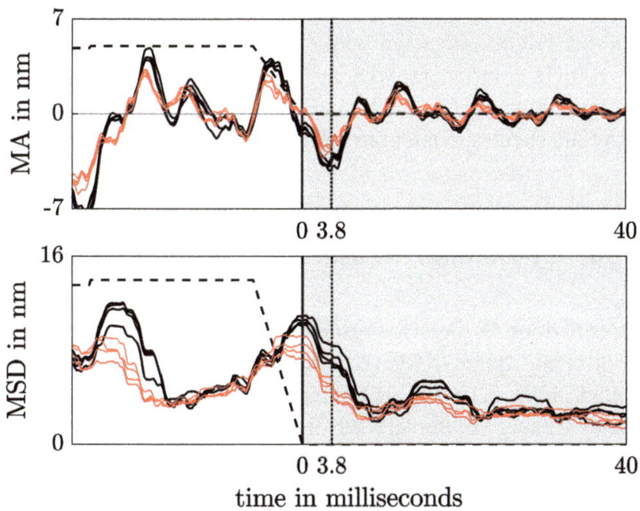

(b) Measurements at die three (top right die).

Fig. 8.17 Four consecutive time-series measurements of the MA- and MSD-filtered servo errors in scanning y-direction at die one and die three obtained with linear control (black) and HIGS-based control (red). The scanning phase is indicated by the grey interval, and the acceleration profile (scaled) is shown by the dashed black line. The start of relevant computations for scanning performance is indicated by the dotted line and MSD-values during the relevant scanning interval. In particular, the peak values of the MA-filtered error at die one and die three, occurring at $t = 3.8$ milliseconds decrease from, respectively, 3.59 and 4.05 nm in the linear case to 1.56 and 2.37 nm in the HIGS controlled case, thus demonstrating a significant improvement of 40–55%. Note that there is a large transient effect present at the start of the scan, which results from a non-ideal feedforward controller in the absence of a settling phase. It can be seen that the HIGS-based controller is able to better deal with these transient effects in terms of reduced overshoot and settling times.

Improvements are also observed for the MSD-filtered error as the peak values at die one and die three reduce from 8.05 nm, and 8.45 nm in the linear case to 5.66 and 5.41 nm in the HIGS controlled case. Note that simultaneous improvements in both MA and MSD is generally not trivial due to (i) the waterbed effect, and (ii) the generation of non-smooth control signals which potentially induce more high-frequency distortion.

The observed improvements in scanning performance are made possible by well-balanced switching between integrator mode and gain mode of both HIGS elements, which is illustrated for a particular measurement result in Fig. 8.18. The switching sequences are depicted in grey: non-zero values correspond to HIGS operating in gain mode, whereas zero values correspond to HIGS operating in integrator mode. It can be seen that HIGS associated with $C_i\{\mathcal{H}\}$ roughly switches on the basis of a dominating 100 Hz component, whereas HIGS associated with $C_{lp}\{\mathcal{H}\}$ roughly switches between 500–600 Hz, thereby indicating that both elements operate on the basis of appropriate spectral content in sufficiently separated frequency intervals.

8.6.3 Frequency-Domain Measurement Results

Harmonic contributions in the error response are studied in more detail through a cumulative root-mean-square (RMS) representation of the measured servo error signals as presented in Fig. 8.19. In this figure, the values of 3.2 and 4.2 nm provide a measure of the total energy of the servo error signals in the scanning interval. The particular values of 1.4 and 2.8 nm are associated with the critical frequency $f_c = 148$ Hz, which is interpreted as the maximum frequency below which the *linear* closed-loop system may benefit from feedback control, i.e., the point at which disturbance suppression is obtained without inducing any amplifications. Since the linear controller is designed to maximize this point (by maximizing the bandwidth ω_b), it provides an objective measure to quantify improvements in the suppression of low-frequency disturbances with HIGS. It can be seen that HIGS-based control is able to reduce the RMS value at the critical frequency by a factor of two, thereby demonstrating better low-frequency disturbance rejection properties. This factor is in good agreement with the expectations raised by the sensitivity characteristics in Fig. 8.16b. Note that more frequency contributions are present in the nonlinear system for the range 400–900 Hz compared to the linear system. Within the range 400–700 Hz this is in line with the expectations raised by the describing function-based sensitivity

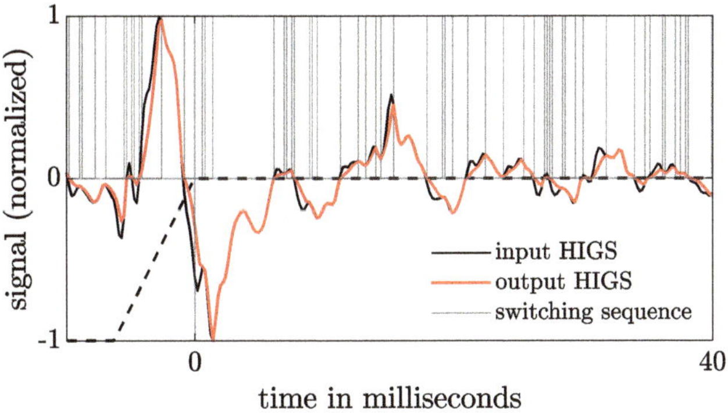

(a) Switching of HIGS in the integrator C_i $\{\mathcal{H}\}$.

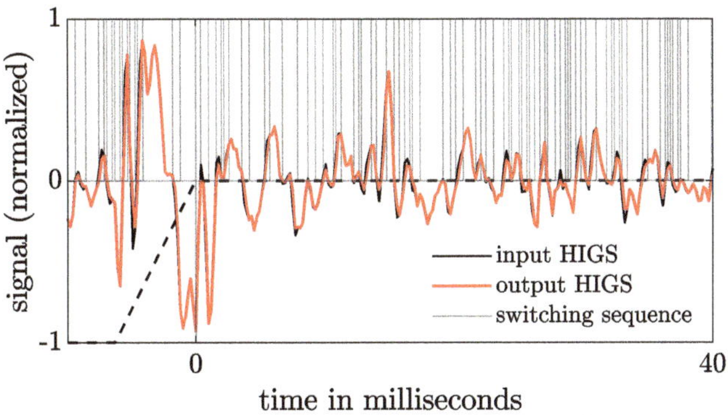

(b) Switching of HIGS in the low-pass filter C_{lp} $\{\mathcal{H}\}$.

Fig. 8.18 Switching characteristics of HIGS associated with the integrator C_i $\{\mathcal{H}\}$ (top figure) and low-pass filter C_{lp} $\{\mathcal{H}\}$ (bottom figure). The switching sequences are indicated in grey: non-zero values correspond to the gain mode, and zero values correspond to the integrator mode

characteristics in Fig. 8.16b. For the range 700–900 Hz, an increased content may be the result of the generation of higher harmonics, as well as an increased sensitivity to high frequencies that does not appear from the characteristics of the describing function, but results from the switching behaviour of HIGS as discussed previously in Sect. 8.5 (see also Chap. 7). This increased sensitivity is emphasized by resonance frequencies in the plant, which for the stage in y-direction occur around 700–900 Hz (recall the FRFs in Fig. 8.6). The design is chosen to be sufficiently robust against these effects in the sense that the influence on performance (in terms of MA- and MSD-values) remains fairly small.

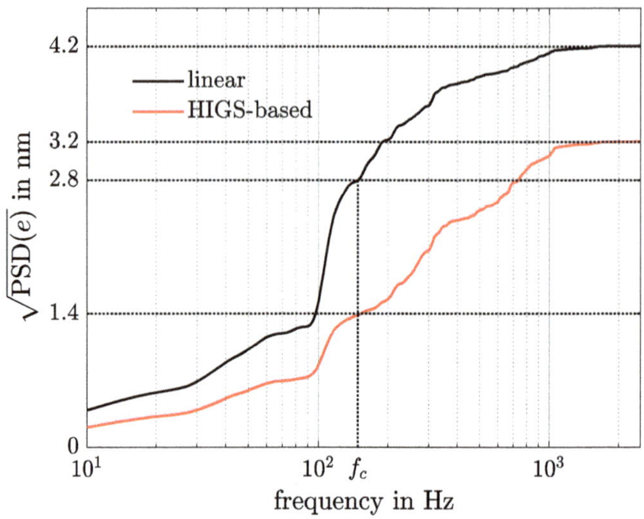

Fig. 8.19 Cumulative root-mean-square (RMS) representation of the measured servo error. The critical frequency f_c indicates the point below which the linear system benefits from feedback control

8.7 Takeaways

This chapter presents a case-study specific to wafer scanners, but many of the design principles presented, insights from measurements, and learned lessons extend well beyond this specific wafer scanning application domain. Readers unfamiliar with wafer scanning applications may still find valuable insights in the way the HIGS-based controller is implemented, and the potential pitfalls associated with HIGS-based control design as discussed in Sect. 8.5. The main takeaways of this chapter for practical HIGS-based design are as follows.

(1) Filter sequence matters: LTI integrators should follow the HIGS elements. When using multiple HIGS elements, separate them with low-pass filters.
(2) Frequency separation is essential: each HIGS element should operate in a distinct frequency band to avoid interaction. Tune the controller parameters to achieve this.
(3) Pre-filter tuning should match intended HIGS behaviour: pre-filters should amplify the target frequency range in the input where the HIGS element should operate.
(4) Account for describing function uncertainty: factor this uncertainty in a frequency-domain loop-shaping procedure, but avoid overly conservative design.

In addition, this chapter provides the ingredients for setting up a simulation environment that can aid in further exploration of HIGS-based control. Specifically, this chapter provides (i) relevant setpoints in Sect. 8.2.1, (ii) a representative dynamic model in Sect. 8.3, (iii) an LTI and HIGS-based control scheme in Sect. 8.4, and (iv) a comparative tuning in Sect. 8.6.

8.8 Summary

In this chapter, the performance improvement potential of HIGS-based PID control is demonstrated on a wafer-stage system of a state-of-practice industrial wafer scanner. The reduced phase characteristics associated with this specific controller structure are exploited in a robust quasi-linear loop-shaping procedure to achieve increased bandwidths, enabling improved low-frequency disturbance rejection properties when compared to a linear equivalent design. The procedure is complemented with knowledge of relevant signal content that may affect switching of HIGS, thereby leading to an effective closed-loop controller design in practice. Measurement results with the HIGS-based controller obtained from a wafer stage show a factor-two improvement in disturbance rejection properties of the system as compared to the baseline linear design, without excessive transmission of high-frequency content. These results support the potential of HIGS-based control for high-precision motion systems.

References

1. Arcak, M., Meissen, C., Packard, A.: Networks of Dissipative Systems. Springer (2016)
2. Beerens, R., Bisoffi, A., Zaccarian, L., Heemels, W., Nijmeijer, H., van de Wouw, N.: Reset integral control for improved settling of PID-based motion systems with friction. Automatica **107**, 483–492 (2019)
3. Beerens, R., Konings, P., Heertjes, M., van den Eijnden, S.: Stability and performance assessment of a MIMO higs-based controller design. In: 2024 European Control Conference (ECC), pp. 1723–1728 (2024)
4. Beijen, M., op Smeijers, T.T., Boerrigter, G., van den Eijnden, S.: Hybrid control of a variable-speed peristaltic pump. In: 2024 American Control Conference (ACC), pp. 4296–4301 (2024)
5. Boerlage, M., Tousain, R., Steinbuch, M.: Jerk derivative feedforward control for motion systems. In: Proceedings of the 2004 American Control Conference, vol. 5, pp. 4843–4848 (2004)
6. Boettcher, U., Fetzer, D., Li, H., de Callafon, R.A., Talke, F.E.: Reference signal shaping for closed-loop systems with application to seeking in hard disk drives. IEEE Trans. Control Syst. Technol. **20**(2), 335–345 (2012)
7. Butler, H.: Position control in lithographic equipment. Control. Syst., IEEE **31**, 28–47 (2011)
8. Butler, H.: Adaptive feedforward for a wafer stage in a lithographic tool. IEEE Trans. Control Syst. Technol. **21**(3), 875–881 (2013)
9. Cai, C., Dastjerdi, A.A., Saikumar, N., HosseinNia, S.: The optimal sequence for reset controllers. In: 2020 European Control Conference (ECC), pp. 1826–1833 (2020)
10. Cutforth, C.F., Pao, L.Y.: Adaptive input shaping for maneuvering flexible structures. Automatica **40**(4), 685–693 (2004)
11. van Eijk, L.F., Beer, S., van Es, R.M.J., Kostić, D., Nijmeijer, H.: Frequency-domain properties of the hybrid integrator-gain system and its application as a nonlinear lag filter. IEEE Trans. Control Syst. Technol. **31**(2), 905–912 (2023)

12. Gelb, A., Vander Velde, W.: Multiple-input Describing Functions and Nonlinear Systems Design. McGraw Hill, New York (1968)
13. Guo, Y., Wang, Y., Xie, L.: Frequency-domain properties of reset systems with application in hard-disk-drive systems. IEEE Trans. Control Syst. Technol. **17**(6), 1446–1453 (2009)
14. Heemels, W.P.M.H., Tanwani, A.: Existence and completeness of solutions to extended projected dynamical systems and sector-bounded projection-based controllers. IEEE Control Syst. Lett. **7**, 1590–1595 (2023)
15. Heertjes, M., van den Eijnden, S., Sharif, B., Heemels, M., Nijmeijer, H.: Hybrid integrator-gain system for active vibration isolation with improved transient response. IFAC-PapersOnLine **52**(15), 454–459, 8th IFAC Symposium on Mechatronic Systems MECHATRONICS 2019 (2019)
16. Heertjes, M., Butler, H., Dirkx, N., van der Meulen, S., Ahlawat, R., O'Brien, K., Simonelli, J., Teng, K.T., Zhao, Y.: Control of wafer scanners: methods and developments. In: 2020 American Control Conference (ACC), pp. 3686–3703 (2020)
17. Heertjes, M.M., Hennekens, D.D., Steinbuch, M.: MIMO feed-forward design in wafer scanners using a gradient approximation-based algorithm. Control Eng. Pract. **18**, 495–506 (2010)
18. Karybakas, C.A.: Nonlinear integrator with zero phase shift. IEEE Trans. Ind. Electron. Control Instrum. **IECI-24**(2), 150–152 (1977)
19. Kontaras, N., Heertjes, M., Zwart, H.: Continuous compliance compensation of position-dependent flexible structures. IFAC-PapersOnLine **49**(13), 76–81, 12th IFAC Workshop on Adaptation and Learning in Control and Signal Processing ALCOSP 2016 (2016)
20. Li, Y., Guo, G., Wang, Y.: Reset control for midfrequency narrowband disturbance rejection with an application in hard disk drives. IEEE Trans. Control Syst. Technol. **19**(6), 1339–1348 (2011)
21. Martinez, V., Edgar, T.F.: Control of lithography in semiconductor manufacturing. IEEE Control Syst. Mag. **26**(6), 46–55 (2006)
22. Oomen, T., van Herpen, R., Quist, S., van de Wal, M.M.J., Bosgra, O.H., Steinbuch, M.: Next-generation wafer stage motion control: connecting system identification and robust control. In: 2012 American Control Conference (ACC), pp. 2455–2460 (2012)
23. Oomen, T., van Herpen, R., Quist, S., van de Wal, M., Bosgra, O., Steinbuch, M.: Connecting system identification and robust control for next-generation motion control of a wafer stage. IEEE Trans. Control Syst. Technol. **22**(1), 102–118 (2014)
24. Potsaid, B., Wen, J.: High performance motion tracking control. In: Proceedings of the 2004 IEEE International Conference on Control Applications, vol. 1, pp. 718–723 (2004)
25. Saikumar, N., Sinha, R.K., HosseinNia, S.H.: "Constant in gain lead in phase" element–application in precision motion control. IEEE/ASME Trans. Mechatron. **24**(3), 1176–1185 (2019)
26. Seron, M.M., Braslavsky, J.H., Goodwin, G.C.: Fundamental Limitations in Filtering and Control. Springer, Berlin (1997)
27. Shi, K., Nikooienejad, N., Petersen, I.R., Moheimani, S.O.R.: Negative imaginary control using hybrid integrator-gain systems: application to mems nanopositioner. IEEE Trans. Control Syst. Technol. **32**(3), 718–730 (2024)
28. Skogestad, S., Postlethwaite, I.: Multivariable Feedback Control: Analysis and Design, 2nd edn. Wiley and Sons (2010)
29. Steinbuch, M., Norg, M.: Advanced motion control: an industrial perspective. Eur. J. Control **4**(4), 278–293 (1998)
30. Trentelman, H.L., Stoorvogel, A.A., Hautus, M.: Control Theory for Linear Systems. Springer, London (2001)
31. van den Eijnden, S., Francke, M., Nijmeijer, H., Heertjes, M.: Improving wafer stage performance with multiple hybrid integrator-gain systems. IFAC-PapersOnLine **53**(2), 8321–8326, 21st IFAC World Congress (2020)
32. van den Eijnden, S., Heertjes, M., Nijmeijer, H.: Experimental demonstration of a nonlinear PID-based control design using multiple hybrid integrator-gain elements. In: 2020 American Control Conference (ACC), pp. 4307–4312 (2020)

Chapter 9
Conclusion and Outlook

9.1 Conclusion

The *hybrid integrator–gain system*, abbreviated as HIGS, has been considered as a possible means to deal with the fundamental performance limitations of linear time-invariant (LTI) control for LTI (motion) systems. In terms of performance, HIGS may offer comparable benefits as other hybrid control concepts such as reset controllers that originate from the Clegg integrator, however, with the additional advantage of HIGS avoiding the need for hard state resets and exploiting continuous control. The latter opens up new opportunities in both application-based and control-theoretic directions.

This book provides a comprehensive overview of both the theory and practice of HIGS-based control systems. The exposition begins by demonstrating the possibility to overcome fundamental performance limitations of LTI control for LTI systems with HIGS. The key conclusion that follows from this result is that performance improvements do not simply come from blind application of HIGS, but arise from a careful interplay between HIGS and the linear filters in the controller. The sequence in which these filters appear with respect to HIGS is of particular importance in the design.

Tools for stability analysis of HIGS-based control systems are developed along two main directions: frequency-domain and time-domain conditions. The frequency-domain approach allows for graphical verification and use of nonparametric plant models, and it extends naturally to robustness analysis. However, it introduces conservatism due to the reliance on a common quadratic Lyapunov function for an essentially piecewise linear system. In contrast, the time-domain conditions reduce conservatism by employing piecewise quadratic Lyapunov functions and are formulated as numerically tractable linear matrix inequalities (LMIs). The latter are naturally extended to include steady-state and transient performance measures in terms of the $\mathcal{L}_2$-gain and $\mathcal{H}_2$-norm. For some examples, the time-domain conditions are found

S. van den Eijnden et al., *Hybrid Integrator-Gain Systems*, Lecture Notes in Control and Information Sciences 499, https://doi.org/10.1007/978-3-032-18101-5_9

to closely resemble what is expected from a necessary condition for closed-loop stability, and tight performance estimates are obtained. However, reduced conservatism may come at the cost of an increased computational burden. Given the different trade-offs between conservatism and complexity in the conditions, both approaches are valuable and complementary.

In addition to the $\mathcal{L}_2$-gain and $\mathcal{H}_2$-norm, which provide generic performance measures, steady-state behaviour under specific inputs is analysed using the framework of convergent systems. A convergent system guarantees a unique and bounded steady-state response for every bounded (possibly time-varying) input. Incremental passivity and incremental gain properties of HIGS are derived and exploited to formulate time- and frequency-domain conditions for convergence. The results enable a precise characterization of performance using nonlinear Bode plots, which offer new insights in, e.g., the sensitivity and disturbance rejection properties of a HIGS-controlled system.

The design of HIGS-based controllers is pursued through two different approaches. The first approach leverages previously developed stability conditions directly for controller synthesis. This method yields HIGS-based controllers with strong theoretical guarantees for robust stability and performance, but often lacks clear physical intuition for parameter tuning and robust controller redesign. The latter may sometimes be perceived as less desirable in practical applications. The second approach exploits the describing function of HIGS within a robust loop-shaping framework, combined with both time- and frequency-domain design guidelines. Although the approach does not come with formal guarantees for stability and performance, it still serves as a practically valuable and insightful design method. This is also demonstrated in the application of HIGS-based feedback control in a state-of-the-art industrial wafer-stage system. Compared to the state-of-practice linear control solution, the HIGS-based design enables a substantial increase in low-frequency disturbance rejection properties without compromising high-frequency noise amplification, which is attributed to its phase benefits, thereby supporting also the potential of HIGS-based control for industrial applications.

9.2 Outlook

HIGS represents a promising but still relatively new control approach. This book demonstrates that HIGS offers compelling advantages in performance, particularly in high-precision industrial applications. However, the theoretical foundations and practical applications of HIGS-based control open up broad directions for future research.

One promising direction lies in generalizing the specific philosophy underlying HIGS, potentially through the framework of extended projected dynamical systems (ePDS) [5, 8, 11]. This direction could enable the development of HIGS-like and projection-based strategies for multivariable and nonlinear systems, extending the applicability of the approach and improving performance capabilities. It is envisioned that careful design of the logic triggering a switch or projection will be key to achieving, e.g., improved phase characteristics; see also the preliminary results in [6].

To bring nonlinear and hybrid control strategies into practice, the development of systematic, easy-to-use, and accurate tools for analysis and design is indispensable. In this book, new graphical analysis tools based on frequency-domain methods have been introduced. These tools are valuable from a practical viewpoint, but reducing their conservatism remains an open issue. Bridging the less conservative LMI-based conditions and practical frequency-domain approaches could lead to improved tools. A starting point in this direction may come from exploiting the ideas in [13, Theorem 1] to reduce a number of (less conservative) LMIs into a single equivalent LMI, the latter which may then be transformed into an equivalent frequency-domain inequality by means of the Kalman–Yakubovich–Popov lemma.

Frequency-domain conditions exploit the linear structure in the system. To go beyond the classical Lur'e setting of a linear part in feedback with a nonlinearity, input–output analysis tools based on integral quadratic constraints (IQCs) [10] and scaled relative graphs (SRGs) [3, 4, 14] may offer novel routes for developing graphical tools for stability analysis. Additionally, although the describing function approach lacks a rigorous theoretical foundation for HIGS at this stage, it has shown surprisingly good predictive value in practice and deserves further exploration. Although this is acknowledged to be a challenging problem, revisiting the line of work in [1, 9] with modern tools might provide a promising starting point.

The conditions presented for incremental stability and convergence of HIGS-controlled systems are conservative at this point. To improve on this aspect, it may be fruitful to explore the ideas and results outlined in [15, 16]. Furthermore, recent results in contraction theory, see, e.g., [2], could inspire novel approaches. Note that incremental stability/convergence analysis of discontinuous dynamical systems in general is a relevant and challenging topic that seems to have received only limited attention in the literature but deserves further exploration.

Regarding the tuning and synthesis of HIGS-based controllers, developments in machine learning and data-based control methods are expected to guide novel approaches to tuning and learning the optimal (HIGS-based) controller structure [7]. In addition, the aforementioned advances in describing function analysis should aid the development of an accurate and reliable nonlinear loop-shaping framework that strongly connects to the current industrial controller design practice.

Finally, in this book, a continuous-time multi-loop single-input single-output approach is pursued for the design of a HIGS-based controller for a wafer-stage system. Although this approach was motivated by the high sampling rate and the largely decoupled wafer-stage dynamics, the discrete-time and multivariate design aspects of HIGS-based control systems are important and largely untouched [12]. Particularly in industrial high-precision applications, these aspects play an essential role, and therefore additional research in this direction is needed.

9.3 Final Thoughts

The development of hybrid control strategies is partly driven by the need of the high-tech industry to realize machine performance beyond what is currently possible with linear control. Despite their promising potential, it is surprising to see that hybrid controllers are still far from being completely adopted in industrial applications. This seems attributed to the fact that hybrid controllers are generally more difficult to analyse and design than their linear counterparts and require advanced training of control engineers. Broader acceptance in industry can be accelerated by focusing on the development of accessible, intuitive, and systematic tools for analysis and design, and by providing experimental proof-of-concepts. This book contributes to these important areas for a special hybrid control strategy known as HIGS. With a suitable design, HIGS-based controllers have the ability to outperform linear controllers, thereby opening up new possibilities to further push the performance of linear systems. It is shown that such possibilities not only exist on paper, but can be realized in practical applications as well. The authors conclude with the hope that the findings in this book merit further developments in the appealing and challenging area of nonlinear and hybrid control systems.

References

1. Bergen, A., Chua, L., Mees, A., Szeto, E.: Error bounds for general describing function problems. IEEE Trans. Circuits Syst. **29**(6), 345–354 (1982)
2. Bullo, F.: Contraction Theory for Dynamical Systems. Francesco Bullo (2022)
3. Chaffey, T.: A rolled-off passivity theorem. Syst. Control Lett. **162**, 105198 (2022)
4. Chaffey, T., Forni, F., Sepulchre, R.: Graphical nonlinear system analysis. IEEE Trans. Autom. Control **68**(10), 6067–6081 (2023)
5. Chu, H., van den Eijnden, S., Heemels, W.: Projection-based controllers with inherent dissipativity properties. In: 2023 62nd IEEE Conference on Decision and Control (CDC), pp. 5664–5669 (2023)
6. Chu, H., van den Eijnden, S., Heertjes, M., Heemels, W.: Filtering in projection-based integrators for improved phase characteristics. In: 2024 IEEE 63rd Conference on Decision and Control (CDC), pp. 2391–2396 (2024)
7. Ferrarotti, L., Bemporad, A.: Learning optimal switching feedback controllers from data. IFAC-PapersOnLine **53**(2), 1602–1607 (2020). (21st IFAC World Congress)
8. Heemels, W.P.M.H., Tanwani, A.: Existence and completeness of solutions to extended projected dynamical systems and sector-bounded projection-based controllers. IEEE Control Syst. Lett. **7**, 1590–1595 (2023)
9. Mees, A.I.: Describing functions: ten years on. IMA J. Appl. Math. **32**(1–3), 221–233 (1984)
10. Megretski, A., Rantzer, A.: System analysis via integral quadratic constraints. IEEE Trans. Autom. Control **42**(6), 819–830 (1997)
11. Sharif, B., Heertjes, M.F., Heemels, W.P.M.H.: Extended projected dynamical systems with applications to hybrid integrator-gain systems. In: IEEE 58th Conference on Decision and Control (CDC'19), pp. 5773–5778 (2019)
12. Sharif, B., Alferink, D., Heertjes, M., Nijmeijer, H., Heemels, M.: Analysis of sampled-data hybrid integrator-gain systems: a discrete-time approach. Automatica **167**, 111765 (2024)

13. Skorodinskii, V.I.: Absolute stability and absolute instability of control systems with two nonlinear nonstationary elements. Autom. Remote. Control. **42**(9), 1149–1157 (1981)
14. van den Eijnden, S., Chaffey, T., Oomen, T., (Maurice) Heemels, W.: Scaled graphs for reset control system analysis. Eur. J. Control. **80**, 101050 (2024)
15. Waitman, S., Massioni, P., Bako, L., Scorletti, G., Fromion, V.: Incremental l_2-gain analysis of piecewise-affine systems using piecewise quadratic storage functions. In: 2016 IEEE 55th Conference on Decision and Control (CDC), pp. 1334–1339 (2016)
16. Waitman, S., Massioni, P., Bako, L., Scorletti, G.: Incremental l2-gain stability of piecewise-affine systems with piecewise-polynomial storage functions. Automatica **107**, 224–230 (2019)

Appendix A
Proofs

A.1 Preliminary Results

Before stating the proofs of several theorems presented in the main text, some useful technical results are given first.

The following general version of the Kalman–Yakubovich–Popov (KYP)-lemma, which is free of any hypothesis on minimality of the system, is provided. This result will play a central role in the upcoming proofs of Chaps. 4 and 6 (see also [6, 7] for other versions of the KYP-lemma).

Theorem A.1 ([8, Theorem 1]) *Given* $A \in \mathbb{R}^{n\times n}$, $B \in \mathbb{R}^{n\times m}$, $Q = Q^\top \in \mathbb{R}^{(n+m)\times(n+m)}$ *with* $\det(j\omega I - A) \neq 0$ *for all* $\omega \in \mathbb{R}$. *The following two statements are equivalent:*

1. *For all* $\omega \in \mathbb{R} \cup \{\infty\}$ *the following inequality is satisfied*

$$\begin{bmatrix} (j\omega I - A)^{-1}B \\ I \end{bmatrix}^* Q \begin{bmatrix} (j\omega I - A)^{-1}B \\ I \end{bmatrix} \prec 0. \tag{A.1}$$

2. *There exists a symmetric matrix* $P = P^\top \in \mathbb{R}^{n\times n}$ *such that*

$$\begin{bmatrix} A^\top P + PA & PB \\ B^\top P & 0 \end{bmatrix} + Q \prec 0. \tag{A.2}$$

The corresponding equivalence holds for non-strict inequalities in both (A.1) and (A.2), if the pair (A, B) *is controllable.*

The following result for the scalar case ($m = 1$) is a special case of Theorem A.1.

S. van den Eijnden et al., *Hybrid Integrator-Gain Systems*, Lecture Notes in Control and Information Sciences 499, https://doi.org/10.1007/978-3-032-18101-5

Lemma A.1 *Given a Hurwitz matrix A, a vector $B \in \mathbb{R}^n$ and a matrix*

$$Q = \begin{bmatrix} 0 & -Q_{12} \\ -Q_{12}^\top & q_{22} \end{bmatrix} \tag{A.3}$$

with $q_{22} < 0$ and arbitrary $Q_{12} \in \mathbb{R}^n$ for which the conditions in Theorem A.1 are satisfied. Then there exist a positive definite matrix $P = P^\top \in \mathbb{R}^{n\times n}$, a positive semi-definite matrix $N = N^\top \in \mathbb{R}^{n\times n}$, a vector $L \in \mathbb{R}^n$, and a number $\varepsilon > 0$ that satisfy

$$A^\top P + PA = -L^\top L - N - \varepsilon P \tag{A.4}$$

$$PB = Q_{12} - \sqrt{|q_{22}|}L^\top. \tag{A.5}$$

Proof Note that since A is Hurwitz, $\det(j\omega I - A) \neq 0$ for all $\omega \in \mathbb{R}$, such that Theorem A.1 can be applied for A, B, and Q as in (A.3). Since $q_{22} < 0$ it follows by virtue of the Schur complement that (A.2) is equivalent to

$$A^\top P + PA + \frac{1}{|q_{22}|}(PB - Q_{12})(PB - Q_{12})^\top \prec 0, \tag{A.6}$$

which implies the existence of a matrix $\hat{N} = \hat{N}^\top \succ 0$ and a number $\varepsilon > 0$ such that

$$A^\top P + PA + \frac{1}{|q_{22}|}(PB - Q_{12})(PB - Q_{12})^\top = -(\hat{N} - \varepsilon P) - \varepsilon P. \tag{A.7}$$

By choosing ε sufficiently small, one can always ensure the matrix $N := \hat{N} - \varepsilon P$ to be positive (semi-)definite. Define

$$L^\top = \frac{1}{\sqrt{|q_{22}|}}(PB - Q_{12}) \tag{A.8}$$

and substitute this identity in (A.7) to obtain (A.4) and (A.5). It remains to show that P is positive definite. Since (A.4) implies $A^\top P + PA \prec 0$ and A is Hurwitz, the result is immediate (see, e.g., [6, Lemma 1.10.1]). □

The next result from [4] is particularly useful for reducing a number of matrix inequalities to a single matrix inequality. The result can be used in a recursive manner, and is used in various proofs of Chaps. 4 and 6.

Lemma A.2 ([4]) *For satisfying a system of two matrix inequalities*

$$I_1 \prec 0, \quad I_2 \prec 0, \tag{A.9}$$

for which the difference can be written as $I_2 - I_1 = Q = pq^\top + qp^\top$, *for some vectors* $p, q \in \mathbb{R}^n$, *it is necessary and sufficient that there exists a number* $\varepsilon > 0$ *such that one inequality*

$$I_1 + Q^+(\varepsilon) = I_2 + Q^-(\varepsilon) \prec 0, \tag{A.10}$$

with

$$Q^\pm(\varepsilon) = \left(\frac{\varepsilon}{\sqrt{2}}p \pm \frac{1}{\varepsilon\sqrt{2}}q\right)\left(\frac{\varepsilon}{\sqrt{2}}p \pm \frac{1}{\varepsilon\sqrt{2}}q\right)^\top \tag{A.11}$$

is satisfied.

Proof The proof can be found in [4, Sect. 3]. □

A.2 Proofs for Chap. 2

A.2.1 Proof of Theorem 2.1

The proof of Theorem 2.1 exploits some of the machinery introduced in Sect. 2.3. Consider the quadratic storage function candidate $V(x_h) = kx_h^2$ with $k = \lambda/(2\omega_h)$ and $\lambda \in (0, 1)$. This function is Lipschitz continuous and positive definite. Since x_h is a locally absolutely continuous solution to $\dot{x}_h(t) \in \{\omega_h z(t), k_h \dot{z}(t)\}$ for some continuously differentiable z, it follows from Theorem 2.7 that in integrator mode

$$\dot{V} = \nabla V(x_h)\omega_h z = \lambda z x_h. \tag{A.12}$$

On the other hand, in gain mode one finds

$$\dot{V} = \nabla V(x_h)k_h\dot{z} = \lambda\frac{k_h}{\omega_h}x_h\dot{z} < \lambda z x_h, \tag{A.13}$$

where for the inequality use is made of the fact that in gain mode one has $x_h = k_h z$ and $\omega_h z^2 > k_h z\dot{z}$, see also the definition of the gain mode set in (2.4b). As such, one finds as a common upper-bound $\dot{V} \leq \lambda z x_h = \lambda z u$, which shows that HIGS is passive. To show *strict* passivity, observe that

$$\dot{V} \leq -(1-\lambda)z x_h + zu \leq -(1-\lambda)\frac{1}{k_h}x_h^2 + zu, \tag{A.14}$$

where for obtaining the second inequality use is made of the sector condition $x_h^2 \leq k_h z x_h$. Denote $c = (1-\lambda)/k_h$ and observe that since $\lambda \in (0, 1)$, one finds $c \in (0, 1/k_h)$.

A.2.2 Proof of Theorem 2.2

Consider the function $V(x_h) = \frac{1}{2k_h}x_h^2$, which is Lipschitz continuous and positive definite since $k_h > 0$. Since x_h is a locally absolutely continuous solution to $\dot{x}_h(t) \in \{\omega_h z(t), k_h \dot{z}(t)\}$ for some continuous $z \in C^1$, it follows from Theorem 2.7 that in integrator mode

$$\dot{V} = \nabla V(x_h)\omega_h z = \frac{\omega_h}{k_h} uz \leq \omega_h z^2 = z\dot{u}, \tag{A.15}$$

where the sector condition $zu \leq k_h z^2$ is used. Similarly, in gain mode one has

$$\dot{V} = \nabla V(x_h)\dot{x}_h = \frac{1}{k_h} k_h z \dot{x}_h = z\dot{u}, \tag{A.16}$$

where use is made of the equality $x_h = k_h z$ in gain mode. Combining the above (in)equalities leads to the common upper bound $\dot{V} \leq z\dot{u}$, which shows that HIGS is negative imaginary.

A.2.3 Proof of Theorem 2.3

Consider the difference between two solutions starting from different consistent initial conditions $x_h'(0), x_h''(0) \in \mathbb{R}$ as $\delta x_h(t) := x_h'(t, x_h'(0), z(t)) - x_h''(t, x_h''(0), z(t))$ and define an incremental Lyapunov function candidate as

$$V(\delta x_h) = \frac{1}{2}(\delta x_h)^2. \tag{A.17}$$

Clearly, (A.17) is positive definite. Moreover, since solutions to HIGS in (2.1) are locally absolutely continuous functions, $t \mapsto V(\delta x_h(t))$ is a locally absolutely continuous function as well, and thus $dV(\delta x_h(t))/dt$ exists almost everywhere.

Suppose two trajectories starting at different initial conditions are both in integrator mode. In this case, one finds the time-derivative along the solutions of (2.1) to satisfy $\dot{V} = 0$, such that V itself remains constant. A similar observation can be made when both trajectories are in gain mode, in which case it holds that $\dot{V} = 0$ and $V = 0$. Note that the latter case implies $\delta x_h = 0$.

Next, consider the case where both trajectories are in different modes, for which one finds

$$\dot{V} = \begin{cases} (x_h'(t) - k_h z(t))(\omega_h z(t) - k_h \dot{z}(t)), & \text{if } (x_h'(t), x_h''(t)) \in \mathcal{F}_1 \times \mathcal{F}_2, \\ (k_h z(t) - x_h''(t))(k_h \dot{z}(t) - \omega_h z(t)), & \text{if } (x_h'(t), x_h''(t)) \in \mathcal{F}_2 \times \mathcal{F}_1, \end{cases} \tag{A.18}$$

where $x'_h(t) = x_h(t, x'_h(0), z(t))$ and $x''_h(t) = x_h(t, x''_h(0), z(t))$. In the remainder, dependency on time is omitted for brevity. Suppose x''_h resides in gain mode. In that case, it holds that $\omega_h z^2 > k_h \dot{z} z$ (see the definition of the gain mode set $\mathcal{F}_2$ in (2.4b)). Moreover, since $z \neq 0$ and $k_h z - x'_h \neq 0$ (x'_h cannot be on the sector-boundary otherwise it must be in gain mode from the fact that $\omega_h z^2 > k_h \dot{z} z$), it follows that $(k_h z - x'_h)z > 0$. Hence, $x'_h - k_h z$ and $\omega_h z - k_h \dot{z}$ have opposite signs and are both non-zero, such that $\dot{V} < 0$. A similar reasoning applies for the case when x'_h is in gain mode. At this point, it is known that $\dot{V} \leq 0$, which implies that $i)$ HIGS in (2.1) is incrementally stable, i.e., $\|x'_h(t, x'_{h,0}, z) - x''_h(t, x''_{h,0}, z)\| \leq \|x'_{h,0} - x''_{h,0}\|$, and $ii)$ solutions to HIGS are unique. However, this does not yet imply the contraction property for $t \to \infty$. To show this property, first note that there are three "trouble-making" situations in which $\dot{V} = 0$, and thus $V \neq 0$ can get stuck after some time $t \geq T$ with $T \geq 0$:

(1) The dynamics governing x'_h and x''_h both remain in integrator mode for all $t \geq T$.
(2) The dynamics governing x'_h and x''_h both remain in gain mode for all $t \geq T$.
(3) The dynamics governing x'_h and x''_h synchronously switch between integrator mode and gain mode for all $t \geq T$.

The last two situations can be easily ruled out. Namely, when both dynamics switch to gain mode at some finite time instance $t^* \geq 0$, one finds $x'_h(t^*) = x''_h(t^*) = k_h z(t^*)$. Since $\dot{V} \leq 0$ one finds $V(\delta x_h(t^*)) = 0$ and thus $x'_h(t^*) = x''_h(t^*)$ for all $t \geq t^*$. Hence, this leaves the case that both dynamics get stuck in integrator mode and remain therein forever. However, due to the assumption that z is bounded this cannot happen as will be shown next. First observe that in integrator mode one has

$$(x'_h(t))^2 - k_h^2 z^2(t) = \left(\int_{t_k}^{t} \omega_h z(\tau) d\tau + x'_h(t_k) \right)^2 - k_h^2 z^2(t), \tag{A.19}$$

where $t_k \geq 0$ denotes the time from which solutions evolve according to the integrator mode. A lower-bound for (A.19) is given by

$$\left(\int_{t_k}^{t} \omega_h z(\tau) d\tau + x'_h(t_k) \right)^2 - k_h^2 z^2(t) \geq \left(\int_{t_k}^{t} \omega_h z(\tau) d\tau \right)^2 - k_h^2 z^2(t), \tag{A.20}$$

in which it is used that $x'_h(t_k) \int_{t_k}^{t} z(\tau) d\tau \geq 0$ due to the sector-condition. Next, suppose that $z(t)$ has no zero-crossings in finite time. This is the relevant case to consider, because a zero-crossing $z(t) = 0$ at $t = t^*$ implies $x'_h(t^*) = x''_h(t^*) = 0$ due to the sector condition, and, in turn, this implies $V(\delta x_h(t^*)) = 0$. Since $\dot{V} \leq 0$ this implies $V(\delta x_h(t)) = 0$ for all $t \geq t^*$. In the absence of zero-crossings in z and thus $z(t) \neq 0$ for all $t \geq 0$, by continuity and boundedness of z one finds $0 < c_1 \leq |z| \leq c_2$ such that

$$(x_h'(t))^2 - k_h z^2(t) \geq \left(\int_{t_k}^{t} \omega_h z(\tau)d\tau\right)^2 - k_h^2 z^2(t) \geq (\omega_h c_1(t-t_k))^2 - (k_h c_2)^2. \tag{A.21}$$

When $t - t_k > T(c_1, c_2) := (k_h c_2)/(\omega_h c_1)$, it follows that $(x_h')^2 - k_h z^2 > 0$, which violates the sector-condition in integrator mode and thus a switch to gain mode must have occurred at a time $t < t_k + T(c_1, c_2)$. As such, $T(c_1, c_2) < \infty$ provides a finite upper-bound on the time-interval in which trajectories can both reside in integrator mode and thus $\dot{V} = 0$. It remains to be shown that V cannot converge to a non-zero constant. Hereto, the following claim is proven first.

Claim: In the absence of zero crossings, i.e., $z(t) \neq 0$ for all time $t \geq 0$, there exists a time $\tau \leq T(c_1, c_2)$ such that $x_h'(\tau) = x_h''(\tau)$.

Proof of the Claim: First note that without zero crossings, z retains a similar sign for all $t \geq 0$. It is sufficient to consider the case $z > 0$ (the case $z < 0$ follows analogously). Without loss of generality, suppose $x_h'(0) \geq x_h''(0)$, and assume that x_h'' starts in integrator mode (starting in gain mode leads to a trivial situation). As shown before, x_h'' must switch to gain mode after some finite time $\tau \leq T(c_1, c_2)$. Suppose that $x_h'(t) \geq x_h''(t)$ for all $t \leq \tau$. This is the relevant case to consider, because when $x_h'(t) \leq x_h''(t)$ there must have been a time $\tau^* \leq \tau$ for which $x_h'(\tau^*) = x_h''(\tau^*)$ (since initially $x_h'(0) \geq x_h''(0)$). Since $z > 0$, it holds that $k_h z(t) \geq x_h'(t)$ for all time $t \geq 0$. Then, at time $t = \tau$ one finds $k_h z(\tau) \geq x_h'(\tau) \geq x_h''(\tau) = k_h z(\tau)$, which implies $x_h'(\tau) = x_h''(\tau)$, thereby proving the claim.

Given the previous claim, it follows that at some time $\tau \leq T(c_1, c_2)$ one has $V(\delta x_h(\tau)) = 0$. In turn, since $\dot{V} \leq 0$, it follows that $V(\delta x_h(t)) = 0$ for all $t \geq \tau$. Thus, one finds $V(\delta x_h(t)) \leq V(\delta x_h(0))$ for $0 \leq t \leq \tau$, and $V(\delta x_h(t)) = 0$ for $t \geq \tau$. Based on this knowledge, one can construct the upper-bound

$$V(\delta x_h(t)) \leq e^{-(t-T^*)} V(\delta x_h(0)), \tag{A.22}$$

in which $T^* = \min\{T_0, T(c_1, c_2)\} \geq \tau$, and where T_0 denotes the time at which the first zero crossing of z occurs. In the absence of zero crossings, one finds $T_0 = \infty$. Based on the bound in (A.22), one finds $\|\delta x_h(t)\| \to 0$ for $t \to \infty$, thereby demonstrating the attractivity property.

A.3 Proofs for Chap. 4

A.3.1 Summary of Relevant Matrices

For readability of the proofs of Chap. 4, relevant system matrices defined in the main text are briefly summarized here, as the upcoming proofs heavily exploit structure in these system matrices. Matrices related to the closed-loop system in (4.6) are given by

$$A_1 = \begin{bmatrix} A_g & B_g \\ -\omega_h C_g & 0 \end{bmatrix}, \quad A_2 = \begin{bmatrix} A_g & B_g \\ -k_h C_g A_g & 0 \end{bmatrix}, \quad B = \begin{bmatrix} B_w \\ 0 \end{bmatrix}, \tag{A.23}$$

and $C = \begin{bmatrix} C_g & 0 \end{bmatrix}$, with matrices (A_g, B_g, B_w, C_g) given in (4.1). Furthermore, the extended system matrix in gain mode is given by $\mathcal{A}_2 = A_2 + k\frac{\omega_h}{k_h} bH$ with

$$b = \begin{bmatrix} 0 & 1 \end{bmatrix}^\top, \quad \text{and} \quad H = \begin{bmatrix} -k_h C_g & -1 \end{bmatrix} = k_h C_z - C_u, \tag{A.24}$$

and in which

$$C_z = -C = \begin{bmatrix} -C_g & 0 \end{bmatrix}, \quad \text{and} \quad C_u = b^\top = \begin{bmatrix} 0 & 1 \end{bmatrix}. \tag{A.25}$$

Additional S-procedure matrices related to the set of LMIs in Theorem 4.1 are given by $\Theta^\top = [I \;\; -k_h C_g^\top]$, and

$$S = \begin{bmatrix} C_u \\ k_h C_z - C_u \end{bmatrix}, \quad T = \begin{bmatrix} C_u \\ C_z(\omega_h I - k_h A_2) \end{bmatrix}, \quad J = \begin{bmatrix} 0 & 1 \\ 1 & 0 \end{bmatrix}. \tag{A.26}$$

Other matrices relevant in the proofs will be specified in detail when needed.

A.3.2 *Proof of Theorem 4.2*

$(i) \implies (ii)$. First observe that by Assumption 2.5 one has

$$\lim_{\omega \to \infty} \left(1 + \mathrm{Re}\left\{k_h C_g (j\omega I - A_g)^{-1} B_g\right\}\right) > 0.$$

Indeed, due to the assumption on the relative degree of $G_{yu}(s) = C_g(sI - A_g)^{-1}B_g$ one has $\lim_{|\omega|\to\infty} G_{yu}(j\omega) = 0$. Remark that satisfaction of the frequency-domain inequality (4.14) implies that $1 - \epsilon + \mathrm{Re}\left\{k_h G_{yu}(j\omega)\right\} > 0$ for all $\omega \in \mathbb{R} \cup \{\infty\}$ must be satisfied for some sufficiently small $0 < \epsilon < 1$. Since A_g is Hurwitz, one finds $\det(j\omega I - A_g) \neq 0$, such that (A.1) in item 1 of Theorem A.1 is satisfied for

$$Q = \begin{bmatrix} 0 & k_h C_g^\top \\ k_h C_g & -2(1-\epsilon) \end{bmatrix}.$$

Furthermore, by Lemma A.1 there exist matrices $M = M^\top \succ 0$, $N = N^\top \succeq 0$ and L, and a constant $\varepsilon > 0$ such that

$$A_g^\top M + M A_g = -L^\top L - \varepsilon M - N, \tag{A.27a}$$

$$M B_g = k_h C_g^\top - \sqrt{2(1-\epsilon)} L^\top. \tag{A.27b}$$

Take $P = \operatorname{diag}(M, m)$ with $m = \mu/\omega_h$ and $\mu > 0$ such that $P \succ 0$. As such, (4.10a) is satisfied with $\tau_1 = 0$. Next, observe that with this choice for P together with the equalities in (A.27), one finds

$$\begin{aligned} A_1^\top P + P A_1 &= \begin{bmatrix} A_g^\top M + M A_g & M B_g - \mu C_g^\top \\ B_g^\top M - \mu C_g & 0 \end{bmatrix} \\ &= -U^\top U + \begin{bmatrix} -\varepsilon M - N & (k_h - \mu) C_g^\top \\ (k_h - \mu) C_g & 2(1-\epsilon) \end{bmatrix}, \end{aligned} \tag{A.28}$$

with $U = [L, \sqrt{2(1-\epsilon)}]$. Next, note that

$$S^\top J S = \begin{bmatrix} 0 & -k_h C_g^\top \\ -k_h C_g & -2 \end{bmatrix}, \tag{A.29}$$

such that one finds

$$A_1^\top P + P A_1 + S^\top J S = -U^\top U + \begin{bmatrix} -\varepsilon M - N & -\mu C_g^\top \\ -\mu C_g & -2\epsilon \end{bmatrix}. \tag{A.30}$$

Via the Schur complement, it follows that the right-hand side of (A.30) is negative definite if $-\varepsilon M + \frac{\mu^2}{2\epsilon} C_g^\top C_g \prec 0$. Since M is positive definite, one finds

$$-\varepsilon M + \frac{\mu^2}{2\epsilon} C_g^\top C_g \preceq \left(-\varepsilon \lambda_{\min}(M) + \frac{\mu^2}{2\epsilon} \lambda_{\max}(C_g^\top C_g) \right) I, \tag{A.31}$$

which by choosing $0 < \mu < \sqrt{2\varepsilon \lambda_{\min}(M) \epsilon / \lambda_{\max}(C_g^\top C_g)}$ ensures (A.30) to be negative definite. Hence, (4.10b) is satisfied with $\tau_2 = 1$.

In a similar manner, observe that the above choice for P together with the equalities in (A.27) leads to

$$\begin{aligned} A_2^\top P + P A_2 &= \begin{bmatrix} A_g^\top M + M A_g & M B_g - \mu \frac{k_h}{\omega_h} (C_g A_g)^\top \\ B_g^\top M - \mu \frac{k_h}{\omega_h} C_g A_g & 0 \end{bmatrix} \\ &= -U^\top U + \begin{bmatrix} -\varepsilon M - N & k_h C_g^\top - \mu \frac{k_h}{\omega_h} (C_g A_g)^\top \\ k_h C_g - \mu \frac{k_h}{\omega_h} C_g A_g & 2(1-\epsilon) \end{bmatrix}, \end{aligned} \tag{A.32}$$

and note that

$$T^\top J T = \begin{bmatrix} 0 & (k_h A_g - \omega_h I)^\top C_g^\top \\ C_g (k_h A_g - \omega_h I) & 0 \end{bmatrix}. \tag{A.33}$$

Now consider $\Gamma = [0, 1]$ and $m = \mu/\omega_h$, and observe that one then finds

$$A_2^\top P + PA_2 + mT^\top JT + \Gamma^\top H + H^\top \Gamma = -U^\top U + \begin{bmatrix} -\varepsilon M - N & -\mu C_g^\top \\ -\mu C_g & -2\epsilon \end{bmatrix}. \tag{A.34}$$

A similar choice for μ as above makes the right-hand side of (A.34) negative definite. Since $H\Theta = 0$ one finds

$$\Theta^\top \left(A_2^\top P + PA_2 + mT^\top JT\right)\Theta \prec 0, \tag{A.35}$$

such that (4.10c) is satisfied with $\tau_3 = m = \mu/\omega_h$.
$(ii) \implies (i)$. Suppose the LMIs (4.10) are feasible with $P = \text{diag}(M, m) \succ 0, \tau_1 = 0, \tau_2, \tau_3 \geq 0$. Then, one finds from (4.10b)

$$A_1^\top P + PA_1 + \tau_2 S^\top JS = \begin{bmatrix} A_g^\top M + MA_g & MB_g - \gamma k_h C_g^\top \\ B_g^\top M - \gamma k_h C_g & -2\tau_2 \end{bmatrix} \prec 0, \tag{A.36}$$

with $\gamma = \tau_2 + m\omega_h/k_h > 0$. From this inequality it immediately follows that

$$\begin{bmatrix} A_g^\top L + LA_g & LB_g - k_h C_g^\top \\ B_g^\top L - k_h C_g & -2 \end{bmatrix} \prec 0, \tag{A.37}$$

with $L = M/\gamma$. By the KYP-lemma (Theorem A.1) this implies

$$1 + \text{Re}\left\{k_h C_g (j\omega I - A_g)^{-1} B_g\right\} > 0 \tag{A.38}$$

for all $\omega \in \mathbb{R} \cup \{\infty\}$, which completes the proof.

A.3.3 *Proof of Theorem 4.3*

$(i) \implies (ii)$. First, observe that the transfer function defined by $W(j\omega) = c(j\omega I - A_1)^{-1}b$ has a relative degree of one such that $\lim_{|\omega|\to\infty} \text{Re}\{W(j\omega)\} = 0$. Since by assumption A_1 is Hurwitz, one finds $\det(j\omega I - A_1) \neq 0$ for all $\omega \in \mathbb{R}$. Together with satisfaction of the frequency-domain inequality (4.15) this implies the transfer function $1 + W(j\omega)$ to satisfy the frequency-domain inequality in Theorem A.1 with

$$Q = \begin{bmatrix} 0 & c^\top \\ c & -2 \end{bmatrix}, \tag{A.39}$$

in which $c = C_z(\omega_h I - k_h A_2) + \alpha_1 C_z + \alpha_2 H$. By Lemma A.1 it follows that this is equivalent to the existence of matrices $P = P^\top \succ 0$, $N = N^\top \succeq 0$ and L, and a constant $\varepsilon > 0$ such that

$$A_1^\top P + PA_1 = -L^\top L - \varepsilon P - N \tag{A.40a}$$
$$Pb = c^\top - \sqrt{2}L^\top. \tag{A.40b}$$

It immediately follows that (4.10a) and (4.10b) are satisfied with $\tau_1 = \tau_2 = 0$. Next, observe that $A_2 = A_1 - bK$ with $b = [0, 1]$ and $K = C_z(\omega_h I - k_h A_2)$, and consider

$$\begin{aligned} A_2^\top P + PA_2 &= A_1^\top P + PA_1 - PbK - (bK)^\top P \\ &= -L^\top L - \varepsilon P - N - \mathrm{He}\left(c^\top K - \sqrt{2}L^\top K\right) \\ &\preceq -(L - \sqrt{2}K)^\top (L - \sqrt{2}K) - \mathrm{He}\left((\alpha_1 C_z + \alpha_2 H)^\top K\right) \end{aligned} \tag{A.41}$$

where $H = k_h C_z - C_u$ and $\mathrm{He}(X) = X^\top + X$. From this, one can find

$$\begin{aligned} &A_2^\top P + PA_2 + (\alpha_1 + \alpha_2)(H^\top K + K^\top H) + \tau_3 T^\top J T \\ &\preceq -\varepsilon P - (L - \sqrt{2}K)^\top (L - \sqrt{2}K) + \left(\tau_3 - \frac{\alpha_1}{k_h}\right)(C_u^\top K + K^\top C_u), \end{aligned} \tag{A.42}$$

such that the choice $\tau_3 = \alpha_1/k_h$ makes (A.42) negative definite. Since $H\Theta = 0$, one finds that negative definiteness of the matrix in the left-hand side of (A.42) implies

$$\Theta^\top \left(A_2^\top P + PA_2 + \tau_3 T^\top J T\right) \Theta \prec 0. \tag{A.43}$$

Hence, (4.10c) is satisfied with $\tau_3 = \alpha_1/k_h \geq 0$.
$(ii) \implies (i)$. Suppose that the LMIs in (4.10) are feasible for some $P \succ 0$, $\tau_1 = \tau_2 = 0$ and $\tau_3 \geq 0$. From Finsler's lemma it follows that there exists a constant $\gamma \in \mathbb{R}$ such that the following set of LMIs is satisfied

$$A_1^\top P + PA_1 \prec 0, \tag{A.44a}$$
$$A_2^\top P + PA_2 + \tau_3 T^\top J T + \gamma H^\top H \prec 0. \tag{A.44b}$$

Two possible cases arise: $\gamma \geq 0$ and $\gamma < 0$. In both cases it still holds that $A_1^\top P + PA_1 - |\gamma| H^\top H \prec 0$ and $A_2^\top P + PA_2 + \tau_3 T^\top J T - |\gamma| H^\top H \prec 0$. Using the identities $A_2 = A_1 - bK$, $T^\top J T = C_u^\top K + K^\top C_u$, and $b = C_u^\top$, the second inequality is written as

$$A_1^\top P + PA_1 - (P - \tau_3 I)bK - (bK)^\top (P - \tau_3 I) - |\gamma| H^\top H \prec 0. \tag{A.45}$$

As the difference between (A.45) and the inequality $A_1^\top P + PA_1 - |\gamma| H^\top H \prec 0$ is given by $(Pb - \tau_3 b)K - K^\top (b^\top P + \tau_3 b^\top)$, it immediately follows from Lemma A.2 that there exists a number $\varepsilon_1 > 0$ such that

$$A_1^\top P + PA_1 - |\gamma| H^\top H + \left(\frac{\varepsilon_1}{\sqrt{2}} Pb - \frac{1}{\varepsilon_1\sqrt{2}}(\tau C_u + K)^\top\right)\left(\frac{\varepsilon_1}{\sqrt{2}} Pb - \frac{1}{\varepsilon_1\sqrt{2}}(\tau C_u + K)^\top\right)^\top \prec 0, \tag{A.46}$$

where $\tau = \tau_3\varepsilon_1^2$. By using the vector property $aa^\top - bb^\top = \frac{1}{2}(a+b)(a-b)^\top + \frac{1}{2}(a-b)(a+b)^\top$ together with (A.46) and the fact that $A_1^\top P + PA_1 \prec 0$, it again follows from Lemma A.2 that there exists a number $\varepsilon_2 > 0$ such that

$$A_1^\top P + PA_1 + \left(\frac{\varepsilon_2}{\sqrt{2}} p + \frac{1}{\varepsilon_2\sqrt{2}} q\right)\left(\frac{\varepsilon_2}{\sqrt{2}} p + \frac{1}{\varepsilon_2\sqrt{2}} q\right)^\top \prec 0, \tag{A.47}$$

with

$$p = \left(\frac{\varepsilon_1}{2} Pb - \frac{1}{2\varepsilon_1}(\tau C_u + K)^\top\right) + \sqrt{\frac{|\gamma|}{2}} H^\top, \tag{A.48a}$$

$$q = \left(\frac{\varepsilon_1}{2} Pb - \frac{1}{2\varepsilon_1}(\tau C_u + K)^\top\right) - \sqrt{\frac{|\gamma|}{2}} H^\top. \tag{A.48b}$$

Denote $\bar{b} = \frac{\varepsilon_1}{2} b$, $\bar{c} = \frac{1}{2\varepsilon_1}(\tau C_u + K)^\top$, and $\bar{h} = \sqrt{\frac{|\gamma|}{2}} H^\top$ such that $p = P\bar{b} - \bar{c} + \bar{h}$ and $q = P\bar{b} - \bar{c} - \bar{h}$, and (A.47) can be written as

$$A_1^\top P + PA_1 + \rho^2 \left(P\bar{b} - \bar{c} + \kappa\bar{h}\right)\left(P\bar{b} - \bar{c} + \kappa\bar{h}\right)^\top \prec 0, \tag{A.49}$$

where $\rho = \frac{\varepsilon_2}{\sqrt{2}} + \frac{1}{\varepsilon_2\sqrt{2}}$ and $\kappa = 1 - 2/(\rho\varepsilon_2\sqrt{2}) = 1 - 2/(\varepsilon_2^2 + 1)$. Note that $-1 < \kappa < 1$ and $\rho \geq \sqrt{2}$ for all $\varepsilon_2 > 0$, and thus (A.49) implies that

$$A_1^\top P + PA_1 + 2\left(P\bar{b} - \bar{c} + \kappa\bar{h}\right)\left(P\bar{b} - \bar{c} + \kappa\bar{h}\right)^\top \prec 0 \tag{A.50}$$

also holds. By applying the Schur complement to (A.50) one finds

$$\begin{bmatrix} A_1^\top P + PA_1 & \sqrt{2}(P\bar{b} - \bar{c} + \kappa\bar{h}) \\ \sqrt{2}(P\bar{b} - \bar{c} + \kappa\bar{h})^\top & -1 \end{bmatrix} \prec 0, \tag{A.51}$$

which by the KYP-lemma (Theorem A.1) is equivalent to the frequency-domain inequality for all $\omega \in \mathbb{R} \cup \{\infty\}$

$$\begin{aligned} &1 + 4\mathrm{Re}\left\{(\bar{c} - \kappa\bar{h})(j\omega I - A_1)^{-1}\bar{b}\right\} \\ &\quad = 1 + \mathrm{Re}\left\{(F + \beta_1 C_u + \beta_2 H(j\omega I - A_1)^{-1} b\right\} > 0, \end{aligned} \tag{A.52}$$

with $\beta_1 = \tau$ and $\beta_2 = -\kappa\varepsilon_1\sqrt{2|\gamma|}$. This corresponds to the frequency-domain inequality (4.15) with $\alpha_1 = k_h\beta_1 \geq 0$ and $\alpha_2 = \beta_2 - \beta_1 \in \mathbb{R}$. This completes the proof.

A.3.4 Proof of Theorem 4.4

First note that from the assumption that $A_g - k_h B_g C_g$ is Hurwitz and $k \geq 1$, it follows that the matrix $\mathcal{A}_2 = A_2 + k\frac{\omega_h}{k_h} bH$ is Hurwitz. To see this, consider the similarity transformation $T\mathcal{A}_2T^{-1}$ with

$$T = \begin{bmatrix} I & 0 \\ k_h C_g & 1 \end{bmatrix}, \quad \text{and} \quad T^{-1} = \begin{bmatrix} I & 0 \\ -k_h C_g & 1 \end{bmatrix}, \tag{A.53}$$

which leads to

$$\bar{\mathcal{A}}_2 := T\mathcal{A}_2T^{-1} = \begin{bmatrix} A_g - k_h B_g C_g & B_g \\ 0 & -k\frac{\omega_h}{k_h} \end{bmatrix}. \tag{A.54}$$

Due to its upper triangular structure, the eigenvalues of the matrix $\bar{\mathcal{A}}_2$ in (A.54) are given by the eigenvalues of $A_g - k_h B_g C_g$ and $-k\omega_h/k_h$. Therefore, $\bar{\mathcal{A}}_2$ is Hurwitz, and thus also $\mathcal{A}_2$ is Hurwitz.

Next, observe that the transfer function

$$W(j\omega) = \left(\lambda k C_u + F + \left(\lambda + k\frac{\omega_h}{k_h}\right) H\right)(j\omega I - \mathcal{A}_2)^{-1} b$$

with $F = C_z(k_h A_1 - \omega_h I)$ has a relative degree of at least one. Hence, one finds $\lim_{|\omega|\to\infty} \mathrm{Re}\{W(j\omega)\} = 0$. Since $\mathcal{A}_2$ is Hurwitz, it follows that $\det(j\omega I - \mathcal{A}_2) \neq 0$ for all $\omega \in \mathbb{R}$. Together with satisfaction of the frequency-domain inequality in (4.16), this implies the transfer function $1 + W(j\omega)$ to satisfy the frequency-domain inequality in Theorem A.1 with

$$Q = \begin{bmatrix} 0 & \left(\lambda k C_u + F + \left(\lambda + k\frac{\omega_h}{k_h}\right) H\right)^\top \\ \lambda k C_u + F + \left(\lambda + k\frac{\omega_h}{k_h}\right) H & -2 \end{bmatrix}. \tag{A.55}$$

By virtue of Lemma A.1 this implies the existence of matrices $M = M^\top \succ 0$, $N = N^\top \succeq 0$ and L, and a constant $\varepsilon > 0$ such that

$$\mathcal{A}_2^\top M + M\mathcal{A}_2 = -L^\top L - \varepsilon M - N \tag{A.56a}$$

$$Mb = \lambda(kC_u + H)^\top + F^\top + k\frac{\omega_h}{k_h} H^\top - \sqrt{2}L^\top. \tag{A.56b}$$

Since $\mathcal{A}_2 = A_1 + bG$ with $G = F + k\frac{\omega_h}{k_h}H$ the equalities in (A.56) yield

$$\begin{aligned} A_1^\top M + MA_1 &= -G^\top b^\top M - MbG - L^\top L - \varepsilon M - N \\ &= -\mathrm{He}\left(\left(\lambda(kC_u + H)^\top + G^\top - \sqrt{2}L^\top\right)G\right) - L^\top L - \varepsilon M - N \\ &= -\left(L - \sqrt{2}G\right)^\top\left(L - \sqrt{2}G\right) - \mathrm{He}\left(\lambda(kC_u + H)^\top G\right) - \varepsilon M - N \\ &\prec -\lambda\left((kC_u + H)^\top\left(F + k\frac{\omega_h}{k_h}H\right) + \left(F + k\frac{\omega_h}{k_h}H\right)^\top(kC_u + H)\right). \end{aligned} \tag{A.57}$$

Observe that one can write $F = HA_1$ such that

$$(kC_u + H)^\top F = (kC_u + H)^\top HA_1 = (kS^\top JS + H^\top H)A_1 - k\omega_h H^\top C_z, \tag{A.58}$$

where use is made of the identities $S^\top JS = C_u^\top H + H^\top C_u$ and $C_u A_1 = \omega_h C_z$. Substituting (A.58) in (A.57) yields

$$\begin{aligned} A_1^\top M + MA_1 + \lambda\mathrm{He}\Bigg(& (kS^\top JS + H^\top H)A_1 \\ & + \left(k\frac{\omega_h}{k_h}(kC_u + H) - k\omega_h C_z\right)^\top H\Bigg) \prec 0. \end{aligned} \tag{A.59}$$

Choose $P = (M + \lambda H^\top H) + \lambda k S^\top JS$ and observe that since $M \succ 0$, this choice yields $P - \lambda k S^\top JS \succ 0$. Hence, (4.10a) is satisfied with $\tau_1 = \lambda k \geq 0$. Furthermore, with this choice for P, (A.59) evaluates to

$$A_1^\top P + PA_1 + \lambda\frac{\omega_h}{k_h}\mathrm{He}\left((k(kC_u + H) - kk_h C_z)^\top H\right) \prec 0. \tag{A.60}$$

Since $C_z = \frac{1}{k_h}(H + C_u)$ the last term in the left-hand side of (A.60) yields

$$\lambda k(kC_u + H)^\top H - \lambda k k_h C_z^\top H = \lambda k(k-1)C_u^\top H. \tag{A.61}$$

Under the assumption that $\lambda \geq 0, k \geq 1$, and with the identity $S^\top JS = C_u^\top H + H^\top C_u$ this yields

$$A_1^\top P + PA_1 + \tau_2 S^\top JS \prec 0, \tag{A.62}$$

with $\tau_2 = \lambda k\omega_h(k-1)/k_h \geq 0$. As such, (4.10b) is satisfied. Additionally, with the choice for P, one finds from the equality in (A.56a) that

$$\mathcal{A}_2^\top M + M\mathcal{A}_2 = \mathcal{A}_2^\top P + P\mathcal{A}_2 - \lambda\mathrm{He}\left(H^\top\left((H + kC_u)\mathcal{A}_2 - k^2 C_u^\top H\right)\right) \prec 0. \tag{A.63}$$

Since $H\Theta = 0$, and $\Theta^\top \mathcal{A}_2 \Theta = \Theta^\top A_2 \Theta$, it follows that (4.10c) is satisfied with $\tau_3 = 0$. This completes the proof.

A.3.5 Proof of Corollary 4.2

First, note that due to the relative degree assumption on G_{yu} (Assumption 2.5) condition (4.31) is satisfied for $\omega = \infty$. In the remainder of the proof, only all $\omega \in \mathbb{R}$ need to be considered. Observe that condition (4.31) can be written as

$$1 + \text{Re}\{W_1(j\omega)\} - \text{Re}\{W_2(j\omega)\} > 0, \tag{A.64}$$

where

$$W_1(j\omega) = ((\omega_h - k_h j\omega)S_P(j\omega) - \lambda)L(j\omega, k), \tag{A.65a}$$

$$W_2(j\omega) = \left(\frac{\omega_h}{k_h} - \lambda + \lambda k_h S_P(j\omega)\right) kL(j\omega, k), \tag{A.65b}$$

with $S_P = T/k_h$. Satisfaction of the conditions in Theorem 4.2 implies that the matrix $A_g - k_h B_g C_g$ is Hurwitz. In turn, this implies that the H_∞-norm of $S_P(s)$ is finite, i.e., $\|S_P(j\omega)\|_\infty := \mu_1 < \infty$. Moreover, observe that due to the relative degree assumption on G_{yu}, one also finds $j\omega S_P(j\omega)_\infty := \mu_2 < \infty$. Given these observations and the definition of L, one finds that

$$\text{Re}\{W_1\} \geq -\|W_1\| \geq -\left(\frac{\nu}{\sqrt{(k_h\omega)^2 + (k\omega_h)^2}}\right) \tag{A.66}$$

with $\nu = \omega_h \mu_1 + k_h \mu_2 + \lambda$. Choose $\lambda = \omega_h / k_h$ and observe that with this choice $\text{Re}\{W_2\}$ can be written as

$$\text{Re}\{W_2\} = k\omega_h (\text{Re}\{L\}\text{Re}\{S_P\} - \text{Im}\{L\}\text{Im}\{S_P\}).$$

Note that

$$|k\omega_h \text{Im}\{L\}\text{Im}\{S_P\}| \geq -\mu_2 \frac{k\omega_h}{(k_h\omega)^2 + (k\omega_h)^2}, \tag{A.67}$$

where use is made of the fact that S_P has a relative degree of at least one, and thus $|\omega \text{Im}(S_P)| \geq -\mu_2$. Next, note that

$$k\omega_h \text{Re}\{L\}\text{Re}\{S_P\} = k\frac{\omega_h}{k_h}\text{Re}\{L\}(1 - \text{Re}\{S\}), \tag{A.68}$$

where $S = 1/(1 + k_h G_{yu})$. Due to satisfaction of the strict frequency-domain inequality in (4.14) it follows that there exists some $\epsilon > 0$ such that for all $\omega \in \mathbb{R}$

one has $\mathrm{Re}\{S\} \geq \epsilon > 0$. The left-hand side of the inequality in (A.64) then admits a lower-bound given by

$$1 + \mathrm{Re}\{W_1\} - \mathrm{Re}\{W_2\} \geq 1 - \left(\frac{\nu}{\sqrt{(k_h\omega)^2 + (k\omega_h)^2}} \right) - \mu_2 \frac{k\omega_h}{(k_h\omega)^2 + (k\omega_h)^2} + (\epsilon - 1)\frac{(k\omega_h)^2}{(k_h\omega)^2 + (k\omega_h)^2}.$$

Due to the low-pass characteristics, this lower-bound attains its minimum at $\omega = 0$ rad/s, and this minimum is given by $\epsilon - \tau/k$, where $\tau = (\nu + \mu_2)/\omega_h$. Hence, choosing $k > \tau/\epsilon$ makes the right-hand side of (A.64) strictly positive for all $\omega \in \mathbb{R} \cup \{\infty\}$ so that all conditions in Theorem 4.4 are satisfied.

A.3.6 Proof of Corollary 4.3

Choose $\lambda = \frac{\omega_h}{kk_h}$ such that the left-hand side of (4.31) evaluates to

$$1 + X(j\omega, k) - \lambda Y(j\omega, k) = 1 + \omega \mathrm{Im}\{k_h S_P L\} - \frac{\omega_h}{k_h}\left(\frac{k^2 - k + 1}{k}\right)\mathrm{Re}\{L\}. \tag{A.69}$$

Since $\mathrm{Re}\{L(j\omega)\} \leq k_h(k\omega_h)^{-1}$ for all $\omega \in \mathbb{R} \cup \{\infty\}$, and $k(k-1) \geq 0$, one finds that (A.69) is lower bounded by

$$1 + X(j\omega, k) - \lambda Y(j\omega, k) \geq \left(\frac{k-1}{k^2}\right) + \omega \mathrm{Im}\{k_h S_P L\},$$

with $S_P = T/k_h$. From the assumption that the matrix $A_g - k_h B_g C_g$ is Hurwitz, it follows that $\|S_P(j\omega)\|_\infty < \infty$. Furthermore, since by Assumption 2.5 G_{yu} is assumed to have a relative degree of at least two, it follows that $\|\omega S_P(j\omega)\|_\infty < \infty$. Then, one finds $\|\omega k_h S_P(j\omega)\|_\infty = \mu < \infty$, and, therefore, $\|\omega k_h S_P(j\omega) L(j\omega)\|_\infty \leq \|\omega k_h S_P(j\omega)\|_\infty \|L(j\omega)\|_\infty = \mu k_h/(k\omega_h)$. This, in turn, implies

$$|\omega \mathrm{Im}\{k_h S_P(j\omega) L(j\omega)\}| \leq \mu \frac{k_h}{k\omega_h}. \tag{A.70}$$

Choosing $k = 2$ then yields

$$\left(\frac{k-1}{k^2}\right) + \omega \mathrm{Im}\{k_h S_P L\} \geq \frac{1}{4} - \frac{\mu k_h}{2\omega_h}. \tag{A.71}$$

Hence, for $\omega_h > 2\mu k_h$ one finds that (4.31) holds. The conditions of Theorem 4.4 are satisfied, thereby guaranteeing ISS of (4.6).

A.3.7 Proof of Proposition 4.1

The proof is based on showing that, when (4.36) and (4.37) are satisfied, the conditions of Theorem 4.4 hold for all $G_{yu}(s) \in \Pi$. Namely, for all possible plants 1) the gain mode dynamics are stable, and 2) the frequency-domain inequality (4.16) is satisfied.

(1) **Stability of the Gain Mode Dynamics for All** $G_{yu}(s) \in \Pi$.
Consider the open-loop characteristics of the gain mode dynamics, which, under the assumption of multiplicative uncertainties, can be represented as

$$L(s) = k_h G_{yu}(s) = L_0(s)(1 + W(s)\Delta(s)), \tag{A.72}$$

where $L_0(s) = k_h G_0(s)$ denotes the nominal open-loop. From the assumption that the nominal closed-loop gain mode dynamics are stable, it immediately follows from [9, Sect. 7.5] that when (4.36) holds for all $\omega \in \mathbb{R}$, the nominal gain mode dynamics are robustly stable with respect to all allowable perturbations. Therefore, under the hypothesis of the proposition, the gain mode dynamics are stable for all $G_{yu}(s) \in \Pi$.

(2) **Satisfying the Frequency-Domain Inequality in** (4.16) **for All** $G_{yu}(s) \in \Pi$.
Observe that the complementary sensitivity function $T(s)$ can be written as

$$T(s) = \frac{k_h G_{yu}(s)}{1 + k_h G_{yu}(s)} = \frac{L(s)}{1 + L(s)} = T_0(s)(1 + W(s)\Delta_T(s)), \tag{A.73}$$

with

$$T_0(s) = \frac{L_0(s)}{1 + L_0(s)}, \quad \text{and} \quad \Delta_T(s) = \frac{\Delta(s)}{1 + L(s)}. \tag{A.74}$$

Since $\Delta(s) \in \mathcal{RH}_\infty$, it follows that whenever the gain mode is robustly stable, one has $\Delta_T(s) \in \mathcal{RH}_\infty$. The magnitude of the relative uncertainty in $T(s)$ then admits an upper-bound of

$$\|\Delta_T(j\omega)\| \leq \frac{1}{\|1 + L_0(j\omega)\| - \|W(j\omega)L_0(j\omega)\|}, \tag{A.75}$$

which follows from selecting the worst-case perturbation $\|\Delta(j\omega)\| = 1$ for all $\omega \in \mathbb{R}$ such that the complex numbers $1 + L_0(j\omega)$ and $W(j\omega)\Delta(j\omega)L_0(j\omega)$ point in opposite directions. Observe that this corresponds to an upper-bound on the magnitude of all possible sensitivity functions $S(s) = 1 - T(s)$.
From substituting the perturbed system (A.73) in the frequency-domain inequality (4.31), it follows that (4.31) is satisfied for all possible plants $G_{yu}(s) \in \Pi$ if

$$1 + (X_0(\omega, k) + \delta_x(\omega, k)) - \lambda(Y_0(\omega, k) + \delta_y(\omega, k)) > 0. \tag{A.76}$$

Here, the nominal terms are given by

$$X_0(\omega,k) = \frac{\omega_h}{k_h}\mathrm{Re}\left\{(T_0(j\omega)-k)C_l(j\omega)\right\} + \omega\mathrm{Im}\left\{T_0(j\omega)C_l(j\omega)\right\},$$
$$Y_0(\omega,k) = k\mathrm{Re}\left\{T_0(j\omega)C_l(j\omega)\right\} + (1-k)\mathrm{Re}\left\{C_l(j\omega)\right\},$$

and the perturbations are expressed as

$$\delta_x(\omega,k) = \frac{\omega_h}{k_h}\delta_u(\omega) + \omega\delta_v(\omega),$$
$$\delta_y(\omega,k) = k\delta_u(\omega),$$

in which δ_u and δ_v are given by

$$\delta_u(\omega) = \mathrm{Re}\left\{T_0(j\omega)W(j\omega)\Delta_T(j\omega)C_l(j\omega)\right\},$$
$$\delta_v(\omega) = \mathrm{Im}\left\{T_0(j\omega)W(j\omega)\Delta_T(j\omega)C_l(j\omega)\right\},$$

and satisfy the constraint

$$\sqrt{\delta_u^2(\omega)+\delta_v^2(\omega)} \le \frac{\|T_0(j\omega)W(j\omega)C_l(j\omega)\|}{\|1+L_0(j\omega)\| - \|W(j\omega)L_0(j\omega)\|} := \rho(\omega). \tag{A.79}$$

The minimum of (A.76) subject to the constraint in (A.79) can be estimated as follows. Observe that one may write

$$\delta_u(\omega) = \nu(\omega)\cos(\theta), \quad \text{and} \quad \delta_v(\omega) = \nu(\omega)\sin(\theta), \tag{A.80}$$

where $\nu(\omega) \le \rho(\omega)$, and $\theta \in [0, 2\pi]$. Then, using the standard trigonometric identity $a\cos(\phi) + b\sin(\phi) = \sqrt{a^2+b^2}\cos(\phi-\psi)$ with $\psi = \arctan(b/a)$ it follows that (A.76) reads

$$1 + X_0(\omega,k) - \lambda Y_0(\omega,k) + \gamma(\omega)\cos(\theta-\varphi) > 0 \tag{A.81}$$

where

$$\gamma(\omega) = \nu(\omega)\sqrt{\left(\frac{\omega_h}{k_h} - \lambda k\right)^2 + \omega^2}, \quad \text{and} \quad \varphi = \arctan\left(\frac{(k_h\omega)^2}{(\omega_h - \lambda k k_h)^2}\right).$$

The minimum of the left-hand side in (A.81) over all possible perturbations is given by

$$1 + X_0(\omega,k) - \lambda Y_0(\omega,k) - \rho(\omega)\sqrt{\left(\frac{\omega_h}{k_h} - \lambda k\right)^2 + \omega^2} > 0. \tag{A.82}$$

Hence, satisfying (A.82) for some $\lambda \geq 0$ and $k \geq 1$ implies that the inequality in (4.31), and, therefore the one in (4.16) is satisfied for all $G_{yu}(s) \in \Pi$.

It follows that the conditions of Theorem 4.4 are satisfied for all possible plants $G_{yu}(s)$, and therefore the HIGS-based control system (4.6) is robustly ISS for all perturbed plants $G_{yu}(s) \in \Pi$. This completes the proof.

A.4 Proofs for Chap. 5

A.4.1 Proof of Proposition 5.1

First, it is shown that $\xi \in \hat{\mathcal{S}}_i \iff \hat{S}_i\xi \geq 0$ for all $i \in \mathcal{N}$.

- $\xi \in \hat{\mathcal{S}}_i \Rightarrow \hat{S}_i\xi \geq 0$. Since $\hat{\mathcal{S}}_i = \text{pos}\left\{r_i^1, r_i^2, r_i^3\right\}$ is a simplicial cone, each $\xi \in \hat{\mathcal{S}}_i$ can uniquely be written as a positive combination of columns from $R_i = [r_i^1, r_i^2, r_i^3]$, that is,
$$\xi = R_i\lambda, \text{ with } 0 \leq \lambda \in \mathbb{R}^3. \tag{A.83}$$
Invertibility of R_i results in $\lambda = R_i^{-1}\xi \geq 0$, which by the choice $\hat{S}_i = R_i^{-1}$ yields the inequality.
- $\hat{S}_i\xi \geq 0 \Rightarrow \xi \in \hat{\mathcal{S}}_i$. Suppose $\hat{S}_i\xi = \lambda \geq 0$. Invertibility of $\hat{S}_i$ yields $\xi = R_i\lambda$ and thus $\xi \in \text{pos}\left\{r_i^1, r_i^2, r_i^3\right\} = \hat{\mathcal{S}}_i$. The result follows.

Next it is shown that $\xi \in \hat{\mathcal{T}}_j \iff \hat{T}_j\xi \geq 0, h\xi = 0$ for all $j \in \mathcal{M}$.

- $\xi \in \hat{\mathcal{T}}_j \Rightarrow \hat{T}_j\xi \geq 0, h\xi = 0$. The equality $h\xi = 0$ follows immediately from the observation that $\hat{\mathcal{T}}_j \subset \mathcal{F}_2$, and thus each $\xi \in \hat{\mathcal{T}}_j$ is perpendicular to the vector defined by h. Since $\hat{\mathcal{T}}_j = \text{pos}\left\{r_j^1, r_j^2\right\}$ is a polyhedral cone, each $\xi \in \hat{\mathcal{T}}_j$ can be written as the positive combination of columns in $R_j' = [r_j^1, r_j^2]$, that is
$$\xi = R_j'\lambda, \text{ with } 0 \leq \lambda \in \mathbb{R}^2. \tag{A.84}$$
Since R_j' has linearly independent columns, it follows that the matrix product $(R_j')^\top R_j'$ also has linearly independent columns and, therefore, is invertible. Multiplying (A.84) from the left by $((R_j')^\top R_j')^{-1}(R_j')^\top$ yields
$$\left((R_j')^\top R_j'\right)^{-1}(R_j')^\top\xi = \lambda \geq 0.$$
From the choice for $\hat{T}_j$ in (5.15) the result follows.

- $\hat{T}_j\xi \geq 0, h\xi = 0 \Rightarrow \xi \in \hat{\mathcal{T}}_j$. Suppose $\hat{T}_j\xi = \lambda$ with $\lambda \geq 0$. Pre-multiplication of this equality with $(R_j')^\top R_j'$ yields $(R_j')^\top \xi = (R_j')^\top R_j'\lambda$, which, in turn, yields

$$(R_j')^\top \left(\xi - R_j'\lambda\right) = 0. \tag{A.85}$$

Clearly, since $R_j' \in \mathbb{R}^{3\times 2}$ there is no unique solution to this equation. Consider a solution that satisfies $\xi - R_j'\lambda \neq 0$. Since the columns of R_j' both belong to the plane defined by $\ker(h)$, it must hold that the vector $\xi - R_j'\lambda$ is perpendicular to this plane. However, since $R_j'\lambda \in \ker(h)$, this contradicts the assumption that $\xi \in \ker(h)$. As such, the only solution that can satisfy (A.85) when $h\xi = 0$ holds, is $\xi = R_j'\lambda$. For $\lambda \geq 0$ this implies $\xi \in \text{pos}\left\{r_j^1, r_j^2\right\} = \hat{\mathcal{T}}_j$.

A.4.2 *Proof of Proposition 5.2*

The first part of the proof follows similar arguments as in [3, Sect. A.4]. First, the continuity property is shown for all $\xi \in \hat{\mathcal{S}}_i \cap \hat{\mathcal{S}}_j, i, j \in \mathcal{N} \times \mathcal{N}, i \neq j$. Observe that each $\xi \in \hat{\mathcal{S}}_i$ can be written as a linear combination of elements from $\bar{\mathcal{R}}$, such that

$$\xi = \sum_{q=1}^{p} r_q w_{q,i} = \bar{R}\bar{w}_i, \tag{A.86}$$

where $\bar{w}_i = [w_{1,i}, \ldots, w_{p,i}]^\top \in \mathbb{R}^p$ collects the weights $w_{q,i}$. One can set $w_{q,i} = 0$ for all q such that $r_q \notin \hat{\mathcal{S}}_i$ (where no distinction is made between $[0, 0, 1]^\top$ and $[0, 0, -1]^\top$). By doing so, it follows that $\bar{w}_i = E_i E_i^\top \bar{w}_i$, such that (A.86) can equivalently be written as $\xi = \bar{R}E_i E_i^\top \bar{w}_i$. Since E_i extracts those columns from $\bar{R}$ that coincide with the generators of $\hat{\mathcal{S}}_i$, it follows that the columns of the matrix $\bar{R}E_i$ are linearly independent and thus $\bar{R}E_i$ is invertible. From this, one finds the mapping $\bar{w}_i = E_i\left(\bar{R}E_i\right)^{-1}\xi$. On the boundary shared by two simplicial cones $\hat{\mathcal{S}}_i$ and $\hat{\mathcal{S}}_j$, the only non-zero elements of $w_{q,i}, w_{q,j}$ are those that describe this common boundary. Therefore, $w_{q,i}(\xi) = w_{q,j}(\xi)$ for all $\xi \in \hat{\mathcal{S}}_i \cap \hat{\mathcal{S}}_j$. The result trivially follows for $\xi \in -\hat{\mathcal{S}}_i \cap -\hat{\mathcal{S}}_j, i, j \in \mathcal{N} \times \mathcal{N}, i \neq j$.

It remains to show that (5.20) also yields the continuity property for $\xi \in \hat{\mathcal{S}}_i \cap -\hat{\mathcal{S}}_j, i, j \in \mathcal{N} \times \mathcal{N}$ which corresponds to the line $\xi = [0, 0, \dot{z}]^\top$. Since the vector $[0, 0, -1]^\top$ is excluded from $\bar{\mathcal{R}}$, this common boundary is represented by only one column in $\bar{R}$, being the vector $[0, 0, 1]^\top$. Therefore, for each $\xi \in \hat{\mathcal{S}}_i \cap -\hat{\mathcal{S}}_j, i, j \in \mathcal{N} \times \mathcal{N}$, the only non-zero element in $w_{q,i}$ and $w_{q,j}$ describes this common boundary.

A.4.3 Proof of Proposition 5.3

Parts of the proof are inspired by the proofs of [1, Lemma 1], and [1, Theorem 2].

$(i) \implies (ii)$: Suppose V in (5.24) is continuous. Consider the state transformation $\bar{x} = Tx$, with $T = \begin{bmatrix} E^\top & U^\top \end{bmatrix}^\top$, and where $\text{im}U = \ker E$, such that one finds $\bar{x} = [\xi^\top, \hat{x}^\top]^\top$ with $\xi = [z, u, \dot{z}]^\top \in \mathbb{R}^3$ and $\hat{x} \in \mathbb{R}^{n-3}$. Note that T has full rank and thus is invertible. In terms of the transformed coordinates, the function V reads as $V(\bar{x}) = \bar{x}^\top \bar{P}_i \bar{x}^\top$ with $\bar{P}_i = T^{-\top} P_i T^{-1}$. Partition the matrix $\bar{P}_i$, $i \in \mathcal{N}$ as

$$\bar{P}_i = \begin{bmatrix} \bar{P}_i^{11} & \bar{P}_i^{12} \\ \bar{P}_1^{21} & \bar{P}_i^{22} \end{bmatrix}, \tag{A.87}$$

with $\bar{P}_i^{12} = (\bar{P}_i^{21})^\top$, such that

$$V(\bar{x}) = \xi^\top \bar{P}_i^{11} \xi + 2\hat{x}^\top \bar{P}_i^{12} \xi + \hat{x}^\top P_i^{22} \hat{x}. \tag{A.88}$$

Continuity of V then implies for all $\xi \in \hat{\mathcal{S}}_i \cap \hat{\mathcal{S}}_j$, $(i, j) \in \mathcal{N} \times \mathcal{N}$ and all $\hat{x} \in \mathbb{R}^{n-3}$

$$\xi^\top (P_i^{11} - P_j^{11})\xi + 2\hat{x}^\top (P_i^{12} - P_j^{12})\xi + \hat{x}^\top (P_i^{22} - P_j^{22})\hat{x} = 0. \tag{A.89}$$

First consider all $p \in \mathcal{N}$ and all $\xi \in \hat{\mathcal{S}}_p$. Since $\xi = 0 \in \left\{\hat{\mathcal{S}}_i \cap \hat{\mathcal{S}}_j\right\}$ for all $(i, j) \in \mathcal{N} \times \mathcal{N}$, it should hold that $\hat{x}^\top (P_i^{22} - P_j^{22})\hat{x} = 0$ for all $\hat{x} \in \mathbb{R}^{n-3}$. This is only possible if there exists a symmetric matrix $\Phi^{22} \in \mathbb{S}^{(n-3)\times(n-3)}$ such that for all $i \in \mathcal{N}$ one has $P_i^{22} = \Phi^{22}$.

In a similar manner, when $\hat{x} = 0$, it should hold for all $\xi \in \left\{\hat{\mathcal{S}}_i \cap \hat{\mathcal{S}}_j\right\}$, $(i, j) \in \mathcal{N} \times \mathcal{N}$ that $\xi^\top (P_i^{11} - P_j^{11})\xi = 0$. On the basis of [1, Theorem 2], one can conclude that the equalities

$$r_m^\top \bar{P}_i^{11} r_m = r_m^\top \bar{P}_j^{11} r_m = \phi_{mm}, \tag{A.90a}$$
$$r_m^\top \bar{P}_i^{11} r_n = r_m^\top \bar{P}_j^{11} r_n = \phi_{mn} \tag{A.90b}$$

hold true for all $r_m, r_n \in \mathcal{R}_i \cap \mathcal{R}_j$, where $\mathcal{R}_i$ denotes the set of generators of $\hat{\mathcal{S}}_i$. Note that for all $r_m, r_n \in \mathcal{R}_i \cap \mathcal{R}_j$, with $r_n = [0, 0, -1]$ one has

$$r_n^\top \bar{P}_i^{11} r_n = r_n^\top \bar{P}_j^{11} r_n = r_k^\top \bar{P}_i^{11} r_k = r_k^\top \bar{P}_j^{11} r_k = \phi_{nn} = \phi_{kk} \tag{A.91a}$$
$$r_m^\top \bar{P}_i^{11} r_n = r_m^\top \bar{P}_j^{11} r_n = -r_m^\top \bar{P}_i^{11} r_k = -r_m^\top \bar{P}_j^{11} r_k = \phi_{mn} = -\phi_{mk}, \tag{A.91b}$$

with $r_k = -r_n = [0, 0, 1]^\top$. Note that in case r_m and r_k already are generators of a region $\hat{\mathcal{S}}_l$, $l \neq j$, $l \neq i$, then ϕ_{mk} is already defined, and one should consider both r_k, and r_n as separate generators, otherwise there will be overlap in the ϕ-elements. This degenerate situation, however, only arises when the $(z, \dot{z})$ region is divided into two regions. For divisions into more than two regions, there is no need to make a distinction between the vectors $[0, 0, 1]^\top$ and $[0, 0, -1]^\top$ since it is always possible to express the constraints in terms of either one of these vectors without creating overlap.

Considering the non-degenerate cases, by the result in [1, Lemma 1] there exists a symmetric matrix $\Phi^{11} = \{\phi_{pq}\} \in \mathbb{S}^{r\times r}$ for all $p = 1, \ldots, r$ and all $q = 1, \ldots, r$. Note that Φ^{11} can always be constructed by simply collecting the elements in (A.91). Moreover, if for a couple (m, n) and an integer i such that $r_m, r_n \in \mathcal{R}_i$ does not exist, then ϕ_{mn} remains unconstrained.

On a per region basis, one can write

$$R_i^\top \bar{P}_i^{11} R_i = \Phi_i^{11} = E_i^\top \Phi^{11} E_i, \tag{A.92}$$

with R_i the matrix collecting the generators of $\hat{\mathcal{S}}_i$, and Φ_i^{11} the matrix containing the parameters associated with the region $\hat{\mathcal{S}}_i$. The second inequality in (A.92) explicitly follows from the construction of the extraction matrices E_i as in Proposition 5.2. Since R_i is invertible, one finds

$$\bar{P}_i^{11} = R_i^{-\top} \Phi_i R_i^{-1} = (\bar{R}E_i)^{-\top} \Phi_i (\bar{R}E_i)^{-1} = (\bar{R}E_i)^{-\top} E_i^\top \Phi^{11} E_i (\bar{R}E_i)^{-1}, \tag{A.93}$$

with $\bar{R}$ the matrix collecting the distinct generators of the partitioning $\mathcal{P}_{\hat{\mathcal{S}}}$. Due to symmetry, the identity (A.93) also applies to all $\xi \in -\hat{\mathcal{S}}_p$, $p \in \mathcal{N}$. Note that continuity over the boundary defined by the vector $[0, 0, -1]^\top$ is implied by (A.91). From the definition of the continuity matrices as $\hat{F}_i = E_i(\bar{R}E_i)^{-1}$, it follows that

$$\bar{P}_i = \hat{F}_i^\top \Phi^{11} \hat{F}_i. \tag{A.94}$$

Since $\hat{x}^\top (P_i^{22} - P_j^{22})\hat{x} = 0$ for all $\hat{x} \in \mathbb{R}^{n-3}$, and $\xi^\top (P_i^{11} - P_j^{11})\xi = 0$ for all $\xi \in \pm\hat{\mathcal{S}}_i \cap \pm\hat{\mathcal{S}}_j$, $i, j \in \mathcal{N} \times \mathcal{N}$, it must also be true that $\hat{x}^\top (P_i^{12} - P_j^{12})\xi = 0$ for all $(\hat{x}, \xi) \in \{\mathbb{R}^{n-3}\} \times \left\{\pm\hat{\mathcal{S}}_i \cap \pm\hat{\mathcal{S}}_j\right\}$. In a similar manner as before, one finds the equality

$$P_i^{12} r_m = P_j^{12} r_m = \phi_m \tag{A.95}$$

to hold for all $r_m \in \mathcal{R}_i \cap \mathcal{R}_j$. Moreover, for all $r_m \in \mathcal{R}_i \cap \mathcal{R}_j$ with $r_m = [0, 0, -1]^\top$ one finds

$$P_i^{12} r_m = P_j^{12} r_m = \phi_m = -P_i^{12} r_n = -P_j^{12} r_n = -\phi_n, \tag{A.96}$$

where $r_n = -r_m = [0, 0, -1]^\top$. Similar as before, there is no need to make a distinction between $[0, 0, 1]^\top$ and $[0, 0, -1]^\top$. Collect all elements in a matrix $\Phi^{12} = \{\phi_p\} \in \mathbb{R}^{3\times r}$ for all $p = 1, \ldots, r$, and observe that

$$\bar{P}_i^{12} R_i = \Phi_i^{12} = \Phi^{12} E_i. \tag{A.97}$$

Since R_i is invertible, it follows that

$$\bar{P}_i^{12} = \Phi^{12} E_i R_i^{-1} = \Phi^{12} E_i (\bar{R} E_i)^{-1} = \Phi^{12} \hat{F}_i. \tag{A.98}$$

As such, the partitioned matrix in (A.87) is equivalently written as

$$\bar{P}_i = \begin{bmatrix} \hat{F}_i^\top \Phi^{11} \hat{F}_i & \Phi^{12} \hat{F}_i \\ \hat{F}_i^\top \Phi^{21} & \Phi^{22} \end{bmatrix} = \begin{bmatrix} \hat{F}_i & 0 \\ 0 & I \end{bmatrix}^\top \begin{bmatrix} \Phi^{11} & \Phi^{12} \\ \Phi^{21} & \Phi^{22} \end{bmatrix} \begin{bmatrix} \hat{F}_i & 0 \\ 0 & I \end{bmatrix}. \tag{A.99}$$

Pre- and post-multiplying this with $\bar{x} = [\xi^\top, \hat{x}^\top]^\top$ yields

$$\begin{aligned} \bar{x}^\top \bar{P}_i \bar{x} &= \xi^\top \hat{F}_i^\top \Phi^{11} \hat{F}_i \xi + 2\hat{x}^\top \Phi^{12} \hat{F}_i \xi + \hat{x}^\top \Phi^{22} \hat{x} \\ &= \xi^\top \hat{F}_i^\top \Phi^{11} \hat{F}_i \xi + \bar{x}^\top \Gamma \Phi^{12} \hat{F}_i \xi + \xi^\top \hat{F}_i^\top \Phi^{21} \Gamma \bar{x} + \bar{x}^\top \Gamma^\top \Phi^{22} \Gamma \bar{x}, \end{aligned} \tag{A.100}$$

with $\Gamma = [0, I]$. As such, one finds

$$\begin{aligned} \bar{x} \bar{P}_i \bar{x} &= \begin{bmatrix} \xi \\ \bar{x} \end{bmatrix}^\top \begin{bmatrix} \hat{F}_i^\top \Phi^{11} \hat{F}_i & \hat{F}_i^\top \Phi^{12} \Gamma \\ \Gamma^\top \Phi^{21} \hat{F}_i & \Gamma^\top \Phi^{22} \Gamma \end{bmatrix} \begin{bmatrix} \xi \\ \bar{x} \end{bmatrix} \\ &= \begin{bmatrix} \xi \\ \bar{x} \end{bmatrix}^\top \begin{bmatrix} \hat{F}_i & 0 \\ 0 & I \end{bmatrix}^\top \begin{bmatrix} \Phi^{11} & \Phi^{12} \Gamma \\ \Gamma^\top \Phi^{21} & \Gamma^\top \Phi^{22} \Gamma \end{bmatrix} \begin{bmatrix} \hat{F}_i & 0 \\ 0 & I \end{bmatrix} \begin{bmatrix} \xi \\ \bar{x} \end{bmatrix}. \end{aligned} \tag{A.101}$$

Then, using $P_i = T^\top \bar{P}_i T$, and $[\xi^\top, \bar{x}^\top]^\top = [E^\top, T]^\top x$ with $T = [E^\top, U^\top]^\top$, one finds

$$\begin{aligned} P_i &= \begin{bmatrix} \hat{F}_i E \\ T \end{bmatrix}^\top \begin{bmatrix} \Phi^{11} & \Phi^{12} \Gamma \\ \Gamma^\top \Phi^{21} & \Gamma^\top \Phi^{22} \Gamma \end{bmatrix} \begin{bmatrix} \hat{F}_i E \\ T \end{bmatrix} \\ &= \begin{bmatrix} \hat{F}_i E \\ I \end{bmatrix}^\top \begin{bmatrix} \Phi^{11} & \Phi^{12} U \\ U^\top \Phi^{21} & U^\top \Phi^{22} U \end{bmatrix} \begin{bmatrix} \hat{F}_i E \\ I \end{bmatrix} = F_i^\top \Phi F_i. \end{aligned} \tag{A.102}$$

Hence, continuity of V implies the existence of a symmetric matrix Φ such that $P_i = F_i^\top \Phi F_i$.

$(ii) \implies (iii)$. Suppose that there exists a symmetric matrix Φ such that $P_i = F_i^\top \Phi F_i$ for all $i \in \mathcal{N}$, and where $F_i = [(E_i R_i^{-1} E)^\top, I]$. By partitioning Φ as

$$\Phi = \begin{bmatrix} \Phi^{11} & \Phi^{12} \\ \Phi^{21} & \Phi^{22} \end{bmatrix}, \tag{A.103}$$

one can write

$$P_i = \Phi^{22} + (\hat{F}_i E)^\top \Phi^{11} \hat{F}_i E + \Phi^{12} \hat{F}_i E + (\hat{F}_i E)^\top \Phi^{21}, \tag{A.104}$$

with $\hat{F}_i = E_i R_i^{-1}$. Now observe that for all $(i, j) \in \mathcal{N} \times \mathcal{N}$ such that $\mathcal{S}_i \cap \mathcal{S}_j \neq \emptyset$, the boundary between two adjacent regions can be written as

$$H_{ij} = (E_i R_i^{-1} - E_j R_j^{-1})E = (\hat{F}_i - \hat{F}_j)E, \tag{A.105}$$

which clearly by construction of the continuity matrices implies $H_{ij}x = 0$ for all $x \in \mathcal{S}_i \cap \mathcal{S}_j$. In turn, one finds $\hat{F}_i E = H_{ij} + \hat{F}_j E$. Then for all $(i, j) \in \mathcal{N} \times \mathcal{N}$ such that $\mathcal{S}_i \cap \mathcal{S}_j \neq \emptyset$ the difference $P_i - P_j$ after simplification reads

$$P_i - P_j = \left(\Phi^{12} + \Phi^{11} \hat{F}_j E\right)^\top H_{ij} + H_{ij}^\top \left(\Phi^{21} + \Phi^{11} \hat{F}_j E\right). \tag{A.106}$$

Since $\mathrm{im} Z_{ij} = \ker H_{ij}$ it holds true that $H_{ij} Z_{ij} = 0$, from which it immediately follows that $Z_{ij}^\top (P_i - P_j) Z_{ij} = 0$.

$(iii) \implies (i)$. Assume there exist matrices P_i such that $Z_{ij}^\top (P_i - P_j) Z_{ij} = 0$. On the boundary between two regions $\mathcal{S}_i$ and $\mathcal{S}_j$, one can write $x = Z_{ij} v$ with $v \in \mathbb{R}^{n-2}$. Assuming the equality holds, then for any $v \in \mathbb{R}^{n-2}$ one finds

$$v^\top Z_{ij}^\top (P_i - P_j) Z_{ij} v = x^\top (P_i - P_j) x = x^\top P_i x - x^\top P_j x = 0, \tag{A.107}$$

which clearly implies a function of the form $V_i(x) = x^\top P_i x$ if $x \in \mathcal{S}_i \cup -\mathcal{S}_i$ to be continuous. This completes the proof.

A.4.4 Proof of Theorem 5.1

Consider the piecewise quadratic function

$$V(x) = V_i(x) := x^\top P_i x, \ \text{ when } x \in \mathcal{S}_i \cup -\mathcal{S}_i, i \in \mathcal{N}, \tag{A.108}$$

where $P_i = F_i^\top \Phi F_i$. The proof is based on showing that, under the conditions of the theorem, V classifies both as a storage function, as well as a suitable ISS-Lyapunov function for the closed-loop system in (5.5) (see Definition 2.5).

First, observe that solutions to (5.5) are locally absolutely continuous, and V is composed by locally Lipschitz continuous functions. By virtue of Proposition 5.2, V is continuous over the boundaries of the partitioning, making V a (non-smooth) locally Lipschitz continuous function on the set $\mathcal{X} = \{x \in \mathbb{R}^n \mid Ex \in \mathcal{F}\}$.

Positive definiteness of V (and thus satisfaction of condition (2.35) in Definition 2.5) is implied by (5.27). Indeed, from the results in Proposition 5.1 and non-negativity of the elements in W_i, it follows that

$$x^\top S_i^\top W_i S_i x \geq 0, \ \text{ when } x \in \mathcal{S}_i \cup -\mathcal{S}_i, \ i \in \mathcal{N}. \tag{A.109}$$

Application of the S-procedure then shows that

$$V(x) > x^\top S_i^\top W_i S_i x \geq 0 \text{ if } 0 \neq x \in \mathcal{S}_i \cup -\mathcal{S}_i,\ i \in \mathcal{N}. \tag{A.110}$$

The strict inequality implies the existence of some sufficiently small $\alpha_1 > 0$ such that $V(x) \geq \alpha_1 \|x\|^2$. Furthermore, due to the piecewise quadratic construction of V there exists $\alpha_2 > 0$ such that $V(x) \leq \alpha_2 \|x\|^2$. Hence, one finds

$$\alpha_1 \|x\|^2 \leq V(x) \leq \alpha_2 \|x\|^2, \tag{A.111}$$

which yields the corresponding result.

Next, it is shown that the LMIs in (5.28) and (5.29) imply satisfaction of the condition (2.36) in Definition 2.5. Hereto, consider first the LMI condition in (5.28). Using the results from Proposition 5.1, and non-negativity of the elements in U_i, one finds that $x^\top S_i^\top U_i S_i x \geq 0$ when $x \in \mathcal{S}_i \cup -\mathcal{S}_i$. Application of the S-procedure in combination with the Schur complement of (5.28) shows that the strict inequality in (5.28) implies for $i \in \mathcal{N}$ that at all points where the gradient of V is defined, it satisfies

$$\begin{aligned}
\nabla V_i(x)(A_1 x + Bw) &= x^\top \left(A_1^\top P_i + P_i A_1\right) x + 2x^\top P_i B w \\
&\leq -\epsilon \|x\|^2 - (Cx + Dw)^\top (Cx + Dw) + \gamma^2 \|w\|^2 \\
&= -\epsilon \|x\|^2 - \|q\|^2 + \gamma^2 \|w\|^2,
\end{aligned}$$

with $\epsilon > 0$ a small constant. In a similar manner, consider the LMI condition in (5.29). Combining the results of Proposition 5.1 with non-negativity of the elements in K_j shows that $x^\top T_j^\top K_j T_j x \geq 0$ for all $x \in \mathcal{T}_j \cup -\mathcal{T}_j$, $j \in \mathcal{M}$. Application of the S-procedure, Finsler's lemma, and the Schur complement then shows that the inequality in (5.29) at all points where the gradient of V is defined it satisfies

$$\begin{aligned}
\nabla V_j(x)(A_2 x + Bw) &= x^\top \left(A_2^\top P_j + P_j A_2\right) x + 2x^\top P_j B w + 2\mu x^\top P_j (b\Pi) x \\
&\leq -\epsilon \|x\|^2 - (Cx + Dw)^\top (Cx + Dw) + \gamma^2 \|w\|^2 \\
&= -\epsilon \|x\|^2 - \|q\|^2 + \gamma^2 \|w\|^2,
\end{aligned}$$

where the algebraic relation $Hx = 0$ is used, which holds true for all $x \in \mathcal{T}_j \cup -\mathcal{T}_j$, $j \in \mathcal{M}$. From the above inequalities, one can conclude that

$$\max_{(k,l) \in \mathcal{I}} \nabla V_l(x)(A_k x + Bw) \leq -\epsilon \|x\|^2 + \gamma^2 \|w\|^2, \tag{A.112}$$

where $\mathcal{I} := \{\{1\} \times \mathcal{N}\} \cup \{\{2\} \times \mathcal{M}\}$. This implies condition (2.36) to hold, thereby demonstrating that under the hypothesis of the theorem, V classifies as an ISS-Lyapunov function. By virtue of Theorem 2.8 this implies ISS of the closed-loop system in (5.5).

In a similar manner, from the above inequalities one can conclude

$$\max_{(k,l)\in\mathcal{I}} \nabla V_l(x)(A_k x + Bw) \leq -\|q\|^2 + \gamma^2\|w\|^2 \tag{A.113}$$

to hold, such that V classifies as a storage function with supply rate $s(w,q,x) = -\|q\|^2 + \gamma^2\|w\|$. The proof can be finished using standard arguments such as in, e.g., [5, 10] for showing that the closed-loop system in (5.5) admits an $\mathcal{L}_2$-gain smaller than or equal to γ.

A.4.5 *Proof of Proposition 5.4*

For proving infeasibility of the LMIs (5.27), (5.28), and (5.30)–(5.33), it is sufficient to prove that the LMIs

$$P_i - S_i^\top W_i S_i \succ 0, \tag{A.114a}$$

$$A_1^\top P_i + P_i A_1 + S_i^\top U_i S_i \prec 0, \tag{A.114b}$$

$$\mathcal{A}_2^\top P_j + P_j \mathcal{A}_2 + T_j^\top K_j T_j + L_j H + (L_j H)^\top \prec 0, \tag{A.114c}$$

with S_i, T_j, and H defined in (5.22), cannot be feasible for some $i \in \mathcal{N}$ and $j \in \mathcal{M}$ under conditions (5.35) and/or (5.36).

C1. Consider the integrator mode subsystem. When feasible, the conditions in (A.114) imply that for any complex non-zero vector $v \in \mathbb{C}^m$ and its complex conjugate $\bar{v}$ it holds that $\bar{v}^\top\left(P_i - S_i^\top W_i S_i\right)v > 0$ and $\bar{v}^\top\left(A_1^\top P_i + P_i A_1 + S_i^\top U_i S_i\right)v < 0$. Choose v to be an eigenvector of A_1 such that $A_1 v = \lambda v$ with $\lambda \in \mathbb{C}$ the corresponding eigenvalue. Observe the identity

$$Ev \overset{(5.7)}{=} \begin{bmatrix} C_z & 0 \\ 0 & 1 \\ C_z A_g & 0 \end{bmatrix} v = \begin{bmatrix} \frac{1}{\omega_h} C_u A_1 \\ C_u \\ \frac{1}{\omega_h} C_u A_1^2 \end{bmatrix} v = \rho C_u v \tag{A.115}$$

with $C_u := [0_{1\times p}, 1]$, $\rho := \begin{bmatrix} \frac{\lambda}{\omega_h} & 1 & \frac{\lambda^2}{\omega_h} \end{bmatrix}^\top$ and the matrices A_g, $C_z = -C_g$ are given in (5.1) and (5.8), respectively. Note that here use is made of the fact that $C_z B_g = 0$ by virtue of Assumption 2.5. Assume $\lambda \in \mathbb{C}$ such that

$$\rho = \frac{1}{\omega_h}\left(\begin{bmatrix} \mathrm{Re}\{\lambda\} \\ \omega_h \\ \mathrm{Re}\{\lambda\}^2 - \mathrm{Im}\{\lambda\}^2 \end{bmatrix} + j \begin{bmatrix} \mathrm{Im}\{\lambda\} \\ 0 \\ 2\mathrm{Re}\{\lambda\}\,\mathrm{Im}\{\lambda\} \end{bmatrix}\right). \tag{A.116}$$

By construction of S_i as in (5.22), it follows that

$$\begin{aligned} S_i v &= R_i^{-1} E v = R_i^{-1} \rho C_u v \\ &= R_i^{-1} \left(\mathrm{Re}\left\{\rho\right\} + j\mathrm{Im}\left\{\rho\right\}\right) C_u v. \end{aligned} \tag{A.117}$$

Note that, additionally,

$$S_i \bar{v} = R_i^{-1} \left(\mathrm{Re}\left\{\rho\right\} - j\mathrm{Im}\left\{\rho\right\}\right) C_u \bar{v}.$$

If $\mathrm{Re}\left\{\rho\right\}, \mathrm{Im}\left\{\rho\right\} \in \hat{\mathcal{S}}_i$ then $R_i^{-1}\mathrm{Re}\left\{\rho\right\} \geq 0$ and $R_i^{-1}\mathrm{Im}\left\{\rho\right\} \geq 0$ hold entry-wise. Similarly, for $\mathrm{Re}\left\{\rho\right\}, \mathrm{Im}\left\{\rho\right\} \in -\hat{\mathcal{S}}_i$ one has $R_i^{-1}\mathrm{Re}\left\{\rho\right\} \leq 0$ and $R_i^{-1}\mathrm{Im}\left\{\rho\right\} \leq 0$. Moreover, if $\mathrm{Re}\left\{\rho\right\} \in \pm\hat{\mathcal{S}}_i$ and $\mathrm{Im}\left\{\rho\right\} \in \mp\hat{\mathcal{S}}_i$, then $R_i^{-1}\mathrm{Re}\left\{\rho\right\}$ and $R_i^{-1}\mathrm{Im}\left\{\rho\right\}$ have opposite signs. For any symmetric matrix X_i with non-negative elements one finds

$$\bar{v}^\top S_i^\top X_i S_i v = \bar{v}^\top C_u^\top \begin{bmatrix} \mathrm{Re}\left\{\rho\right\} \\ \mathrm{Im}\left\{\rho\right\} \end{bmatrix}^\top Q_i \begin{bmatrix} \mathrm{Re}\left\{\rho\right\} \\ \mathrm{Im}\left\{\rho\right\} \end{bmatrix} C_u v \geq 0, \tag{A.118}$$

with $Q_i = \mathrm{diag}\left(R_i^{-\top} X_i R_i^{-1}, R_i^{-\top} X_i R_i^{-1}\right)$ a block diagonal matrix. Note that the cross product of real and imaginary parts cancels.
Suppose (A.114a) holds. Through (A.118) with $X_i = W_i$, this implies $\bar{v}^\top P_i v > 0$, which, in turn, for $\mathrm{Re}\left\{\lambda\right\} > 0$ implies

$$\begin{aligned} &\bar{v}^\top \left(A_1^\top P_i + P_i A_1\right) v + \bar{v}^\top S_i^\top U_i S_i v \\ &= 2\mathrm{Re}\left\{\lambda\right\} \bar{v}^\top P_i v + \bar{v}^\top S_i^\top U_i S_i v > 0. \end{aligned} \tag{A.119}$$

However, this contradicts (A.114b). Hence, under condition (5.35) the inequalities in (A.114) cannot be satisfied, which shows that the LMIs in Theorems 5.1 and 5.2 are infeasible in this case.
C2. Consider the gain mode subsystem. As these dynamics are defined in a lower-dimensional region of $\mathcal{F}$, namely $\mathcal{F}_2$ in (5.3b), it follows from Finsler's lemma that for any $j \in \mathcal{M}$, the first and third LMI in (A.114) can equivalently be written as

$$\Theta^\top \left(P_j - S_j^\top W_j S_j\right) \Theta \succ 0, \tag{A.120a}$$

$$\Theta^\top \left(A_2^\top P_j + P_j A_2\right) \Theta + \Theta^\top T_j^\top K_j T_j \Theta \prec 0, \tag{A.120b}$$

with $\Theta = [I, k_h C_z^\top]^\top$ and A_2 given in (5.8). Also note that $H\Theta = 0$. Next, observe that

$$A_2 \Theta = \begin{bmatrix} A_g & B_g \\ k_h C_z A_g & 0 \end{bmatrix} \begin{bmatrix} I \\ k_h C_z \end{bmatrix} = \begin{bmatrix} A_g + k_h B_g C_z \\ k_h C_z A_g \end{bmatrix} = \Theta \hat{A}_2,$$

where $\hat{A}_2 := A_g - k_h B_g C_g$ with A_g, B_g, $C_z = -C_g$ given in (5.1) and (5.8), and use is made of the fact that $C_g B_g = 0$. Consider $v \in \mathbb{C}^{m-1}$ to be an eigenvector of $\hat{A}_2$ such that $\hat{A}_2 v = \lambda v$ with $\lambda \in \mathbb{C}$ the corresponding eigenvalue. Then, the following identity holds

$$E\Theta v = \begin{bmatrix} C_z & 0 \\ 0 & 1 \\ C_z A_g & 0 \end{bmatrix} \begin{bmatrix} I \\ k_h C_z \end{bmatrix} v = \begin{bmatrix} C_z \\ k_h C_z \\ C_z \bar{A}_2 \end{bmatrix} v = \rho C_z v, \tag{A.121}$$

with $\rho := \begin{bmatrix} 1 \ k_h \ \lambda \end{bmatrix}^\top$. Assume $\lambda \in \mathbb{C}$ such that

$$\rho = \begin{bmatrix} 1 \\ k_h \\ \operatorname{Re}\{\lambda\} \end{bmatrix} + j \begin{bmatrix} 0 \\ 0 \\ \operatorname{Im}\{\lambda\} \end{bmatrix}. \tag{A.122}$$

The vectors $\operatorname{Re}\{\rho\}$ and $\operatorname{Im}\{\rho\}$ both belong to the plane in $\mathcal{F}$ defined by $h\xi = [k_h, -1, 0]\xi = 0$, $\xi \in \mathbb{R}^3$, i.e., the plane in $\mathcal{F}$ on which the gain mode is defined. Consequently, since $\hat{\mathcal{T}}_j$, $j \in \mathcal{M}$ is a subset of this plane, and $\hat{\mathcal{T}}_j \subset \hat{\mathcal{S}}_j$ for all $j \in \mathcal{M}$, it follows that

$$\left\{ \begin{bmatrix} 1 \\ k_h \\ \operatorname{Re}\{\lambda\} \end{bmatrix}, \begin{bmatrix} 0 \\ 0 \\ \operatorname{Im}\{\lambda\} \end{bmatrix} \right\} \subset \hat{\mathcal{T}}_j \Leftrightarrow \{\operatorname{Re}\{\rho\}, \operatorname{Im}\{\rho\}\} \subset \hat{\mathcal{S}}_j.$$

Using a similar reasoning as before, it follows that if (5.36) is true, then $\bar{v}^\top \Theta S_j^\top W_j S_j \Theta v \geq 0$ and $\bar{v}^\top \Theta T_j^\top K_j T_j \Theta v \geq 0$. Suppose (A.120a) is feasible. Together with the above this implies $\bar{v}^\top \bar{P}_j v > 0$ with $\bar{P}_j := \Theta^\top P_j \Theta$. Consequently it must be true that

$$\begin{aligned} &\bar{v}^\top \left(\hat{A}_2^\top \bar{P}_j + \bar{P}_j \hat{A}_2 \right) v + \bar{v} \Theta T_j^\top K_j T_j \Theta v \\ &= 2\operatorname{Re}\{\lambda\} \bar{v}^\top \bar{P}_j v + \bar{v}^\top \Theta T_j^\top K_j T_j \Theta v > 0. \end{aligned} \tag{A.123}$$

This contradicts (A.120), which implies that (A.114) cannot be feasible when (5.36) is true. In turn, this shows that for this case the LMIs in Theorems 5.1 and 5.2 cannot be feasible. This completes the proof.

A.5 Proofs for Chap. 6

A.5.1 Proof of Property 6.1

In proving Property 6.1, first consider the following observations:

(1) The inequality describing the region Ω_1 in (6.22) is directly equivalent to (6.21a).
(2) When both trajectories are in gain mode, i.e., $x'_h - k_h z'$ and $x''_h = k_h z''$ then also $\delta x_h = k_h \delta z$ and thus the incremental sector-condition is inherently satisfied. Hence $(\delta z, \delta x_h) \in \Omega_1$. Note that in this case, (6.21a) is satisfied with equality.
(3) When trajectories belong to Ω_2, and are both in integrator mode, it follows that

$$(\delta \dot{x}_h)\delta x_h = (f_h(x'_h, z') - f_h(x''_h, z''))\delta x_h,$$

which, by virtue of Assumption 6.2, implies that (6.21b) is satisfied in that case.
(4) The set $\mathcal{F}$ in (6.3) can be written as the union of two polyhedral cones, that is, $\mathcal{F} = \mathcal{K} \cup -\mathcal{K}$, with $\mathcal{K} := \{(z, u, \dot{z}) \in \mathbb{R}^3 \mid u \geq 0, k_h z \geq u\}$ and where $-\mathcal{K}$ indicates the set $\{\xi \in \mathbb{R}^3 \mid -\xi \in \mathcal{K}\}$. From the definition of Ω_2 in (6.22) it follows that all $(\delta z, \delta x_h) \in \Omega_2$ inherently violate the $[0, k_h]$-sector condition in (6.22a). It then follows that the pairs (z', x'_h) and (z'', x''_h) must belong to the same cone $\mathcal{K}$, respectively $-\mathcal{K}$ (otherwise the sector-condition is inherently satisfied), and, consequently, z', z'', x'_h, x''_h all have similar signs and z', z'' are non-zero. The latter follows from the fact that if $z' = 0$ (similar for $z'' = 0$), then also $x'_h = 0$, and consequently $k_h \delta x_h \delta z = k_h x''_h z'' \geq x''^2_h = \delta x_h^2$, such that the $[0, k_h]$-sector condition is satisfied and $(\delta z, \delta x_h)$ should belong to Ω_1. This, however, contradicts the earlier assumption that $(\delta z, \delta x_h) \in \Omega_2$.

Given the above observations, it must be shown that (6.21b) holds in the remaining case when $(\delta z, \delta x_h) \in \Omega_2$ and the trajectories being in *different* modes. To show this, the following claim is proven first.

Claim: for all $(\delta z, \delta x_h) \in \Omega_2$ the next implications hold

(1) if $x'_h = k_h z'$ then $z' \delta x_h > 0$,
(2) if $x''_h = k_h z''$ then $z'' \delta x_h < 0$.

Proof of the Claim: Observe that for all $(\delta z, \delta x_h) \in \Omega_2$, the following inequality is satisfied:

$$\delta x_h^2 > k_h \delta z \delta x_h. \tag{A.124}$$

It will be shown that due to this inequality, the only possible cases when $x'_h = k_h z'$ or $x''_h = k_h z''$ are given by the implications that are stated in the claim. The proof is based on inspecting all possible cases individually.

Suppose $x'_h = k_h z'$ and $\delta x_h < 0$. Dividing both sides in (A.124) by δx_h then yields

$$x'_h - x''_h = k_h z' - x''_h < k_h(z' - z''), \tag{A.125}$$

which, in turn, implies $k_h z'' < x_h''$. Due to the sector condition in (6.2a) it must be true that both $z'' < 0$ and $x_h'' < 0$ by observation 4) above. As for all $(\delta z, \delta x_h) \in \Omega_2$ the signals z', x_h', z'', x_h'' have similar signs and z', z'' are non-zero, one finds $z' < 0$, and thus $z'\delta x_h > 0$.

Suppose $x_h' = k_h z'$ and $\delta x_h > 0$. Dividing both sides in (A.124) by δx_h yields

$$x_h' - x_h'' = k_h z' - x_h'' > k_h(z' - z''), \tag{A.126}$$

which implies $k_h z'' > x_h''$. Due to the sector condition in (6.2a) it must be true in this case that both $z'' > 0$, and $x_h'' > 0$. As for all $(\delta z, \delta x_h) \in \Omega_2$ the signals z', x_h', z'', x_h'' have similar signs and are non-zero, one finds $z' > 0$, and, consequently, $z'\delta x_h > 0$.

Suppose $x_h'' = k_h z''$ and $\delta x_h < 0$. It follows from (A.124) that

$$x_h' - x_h'' = x_h' - k_h z'' =< k_h(z' - z''), \tag{A.127}$$

which, in turn, implies $x_h' < k_h z'$. By the sector condition in (6.2a) it must be true that $z' > 0$, and $x_h' > 0$. Sign-equivalence then yields $z'' > 0$, and thus $z''\delta x_h < 0$.

Suppose $x_h'' = k_h z''$, and $\delta x_h > 0$. As before it follows from (A.124) that

$$x_h' - x_h'' = x_h' - k_h z'' > k_h(z' - z''), \tag{A.128}$$

which implies $x_h' > k_h z'$. Invoking the sector condition in (6.2a) shows that $z' < 0$ and $x_h' < 0$. Sign-equivalence yields $z'' < 0$, and thus $z''\delta x_h < 0$, by which the earlier claim is proven.

The proof of Property 6.1 is now continued. Suppose the trajectories are in different modes. Then, according to (6.19) two possibilities arise

$$(\delta \dot{x}_h)\delta x_h = \left(k_h \dot{z}' - f_h(x_h'', z'')\right)\delta x_h, \tag{A.129a}$$

$$(\delta \dot{x}_h)\delta x_h = \left(f_h(x_h', z') - k_h \dot{z}''\right)\delta x_h. \tag{A.129b}$$

From the previous claim, one finds that whenever the first trajectory is in gain mode, and thus $x_h' = k_h z'$, then z' and δx_h have similar signs and are non-zero. Exploiting the inequality $k_h \dot{z}' z' < f_h(x_h', z')z'$ in gain mode, see (6.2b), yields $k_h \dot{z}'\delta x_h < f_h(x_h', z')\delta x_h$. One finds

$$\left(k_h \dot{z}' - f_h(x_h'', z'')\right)\delta x_h \leq (f_h(x_h', z') - f_h(x_h'', z''))\delta x_h. \tag{A.130}$$

In a similar manner, when the second trajectory is in gain mode, and thus $x_h'' = k_h z''$, then according to the claim z'' and δx_h have opposite signs. By the gain mode condition in (6.2b) this leads to $k_h \dot{z}''\delta x_h > f_h(x_h'', z'')\delta x_h$. Consequently,

$$\left(f_h(x_h', z') - k_h \dot{z}''\right)\delta x_h \leq (f_h(x_h', z') - f_h(x_h'', z''))\delta x_h. \tag{A.131}$$

Invoking Assumption 6.2 finally leads to the inequality in (6.21b).

A.5.2 Proof of Theorem 6.3

Without loss of generality one can assume (possibly after a state transformation) that C_g in (6.5) satisfies $\|C_g\| = 1$, such that $\|\delta z(t)\| = \| - C_g \delta x_g(t)\| \leq \|\delta x_g(t)\|$. Since under the small-gain assumption stated in the theorem, the non-incremental system in (6.7), i.e., the interconnection of the LTI system in (6.5) and $\mathcal{H}$ in (6.1) is ISS, it follows that solutions of (6.7) remain bounded and thus $\|\delta z\| < \infty$ for all $t \in \mathbb{R}_{\geq 0}$. Hence, the result from Lemma 6.2 holds true. The solution of the LTI part of the incremental closed-loop system (i.e., the incremental version of (6.5)) is given by

$$\delta x_g(t) = e^{A_g t}\delta x_g(0) + \int_0^t e^{A_g(t-\tau)} B_g \delta u(\tau) d\tau + \int_0^t e^{A_g(t-\tau)} B_w \delta w(\tau) d\tau. \tag{A.132}$$

Since both $\delta u = \delta x_h$ and δw are bounded, an upper-bound on δz is obtained as

$$\begin{aligned} \|\delta z(t)\| \leq \|C_g \delta x_g(t)\| &\leq k e^{-\lambda t} \|\delta x_g(0)\| \\ &+ \gamma_u \left(\sup_{0 \leq \tau \leq t} \|\delta u(\tau)\| \right) + \gamma_w \left(\sup_{0 \leq \tau \leq t} \|\delta w(\tau)\| \right), \end{aligned} \tag{A.133}$$

with $\gamma_u = \int_0^\infty |C_g e^{A_g \tau} B_g| d\tau < \infty$ and $\gamma_w = \int_0^\infty |C_g e^{A_g \tau} B_w| d\tau < \infty$, where boundedness results from the fact that A_g is Hurwitz (see, e.g., [5, p. 174]). Take ϵ such that $\gamma_z = \max\{k_h, c_2/(|c_1| - \epsilon)\}$ is such that $\gamma_z \gamma_u < 1$, which, under the hypothesis $\bar{\gamma}_z \gamma_u < 1$ in (6.29) is always possible (take, for instance, $0 < \epsilon < |c_1| - \gamma_u c_2 < |c_1|$). Applying the small-gain theorem for ISS systems (or in this case input-to-output stable systems) [2, Theorem 2.1] with $\gamma_z \gamma_u < 1$ then shows that there exist $m, \mu, \tilde{\gamma}_w > 0$ such that

$$\|\delta z(t)\| + \|\delta x_h(t)\| \leq m e^{-\mu t} (\|\delta x_g(0)\| + \|\delta x_h(0)\|) + \tilde{\gamma}_w \|\delta w\|_\infty. \tag{A.134}$$

Define $\delta x(t) = [\delta x_g(t)^\top, \delta x_h(t)]^\top$ and observe from the fact that $\|\delta x\| \leq \|\delta x_g\| + \|\delta x_h\|$ that at this point one has the following two inequalities:

$$\|\delta x(t)\| \leq n e^{-\kappa t} \|\delta x(0)\| + \gamma_h \left(\sup_{0 \leq \tau \leq t} \|\delta x_h(\tau)\| \right) + \hat{\gamma}_w \left(\sup_{0 \leq \tau \leq t} \|\delta w(\tau)\| \right) \tag{A.135}$$

$$\|\delta x_h(t)\| \leq m e^{-\mu t} \|\delta x(0)\| + \tilde{\gamma}_w \left(\sup_{0 \leq \tau \leq t} \|\delta w(\tau)\| \right). \tag{A.136}$$

Following the same arguments as in the proof of [2, Proposition 3.1], then shows that there exist $r, \sigma, \bar{\gamma}_w > 0$ such that

$$\|\delta x(t)\| \leq re^{-\sigma t}\|\delta x(0)\| + \bar{\gamma}_w \left(\sup_{0\leq\tau\leq t} \|\delta w(\tau)\| \right), \tag{A.137}$$

which shows that the system in (6.7) is δISS. Together with the existence of an ISS Lyapunov function for the system in (6.7) (which implies the existence of a forward invariant set), it follows by application of Lemma 6.1 that the system is input-to-state convergent.

A.5.3 Proof of Theorem 6.5

The proof is based on showing that the LMI conditions in (6.40) imply the piecewise quadratic function of the form

$$V(\delta x) = U(\delta x_g) + k \max\left\{\delta x_h^2, k_h \delta z^2\right\} + \max\left\{0, \delta x_h(\delta x_h - k_h \delta z)\right)\}, \tag{A.138}$$

where $U(\delta x_g) := \delta x_g^\top M \delta x_g$ to satisfy all conditions in Theorem 6.4 on continuity, positive definiteness, and negative definiteness of the (generalized) time-derivative.

Before proceeding, recall that V induces a partitioning of the incremental input–output region of (6.19), which is depicted in Fig. 6.3. It is useful to write V explicitly, leading to

$$V(\delta x) = \begin{cases} V_1(\delta x) = U(\delta x_g) + k k_h^2 \delta z^2 & \text{if } (\delta x_h, \delta z) \in \hat{\Omega}_1, \\ V_2(\delta x) = U(\delta x_g) + k \delta x_h^2 + \delta x_h(\delta x_h - k_h \delta z) & \text{if } (\delta x_h, \delta z) \in \hat{\Omega}_2, \\ V_3(\delta x) = U(\delta x_g) + k k_h^2 \delta z^2 + \delta x_h(\delta x_h - k_h \delta z) & \text{if } (\delta x_h, \delta z) \in \hat{\Omega}_3, \end{cases}$$

with regions

$$\hat{\Omega}_1 := \left\{(\delta z, \delta x_h) \in \mathbb{R}^2 \mid k_h \delta z \delta x_h \geq (\delta x_h)^2\right\}, \tag{A.139a}$$

$$\hat{\Omega}_2 := \left\{(\delta z, \delta x_h) \in \mathbb{R}^2 \mid (k_h \delta z)^2 \leq (\delta x_h)^2\right\}, \tag{A.139b}$$

$$\hat{\Omega}_3 := \left\{(\delta z, \delta x_h) \in \mathbb{R}^2 \mid -k_h \delta z \delta x_h \geq (\delta x_h)^2\right\}. \tag{A.139c}$$

Remark that $\hat{\Omega}_1 = \Omega_1$ and $\hat{\Omega}_2 \cup \hat{\Omega}_3 = \Omega_2$ with Ω_1, Ω_2 defined in (6.22) (see also Fig. 6.3 for a visualization of these regions).

Continuity: For showing continuity of V in (A.138), it is sufficient to note that U is a Lipschitz continuous function, and the max-functions are locally Lipschitz continuous. As such, V is a locally Lipschitz continuous function.

Positive Definiteness: Observe that condition (6.40a) implies the existence of numbers $\alpha_1, \alpha_2 > 0$ such that

$$\alpha_1 \|\delta x_g\|^2 \leq V_1(\delta x) = U(\delta x_g) + k k_h^2 \delta z^2 \leq \alpha_2 \|\delta x_g\|^2, \tag{A.140}$$

where specifically $\alpha_1 = \lambda_{\min}(M + kk_h^2 C_g^\top C_g)$ and $\alpha_2 = \lambda_{\max}(M + kk_h^2 C_g^\top C_g)$. For all $(\delta x_h, \delta z) \in \hat{\Omega}_1$ the incremental sector condition $\|\delta x_h\| \leq k_h \|\delta z\| \leq k_h \|C_g\| \|\delta x_g\|$ as in (6.21a) holds true such that

$$\alpha_3 \|\delta x\|^2 \leq V(\delta x) = V_1(\delta x) \leq \alpha_4 \|\delta x\|^2, \tag{A.141}$$

with $\alpha_3 = \alpha_1 \cdot \min\left\{\frac{\mu}{(k_h \|C_g\|)^2}, 1 - \mu\right\}$, $0 < \mu < 1$, and $\alpha_4 = \alpha_2$. For showing the left-hand side inequality in (A.141), specific use is made of the sector condition $\|\delta x_h\|^2 - k_h^2 \|C_g\|^2 \|\delta x_g\|^2 \leq 0$.

For all $(\delta x_h, \delta z) \in \hat{\Omega}_2$ it holds that V can be written as

$$V(\delta x) = V_2(\delta x) = V_1(\delta x) + k(\delta x_h^2 - k_h^2 \delta z^2) + \delta x_h(\delta x_h - k_h \delta z).$$

An upper-bound on V for all $(\delta z, \delta x_h) \in \hat{\Omega}_2$ then follows trivially from $\delta x_h(\delta x_h - k_h \delta z) \leq \left(1 + \frac{k_h}{2}\right) \|\delta x_h\|^2 + \frac{k_h}{2} \|C_g\|^2 \|\delta x_g\|^2$, where use is made of Young's inequality. For deriving a lower-bound, two cases are considered: *(i)* $k \geq 0$ and *(ii)* $-1/2 < k \leq 0$. Observe that in both cases one has for all $(\delta x_h, \delta z) \in \hat{\Omega}_2$ that $\delta x_h(\delta x_h - k_h \delta z) \geq \nu \delta x_h(\delta x_h - k_h \delta z)$ with $0 < \nu \leq 1$ and $\delta x_h^2 \geq k_h^2 \delta z^2$. Then, for $k \geq 0$ this results in

$$\begin{aligned} V(\delta x) &\geq \alpha_1 \|\delta x_g\|^2 + \nu \delta x_h(\delta x_h - k_h \delta z) \\ &\geq (1 - \kappa)\alpha_1 \|\delta x_g\|^2 + \frac{\kappa \alpha_1}{\|C_g\|^2} \delta z^2 + \nu(\delta x_h^2 - k_h \delta x_h \delta z) \\ &= (1 - \kappa)\alpha_1 \|\delta x_g\|^2 + \xi^\top L \xi, \end{aligned} \tag{A.142}$$

where $0 < \kappa < 1$, $\xi := [\delta x_h, \delta z]^\top$, and

$$L := \begin{bmatrix} \nu & -\frac{\nu k_h}{2} \\ -\frac{\nu k_h}{2} & \frac{\kappa \alpha_1}{\|C_g\|^2} \end{bmatrix}. \tag{A.143}$$

Since $\nu > 0$ it follows that the matrix L in (A.143) is positive definite if and only if $\frac{\kappa \alpha_1}{\|C_g\|^2} - \frac{\nu k_h^2}{4} > 0$, which is achieved by choosing $0 < \nu < \min\left\{1, \frac{4\kappa \alpha_1}{k_h^2 \|C_g\|^2}\right\}$. Using $\|\xi\|^2 \geq \|\delta x_h\|^2$ one finds for all $(\delta x_h, \delta z) \in \Omega_2$ and $k \geq 0$ that

$$\alpha_5^+ \|\delta x\|^2 \leq V(\delta x) \leq \alpha_6^+ \|\delta x\|^2, \tag{A.144}$$

with

$$\begin{aligned} \alpha_5^+ &= \min\left\{(1 - \kappa)\alpha_1, \lambda_{\min}(L)\right\}, \\ \alpha_6^+ &= \max\left\{\alpha_2 + \frac{k_h}{2}\|C_g\|^2, 1 + k + \frac{k_h}{2}\right\}. \end{aligned}$$

For the second case where $-1/2 < k \leq 0$ one finds

$$
\begin{aligned}
V(\delta x) &\geq \alpha_1 \|\delta x_g\|^2 + |k|(\delta x_h - k_h \delta z)^2 + (1 - 2|k|)\delta x_h(\delta x_h - k_h \delta z) \\
&\geq (1-\kappa)\alpha_1 \|\delta x_g\|^2 + \frac{\kappa \alpha_1}{\|C_g\|^2}\delta z^2 + \varepsilon(\delta x_h^2 - k_h \delta x_h \delta z) \\
&= (1-\kappa)\alpha_1 \|\delta x_g\|^2 + \xi^\top \hat{L} \xi,
\end{aligned}
\tag{A.145}
$$

where $0 < \kappa < 1$, $0 < \varepsilon < 1 - 2|k|$, $\xi = [\delta x_h, \delta z]^\top$, and

$$
\hat{L} := \begin{bmatrix} \varepsilon & -\frac{\varepsilon k_h}{2} \\ -\frac{\varepsilon k_h}{2} & \frac{\kappa \alpha_1}{\|C_g\|^2} \end{bmatrix}. \tag{A.146}
$$

The matrix $\hat{L}$ is positive definite if $\varepsilon < \frac{4\kappa\alpha_1}{k_h^2 \|C_g\|^2}$. Hence, by choosing $0 < \varepsilon < \min\left\{1 - 2|k|, \frac{4\kappa\alpha_1}{k_h^2\|C_g\|^2}\right\}$ one obtains the bounds

$$
\alpha_5^- \|\delta x\|^2 \leq V(\delta x) \leq \alpha_6^- \|\delta x\|^2, \tag{A.147}
$$

with

$$
\begin{aligned}
\alpha_5^- &= \min\left\{(1-\kappa)\alpha_1, \lambda_{\min}(\hat{L})\right\}, \\
\alpha_6^- &= \max\left\{\alpha_2 + \left(k k_h^2 + \frac{k_h}{2}\right)\|C_g\|^2, 1 + \frac{k_h}{2}\right\}.
\end{aligned}
$$

As such, it follows that for all $(\delta x_h, \delta z) \in \hat{\Omega}_2$ one finds

$$
\alpha_5 \|\delta x\|^2 \leq V(\delta x) \leq \alpha_6 \|\delta x\|^2, \tag{A.148}
$$

with $\alpha_5 = \min\left\{\alpha_5^+, \alpha_5^-\right\}$ and $\alpha_6 = \max\left\{\alpha_6^+, \alpha_6^-\right\}$.

Finally, for all $(\delta x_h, \delta z) \in \hat{\Omega}_3$ it holds that V can be written as

$$
V(\delta x) = V_3(\delta x) = V_1(\delta x) + \delta x_h(\delta x_h - k_h \delta z),
$$

which admits an equivalent lower-bound as in (A.148) (this corresponds to the previous case with $k \geq 0$). As such, one finds for all $(\delta x_h, \delta z) \in \hat{\Omega}_3$ that

$$
\alpha_7 \|\delta x\|^2 \leq V(\delta x) \leq \alpha_8 \|\delta x\|^2, \tag{A.149}
$$

where $\alpha_7 = \alpha_5^+$, $\alpha_8 = \max\left\{1 + \frac{k_h}{2}, \alpha_2 + \frac{k_h}{2}\|C_g\|^2\right\}$. Combining (A.141), (A.148), and (A.149) yields a common upper- and lower-bound for all $\delta x \in \mathbb{R}^n$ as

$$
\alpha_9 \|\delta x\|^2 \leq V(\delta x) \leq \alpha_{10} \|\delta x\|^2, \tag{A.150}
$$

with $\alpha_9 = \min\{\alpha_3, \alpha_5, \alpha_7\}$, and $\alpha_{10} = \max\{\alpha_4, \alpha_6, \alpha_8\}$. Hence, the function V in (A.138) satisfies condition (6.30a) in Theorem 6.4.

Generalized Derivative: For the time being, consider $w'(t) = w''(t)$, i.e., $\delta w(t) \equiv 0$ for all $t \in \mathbb{R}$. Then, for all $(\delta x_h, \delta z) \in \hat{\Omega}_1$ one finds

$$\partial V(\delta x)\delta \dot{x} = \mathrm{He}\left(\delta x_g^\top (A_g^\top (M + kC_g^\top C_g)\delta x_g + \delta x_g^\top M B_g \delta x_h\right). \tag{A.151}$$

By pre- and post multiplication of the LMI in (6.40c) with $\delta x^\top$ and δx, respectively, and noting that for all $(\delta x_h, \delta z) \in \hat{\Omega}_1$ one has $\tau_1 \delta x^\top S_1 \delta x = 2\tau_1 \delta x_h (k_h \delta z - \delta x_h) \geq 0$, it follows by virtue of the S-procedure that (6.40c) implies $\partial V(\delta x)\delta \dot{x} \leq -\epsilon_1 \|\delta x\|^2$, $\epsilon_1 > 0$, for all $(\delta x_h, \delta z) \in \hat{\Omega}_1$.

For all $(\delta x_h, \delta_z) \in \hat{\Omega}_2$, one finds at almost all t and for all points where $\partial V(\delta x)$ exists that

$$\begin{aligned} \partial V(\delta x)\delta \dot{x} &= \partial U(\delta x_g)\delta \dot{x}_g + 2(k+1)\delta x_h \delta \dot{x}_h - k_h(\delta x_h \delta \dot{z} + \delta z \delta \dot{x}_h) \\ &= \delta x^\top (\mathcal{A}^\top Q + Q\mathcal{A})\delta x + 2\delta x^\top Q b \phi, \end{aligned} \tag{A.152}$$

where $\mathcal{A}$ and b are given in (6.33), Q is given by

$$Q = \begin{bmatrix} M & -\frac{k_h}{2} C_z^\top \\ -\frac{k_h}{2} C_z^\top & 1+k \end{bmatrix}, \tag{A.153}$$

and $\phi = \delta \dot{x}_h - c_1 \delta x_h - c_2 \delta z$. By observing that $\mathcal{A}^\top Q + Q\mathcal{A}$ corresponds to the matrix in the left-hand side of the LMI in (6.40b), satisfying the LMI in (6.40b) implies that $\mathcal{A}^\top Q + Q\mathcal{A} \prec 0$. Moreover, since $Qb = (1+k)C_u^\top - \frac{k_h}{2} C_z^\top$ it follows that $\delta x^\top Q b \phi = ((1+k)\delta x_h - \frac{k_h}{2}\delta z)\phi$. By virtue of Properties 6.1 and 6.2 one finds along all possible vector fields of δx_h that $((1+k)\delta x_h - \frac{k_h}{2}\delta z)\phi \leq 0$ for all $(\delta x_h, \delta z) \in \hat{\Omega}_2$. As such, $\partial V(\delta x)\delta \dot{x} \leq -\epsilon_2 \|\delta x\|^2$, $\epsilon_2 > 0$, for all $(\delta x_h, \delta z) \in \hat{\Omega}_2$.

In a similar manner, one finds for all $(\delta x_h, \delta_z) \in \hat{\Omega}_2$ that at those points where $\partial V(\delta x)$ exists, and for almost all t

$$\begin{aligned} \partial V(\delta x)\delta \dot{x} &= \partial U(\delta x_g)\delta \dot{x}_g + 2kk_h^2 \delta z \delta \dot{z} + \delta \dot{x}_h(2\delta x_h - k_h \delta z) - k_h \delta x_h \delta \dot{z} \\ &= \delta x^\top (\mathcal{A}^\top \bar{Q} + \bar{Q}\mathcal{A})\delta x + 2\delta x^\top \bar{Q} b \phi, \end{aligned} \tag{A.154}$$

where $\mathcal{A}$ and b are given in (6.33), $\bar{Q}$ is given by

$$\bar{Q} = \begin{bmatrix} M + kk_h^2 C_z^\top C_z & -\frac{k_h}{2} C_z^\top \\ -\frac{k_h}{2} C_z^\top & 1 \end{bmatrix}, \tag{A.155}$$

and $\phi = \delta \dot{x}_h - c_1 \delta x_h - c_2 \delta z$. From the fact that for all $(\delta x_h, \delta z) \in \hat{\Omega}_3$ one has $\tau_2 \delta x^\top S_2 \delta x = -2\tau_1 \delta x_h (\delta x_h + k_h \delta z) \geq 0$. It follows by virtue of the S-procedure that the LMI in (6.40d) implies $\delta x^\top (\mathcal{A}^\top \bar{Q} + \bar{Q}\mathcal{A})\delta x < 0$ for all $(\delta x_h, \delta z) \in \hat{\Omega}_3$. Moreover, since $\delta x^\top \bar{Q} b\phi = (\delta x_h - \frac{k_h}{2}\delta z)\phi$ it follows by a similar reasoning as before that $\partial V(\delta x)\delta\dot{x} \leq -\epsilon_3\|\delta x\|^2$, $\epsilon_3 > 0$ for all $(\delta x_h, \delta z) \in \hat{\Omega}_3$.

Combining all upper-bounds, by Theorem 2.7 one finds

$$\frac{d}{dt}V(\delta x(t)) \leq \max_{p\subset\partial V(x(t))} \langle p, F(x(t), w(t))\rangle \leq -\epsilon\|\delta x(t)\|^2 \tag{A.156}$$

with $\epsilon = \min\{\epsilon_1, \epsilon_2, \epsilon_3\}$ to hold for almost all times t. For $w' \neq w''$, this result is easily extended as $\delta w = w' - w''$ enters the incremental LTI part of the system (i.e., the incremental version of G in (6.5)) in an affine manner. Besides, due to the assumptions on the relative degree of G (Assumption 2.5), the external inputs w' and w'' do not directly enter in δx_h, δz, and $\delta\dot{z}$. By applying Young's inequality one finds for almost all times $t \in \mathbb{R}_{\geq 0}$ that

$$\frac{d}{dt}V(\delta x(t)) \leq -(\epsilon - \sigma)\|\delta x(t)\|^2 + \beta\|\delta w(t)\|^2, \tag{A.157}$$

where $0 < \sigma < \epsilon$ and $\beta = 2\|MB_w\|/\sigma$. As such, condition (6.30b) is satisfied, so that all conditions in Theorem 6.4 are satisfied and the system is input-to-state convergent.

A.5.4 Proof of Theorem 6.7

The proof consists of two main steps. In the first step it is shown through application of Lemma A.2 that feasibility of the two LMIs in (6.44) is equivalent to feasibility of a single matrix inequality. In the second step, it is shown by the Kalman–Yakubovich–Popov (KYP)-lemma (Theorem A.1) that feasibility of this matrix inequality, in turn, is equivalent to satisfying the frequency-domain inequality in (6.45).

First, it is important to note that for feasibility of the LMIs it is necessary that $c_1 < 0$. Via the Schur complement, the LMIs in (6.44a) and (6.44b) can be reformulated as

$$I_1: \ A_g^\top M + MA_g + \frac{1}{2\tau}\left(MB_g - \tau k_h C_g^\top\right)\left(MB_g - \tau k_h C_g^\top\right)^\top \prec 0, \tag{A.158}$$

$$I_2: \ A_g^\top M + MA_g + \frac{1}{2|c_1|}\mathrm{Pm}\left(MB_g + \left(\lambda I + \frac{k_h}{2}A_g^\top\right)C_g^\top\right) \prec 0, \tag{A.159}$$

with $\mathrm{Pm}\,(X) = XX^\top$. Consider the difference $I_2 - I_1$, which reads

$$\begin{aligned} I_2 - I_1 =& \frac{1}{2|c_1|}\mathrm{Pm}\left(MB_g + \left(\lambda I + \frac{k_h}{2}A_g^\top\right)C_g^\top\right) \\ &- \frac{1}{2\tau}\left(MB_g - \tau k_h C_g^\top\right)\left(MB_g - \tau k_h C_g^\top\right)^\top. \end{aligned} \tag{A.160}$$

Using the vector relation $aa^\top - bb^\top = \frac{1}{2}(a+b)(a-b)^\top + \frac{1}{2}(a-b)(a+b)^\top$ for all $a, b \in \mathbb{R}^n$ it follows that (A.160) is equivalent to

$$I_2 - I_1 = pq^\top + qp^\top \tag{A.161}$$

in which p and q are given by

$$p = \frac{1}{\sqrt{2}}\left(a_1 M B_g + Q_1 C_g^\top\right), \quad \text{and} \quad q = \frac{1}{\sqrt{2}}\left(a_2 M B_g + Q_2 C_g^\top\right), \tag{A.162}$$

where

$$a_1 = \left(\frac{1}{\sqrt{2|c_1|}} + \frac{1}{\sqrt{2\tau}}\right), \tag{A.163a}$$

$$a_2 = \left(\frac{1}{\sqrt{2|c_1|}} - \frac{1}{\sqrt{2\tau}}\right), \tag{A.163b}$$

$$Q_1 = \frac{1}{\sqrt{2|c_1|}}\left(\lambda I + \frac{k_h}{2}A_g^\top\right) - \sqrt{\frac{\tau}{2}}k_h I, \tag{A.163c}$$

$$Q_2 = \frac{1}{\sqrt{2|c_1|}}\left(\lambda I + \frac{k_h}{2}A_g^\top\right) + \sqrt{\frac{\tau}{2}}k_h I. \tag{A.163d}$$

By Lemma A.2, it follows that feasibility of the two inequalities I_1 and I_2 is equivalent to the existence of $\varepsilon > 0$ such that the single inequality

$$\begin{aligned} A_g^\top M + M A_g &+ \frac{1}{2\tau}\left(M B_g - \tau k_h C_g^\top\right)\left(M B_g - \tau k_h C_g^\top\right)^\top \\ &+ \left(\frac{\varepsilon}{\sqrt{2}}p + \frac{1}{\varepsilon\sqrt{2}}q\right)\left(\frac{\varepsilon}{\sqrt{2}}p + \frac{1}{\varepsilon\sqrt{2}}q\right)^\top \prec 0 \end{aligned} \tag{A.164}$$

is feasible. Using p and q as in (A.162), it follows that (A.164) can be written as

$$\begin{aligned} A_g^\top M + M A_g &+ \frac{1}{2\tau}\left(M B_g - \tau k_h C_g^\top\right)\left(M B_g - \tau k_h C_g^\top\right)^\top \\ &+ \frac{1}{4}\mathrm{Pm}\left(\left(\varepsilon a_1 + \frac{1}{\varepsilon}a_2\right) M B_g + \left(\varepsilon Q_1 + \frac{1}{\varepsilon}Q_2\right) C_g^\top\right) \prec 0. \end{aligned} \tag{A.165}$$

Define $\bar{B} = [B_1, B_2]$, $\bar{C} = [C_1, C_2]$ with

$$B_1 = B_g, \tag{A.166a}$$

$$B_2 = \frac{\sqrt{2|c_1|}}{2}\left(\varepsilon a_1 + \frac{1}{\varepsilon}a_2\right) B_g, \tag{A.166b}$$

$$C_1 = -\tau k_h C_g^\top, \tag{A.166c}$$

$$C_2 = \frac{\sqrt{2|c_1|}}{2}\left(\varepsilon Q_1 + \frac{1}{\varepsilon} Q_2\right) C_g^\top. \tag{A.166d}$$

With this choice, the inequality in (A.165) can be written more compactly as

$$A_g^\top M + M A_g + \left(M\bar{B} + \bar{C}^\top\right)\Gamma^{-1}\left(M\bar{B} + \bar{C}^\top\right)^\top \prec 0, \tag{A.167}$$

with $\Gamma^{-1} = \mathrm{diag}(1/(2\tau), 1/(2|c_1|))$. Since Γ is invertible, by virtue of Schur's complement, (A.167) is equivalent to

$$\begin{bmatrix} A_g^\top M + M A_g & M\bar{B} + \bar{C}^\top \\ \bar{B}^\top M + \bar{C}^\top & -\Gamma \end{bmatrix} \prec 0. \tag{A.168}$$

Hence, by Lemma A.2, feasibility of (A.168) is equivalent to feasibility of (6.44). Note that under the assumption that $G_{yu}(s)$ is stable, A_g is Hurwitz and M must necessarily be positive definite (see, e.g., [6, Lemma 1.10.1]).

From the KYP-lemma, see Theorem A.1, it follows that, in turn, feasibility of (A.168) is equivalent to satisfying the frequency-domain inequality given by

$$\begin{bmatrix} W_1(j\omega) & W_2(j\omega) \\ 1 & 0 \\ 0 & 1 \end{bmatrix}^* \begin{bmatrix} 0 & C_1^\top & C_2^\top \\ C_1 & -2\tau & 0 \\ C_2 & 0 & -2|c_1| \end{bmatrix} \begin{bmatrix} W_1(j\omega) & W_2(j\omega) \\ 1 & 0 \\ 0 & 1 \end{bmatrix} \prec 0, \tag{A.169}$$

for all $\omega \in \mathbb{R}$, where $W_1(j\omega) = (j\omega I - A_g)^{-1} B_1 \in \mathbb{C}^n$, and $W_2(j\omega) = (j\omega I - A_g)^{-1} B_2 \in \mathbb{C}^n$. Expanding the left-hand side of (A.169) yields

$$\begin{bmatrix} W_1^*(j\omega)C_1^\top + C_1 W_1(j\omega) - 2\tau & W_1^*(j\omega)C_2^\top + C_1 W_2(j\omega) \\ C_2 W_1(j\omega) + W_2^*(j\omega)C_1^\top & W_2^*(j\omega)C_2^\top + C_2 W_2(j\omega) - 2|c_1| \end{bmatrix} =$$
$$\begin{bmatrix} -2\tau\left(k_h \mathrm{Re}\left\{G_{yu}(j\omega)\right\} + 1\right) & -\tilde{W}(j\omega) \\ -\tilde{W}^*(j\omega) & -|c_1|\left(2 - k_1 \mathrm{Re}\left\{(k_2 + k_3 j\omega)G_{yu}(j\omega)\right\}\right) \end{bmatrix}, \tag{A.170}$$

where

$$\tilde{W}(j\omega) = \frac{\sqrt{2|c_1|}}{2}\left((k_2 - k_3 j\omega)G_{yu}^*(j\omega) - \tau k_h k_1 G_{yu}(j\omega)\right) \tag{A.171}$$

and

$$k_1 = \left(\varepsilon + \frac{1}{\varepsilon}\right)\sqrt{\frac{1}{2|c_1|}} + \left(\varepsilon - \frac{1}{\varepsilon}\right)\sqrt{\frac{\tau}{2}} k_h, \tag{A.172a}$$

$$k_2 = \left(\varepsilon + \frac{1}{\varepsilon}\right)\sqrt{\frac{1}{2|c_1|}}\left(\frac{k_h c_1}{2} - c_2\right) + \left(\frac{1}{\varepsilon} - \varepsilon\right)\sqrt{\frac{\tau}{2}}k_h, \tag{A.172b}$$

$$k_3 = \left(\varepsilon + \frac{1}{\varepsilon}\right)\sqrt{\frac{1}{2|c_1|}}\frac{k_h}{2}. \tag{A.172c}$$

By performing a congruence transformation to (A.170) with the matrix

$$T = \text{diag}(1/\sqrt{2\tau}, 1/\sqrt{2|c_1|})$$

one finds the equivalent frequency-domain inequality

$$\begin{bmatrix} k_h \text{Re}\left\{G_{yu}(j\omega)\right\} + 1 & W(j\omega) \\ W^*(j\omega) & 1 - \frac{k_1}{2}\text{Re}\left\{(k_2 + k_3 j\omega)G_{yu}(j\omega)\right\} \end{bmatrix} \succ 0 \tag{A.173}$$

in which

$$W(j\omega) = \frac{1}{2\sqrt{2\tau}}(k_2 - k_3 j\omega)G^*_{yu}(j\omega) - \frac{k_h k_1}{2}\sqrt{\frac{\tau}{2}}G_{yu}(j\omega). \tag{A.174}$$

Hence, satisfying the frequency-domain inequality in (A.173) is equivalent to satisfying (A.169), which, in turn, is equivalent to satisfying the LMIs in (6.44).

Note that the crux in proving necessity and sufficiency of the frequency-domain conditions for feasibility of the LMIs in (6.44) is the possibility to reformulate the two LMIs in (6.44) into a single matrix inequality.

A.5.5 *Proof of Corollary 6.1*

Consider the conditions in Theorem 6.7 with the choice $\varepsilon = 1$. In that case, one has

$$k_1 = \sqrt{\frac{2}{|c_1|}}, \quad k_2 = k_1\left(\frac{k_h c_1}{2} - c_2\right), \quad \text{and} \quad k_3 = k_1\frac{k_h}{2},$$

and the frequency-domain inequality in (6.45) reduces to

$$\begin{bmatrix} k_h \text{Re}\left\{G_{yu}(j\omega)\right\} + 1 & \frac{k_1}{2\sqrt{2}}\hat{W}(j\omega) \\ \frac{k_1}{2\sqrt{2}}\hat{W}^*(j\omega) & 1 + \frac{1}{c_1}\text{Re}\left\{\hat{U}(j\omega)\right\} \end{bmatrix} \succ 0, \tag{A.175}$$

with $\hat{U}(j\omega) = \left(\frac{k_h c_1}{2} - c_2 + \frac{k_h}{2} j\omega\right) G_{yu}(j\omega)$, $\hat{W}(j\omega) = \frac{1}{\sqrt{\tau}}\hat{U}^*(j\omega) - \sqrt{\tau}k_h G_{yu}(j\omega)$. By invoking a Schur complement, it follows that the necessary and sufficient conditions for the matrix in (A.175) to be positive definite are given by

$$M_{11}(j\omega) > 0, \tag{A.176}$$
$$M_{11}(j\omega)M_{22}(j\omega) - M_{12}(j\omega)M_{12}^*(j\omega) > 0, \tag{A.177}$$

for all $\omega \in \mathbb{R}$, and where

$$\begin{aligned} M_{11}(j\omega) &= k_h \mathrm{Re}\left\{G_{yu}(j\omega)\right\} + 1, \\ M_{22}(j\omega) &= 1 + \frac{\lambda}{c_1}\mathrm{Re}\left\{G_{yu}(j\omega)\right\} - \frac{k_h}{2c_1}\omega \mathrm{Im}\left\{G_{yu}(j\omega)\right\}, \\ M_{12}(j\omega) &= \frac{1}{2\sqrt{|c_1|}}\left(\frac{1}{\sqrt{\tau}}\left(\lambda - \frac{k_h}{2}j\omega\right)G_{yu}^*(j\omega) - \sqrt{\tau}k_h G_{yu}(j\omega)\right), \end{aligned}$$

with $\lambda = \left(\frac{k_h c_1}{2} - c_2\right)$. Expansion of the product $M_{12}(j\omega)M_{12}^*(j\omega)$ yields

$$M_{12}(j\omega)M_{12}^*(j\omega) = F_1(j\omega) + F_2(j\omega)$$

with

$$\begin{aligned} F_1(j\omega) = \frac{k_h}{2c_1}\Big(\lambda\left(\mathrm{Re}\left\{G_{yu}(j\omega)\right\}^2 - \mathrm{Im}\left\{G_{yu}(j\omega)\right\}^2\right) \\ -k_h\omega \mathrm{Re}\left\{G_{yu}(j\omega)\right\}\mathrm{Im}\left\{G_{yu}(j\omega)\right\}\Big), \\ F_2(j\omega) = \frac{1}{4|c_1|}\left(\frac{1}{\tau}\left(\frac{k_h^2}{4}\omega^2 + \lambda^2\right) + \tau k_h^2\right)G_{yu}(j\omega)^2. \end{aligned}$$

It follows that a necessary condition for satisfying (A.176) and (A.177) is that $M_{11}(j\omega) = 1 + k_h\mathrm{Re}\left\{G_{yu}(j\omega)\right\} > 0$, resulting in condition (1) in the corollary.

To satisfy the second condition in (A.177), observe that since positive values for τ are considered, one has $F_2(j\omega) \geq 0$ for all $\omega \in \mathbb{R}$. Hence, a necessary condition for the existence of $\tau > 0$ such that the inequality in (A.177) holds is that for all $\omega \in \mathbb{R}$

$$\begin{aligned} &M_{11}(j\omega)M_{22}(j\omega) - F_1(j\omega) = \\ &1 + \left(k_h + \frac{\lambda}{c_1}\right)\mathrm{Re}\left\{G_{yu}(j\omega)\right\} - \frac{k_h}{2c_1}\omega\mathrm{Im}\left\{G_{yu}(j\omega)\right\} + \frac{k_h}{2}\frac{\lambda}{c_1}G_{yu}(j\omega)^2 > 0. \end{aligned} \tag{A.178}$$

Collecting all terms that contain c_1 in (A.178) leads to the equivalent expression

$$1 + X(j\omega) - \beta Y(j\omega) > 0, \tag{A.179}$$

where $\beta = 1/c_1$ and

$$X(j\omega) = \frac{k_h}{2}\text{Re}\left\{G_{yu}(j\omega)\left(3 + \frac{k_h}{2}G_{yu}(j\omega)\right)\right\}, \tag{A.180a}$$

$$Y(j\omega) = \text{Re}\left\{G_{yu}(j\omega)\left(c_2 - \frac{k_h}{2}j\omega + c_2\frac{k_h}{2}G_{yu}(j\omega)\right)\right\}. \tag{A.180b}$$

Hence, condition (2) in the corollary.

Finally, observe that (A.177) can be written as $1 + X(j\omega) - \beta Y(j\omega) > F_2(j\omega)$. Suppose that condition 2) is satisfied, i.e., for all $\omega \in \mathbb{R}$ it holds that $1 + X(j\omega) - \beta Y(j\omega) > 0$. Then, (A.177) is equivalent to (after division by the left-hand side and taking the square root)

$$\left\| \frac{\sqrt{\tau}}{j\omega + 2\omega_c(\tau)} \right\| > \frac{k_h}{4}\frac{\sqrt{|\beta|}\|G_{yu}(j\omega)\|}{\sqrt{1 + X(j\omega) - \beta Y(j\omega)}}, \tag{A.181}$$

with $\omega_c(\tau) = \sqrt{\left(\frac{1}{2\beta} - \frac{c_2}{k_h}\right)^2 + \tau^2}$. This results in condition (3).

References

[1] Ambrosino, R., Garone, E.: Piecewise quadratic lyapunov functions over conical partitions for robust stability analysis. Int. J. Robust Nonlinear Control **25**(14), 2348–2361 (2015)

[2] Jiang, Z., Teel, A., Praly, L.: Small-gain theorem for iss systems and applications. Math. Control Signals Syst. **7**, 95–120 (1994)

[3] Johansson, M.: Piecewise Linear Control Systems-A Computational Approach. Lecture Notes in Control and Information Sciences, vol. 284 (2002)

[4] Kamenetskiy, V.: Frequency-domain stability conditions for hybrid systems. Autom. Remote Control **78**, 2101–2119 (2017)

[5] Khalil, H.K.: Nonlinear Systems, 3rd ed. Prentice-Hall, Upper Saddle River, New Jersey (2002)

[6] Leonov, G.A., Ponomarenko, D.V., Smirnova, V.B.: Frequency-Domain Methods for Nonlinear Analysis: Theory and Applications. World Scientific (1996)

[7] Narendra, K., Taylor, J.: Frequency Domain Criteria for Absolute Stability. Academic Press (1973)

[8] Rantzer, A.: On the kalman—yakubovich—popov lemma. Syst. Control Lett. **28**(1), 7–10 (1996)

[9] Skogestad, S., Postlethwaite, I.: Multivariable Feedback Control: Analysis and Design, 2nd ed. Wiley and Sons (2010)

[10] van der Schaft, A.: L_2-Gain and Passivity Techniques in Nonlinear Control. Springer International Publishing (2017)

Appendix B
Additional Details

B.1 Additional Details for Chap. 3

Some additional details and numerical results for the example presented in Chap. 3 are discussed. This includes the explicit derivation of the relevant closed-loop system matrices, and numerical solutions to the matrix inequalities in Theorem 3.1.

B.1.1 *State Space Derivation*

Consider the feedback control system with HIGS as depicted in Fig. 3.2 and recall that the closed-loop system dynamics can be written as a switched linear system of the form

$$\Sigma : \begin{cases} \dot{x}(t) = A_i x(t) + B_i w(t), & \text{if } (x(t), w(t)) \in \mathcal{X}_i, \quad i = \{1, 2\}, \\ y(t) = Cx(t) + Dw(t), \end{cases} \tag{B.1}$$

with state vector $x(t) = [x_p^\top(t), x_{c_2}^\top(t), x_h(t), x_{c_1}^\top(t)]^\top \in \mathbb{R}^n$, where x_p denotes the states of the plant $P(s)$, and x_{c_1}, x_{c_2} denote the states of the LTI filters $C_1(s)$ and $C_2(s)$, respectively, and x_h denotes the state of HIGS. The exogenous inputs are denoted by $w(t) \in \mathbb{R}^3$, and $y(t) \in \mathbb{R}$ is the relevant performance output. For the considered set-up, the closed-loop system matrices A_1 and A_2 are given by

S. van den Eijnden et al., *Hybrid Integrator-Gain Systems*, Lecture Notes in Control and Information Sciences 499, https://doi.org/10.1007/978-3-032-18101-5

$$A_1 = \begin{bmatrix} A_p & B_p C_{c_2} & B_p D_{c_2} & 0 \\ 0 & A_{c_2} & B_{c_2} & 0 \\ -\omega_h D_{c_1} C_p & 0 & 0 & \omega_h C_{c_1} \\ -B_{c_1} C_p & 0 & 0 & A_{c1} \end{bmatrix}, \tag{B.2a}$$

$$A_2 = \begin{bmatrix} A_p & B_p C_{c_2} & B_p D_{c_2} & 0 \\ 0 & A_{c_2} & B_{c_2} & 0 \\ -k_h(C_{c_1} B_{c_1} C_p + D_{c_1} C_p A_p) & 0 & 0 & k_h C_{c_1} A_{c_1} \\ -B_{c_1} C_p & 0 & 0 & A_{c1} \end{bmatrix}, \tag{B.2b}$$

in which the matrices (A_p, B_p, C_p, D_p) describe the plant $P(s)$, i.e.,

$$P(s) = C_p(sI - A_p)^{-1} B_p + D_p,$$

and the matrices $(A_{c_1}, B_{c_1}, C_{c_1}, D_{c_1})$, and $(A_{c_2}, B_{c_2}, C_{c_2}, D_{c_2})$ describe the filters $C_1(s)$ and $C_2(s)$, respectively, such that

$$C_1(s) = C_{c_1}(sI - A_{c_1})^{-1} B_{c_1} + D_{c_1},$$
$$C_2(s) = C_{c_2}(sI - A_{c_2})^{-1} B_{c_2} + D_{c_2}.$$

For the considered configuration, the relevant input vector is given by $w = (r, \dot{r}, d)^\top \in \mathbb{R}^3$, and the performance channel y is defined as the output of the plant, such that the input matrices B_1, B_2 are given by

$$B_1 = \begin{bmatrix} 0 & 0 & \omega_h D_{c_1} & B_{c_1} \\ 0 & 0 & 0 & 0 \\ B_p & 0 & 0 & 0 \end{bmatrix}^\top, \quad B_2 = \begin{bmatrix} 0 & 0 & k_h C_{c_1} B_{c_1} & B_{c_1} \\ 0 & 0 & k_h D_{c_1} & 0 \\ B_p & 0 & 0 & 0 \end{bmatrix}^\top, \tag{B.3}$$

and with output matrices $C = \begin{bmatrix} C_p & 0 & 0 & 0 \end{bmatrix}$, and $D = \begin{bmatrix} 0 & 0 & B_p \end{bmatrix}$. Finally, the matrices E and F in the sets $\mathcal{X}_i$, $i = \{1, 2\}$ defined in (3.8) are given by

$$E = \begin{bmatrix} -D_{c_1} C_p & 0 & 0 & C_{c_1} \\ 0 & 0 & 1 & 0 \\ -C_{c_1} B_{c_1} C_p - D_{c_1} C_p A_p & -D_{c_1} C_p B_p & 0 & C_{c_1} A_{c_1} \end{bmatrix}, \quad F = \begin{bmatrix} D_{c_1} & 0 & 0 \\ 0 & 0 & 0 \\ C_{c_1} B_{c_1} & D_{c_1} & 0 \end{bmatrix}. \tag{B.4}$$

B.1.2 Numerical Solutions to the Matrix Inequalities

Recall the example of Sect. 3.4 with LTI plant (3.10) and LTI filters (3.13). The respective matrices are given by

$$\left[\begin{array}{c|c} A_p & B_p \\ \hline C_p & D_p \end{array}\right] = \left[\begin{array}{ccc|c} 0 & 1 & 0 & 0 \\ 0 & 0 & 1 & 0 \\ p\omega^2 & \omega(2p\zeta - \omega) & p - 2\zeta\omega & \omega^2 \\ \hline q & 1 & 0 & 0 \end{array}\right],$$

$$\left[\begin{array}{c|c} A_{c_1} & B_{c_1} \\ \hline C_{c_1} & D_{c_1} \end{array}\right] = \left[\begin{array}{c|c} -\frac{1}{\tau} & \frac{1}{\tau} \\ \hline 1 - \frac{1}{\tau\omega_c} & \frac{1}{\tau\omega_c} \end{array}\right], \quad \text{and} \quad \left[\begin{array}{c|c} A_{c_2} & B_{c_2} \\ \hline C_{c_2} & D_{c_2} \end{array}\right] = \left[\begin{array}{c|c} 0 & \omega_i \\ \hline 1 & \omega_i\tau \end{array}\right],$$

with parameters $\omega = 5 \cdot 2\pi$ rad/s, $\zeta = 1.5$, $p = 1$, $q = 3$, $\omega_i = 10$ rad/s, $k_h = 1$, $\omega_h = 0.8$ rad/s, $\omega_c = 1.29$ rad/s, and $\tau = 10^{-4}$. Using the matrices A_1 and A_2 in (B.2), the matrix inequalities in (3.9) are iteratively solved by means of the MATLAB® toolbox `Yalmip` [4] together with the external solver `MOSEK` [1]. The resulting matrix M is found to be

$$M = \begin{bmatrix} 5268896.19 & 1756629.55 & -1384.5403 & -4474.1597 & 61.1298 & 1757783.13 \\ 1756629.55 & 585690.381 & -459.0564 & -1521.3239 & 20.6351 & 586047.961 \\ -1384.5403 & -459.0564 & 0.5415 & -0.87052 & 0.00006012 & -461.2236 \\ -4474.1597 & -1521.3239 & -0.87052 & 27.6577 & -0.2525 & -1500.5993 \\ 61.1298 & 20.6351 & 0.00006012 & -0.2525 & 0.005326 & 20.4599 \\ 1757783.13 & 586047.961 & -461.2236 & -1500.5993 & 20.4599 & 586425.592 \end{bmatrix},$$

with corresponding eigenvalues

$$\lambda(M) = \left[6440982.17\ 58.1617\ 0.01999\ 0.009841\ 0.003120\ 0.0007173\right],$$

showing M to be positive definite, and with $\mu = 1.5$.

B.2 Additional Details for Chap. 4

Numerical results for the rotating beam example presented in Chap. 4 are discussed. This includes solutions to the matrix inequalities for both the nominal system and two perturbed situations.

B.2.1 Numerical Results for the Rotating Beam Example

By virtue of the KYP-lemma, see Theorem A.1, satisfaction of the frequency-domain inequality (4.16) (or equivalently (4.31)) is equivalent to feasibility of the matrix inequality

$$\begin{bmatrix} \mathcal{A}_2^\top M + M\mathcal{A}_2 & Mb - (\lambda(kC_u + H)^\top + F^\top + k\frac{\omega_h}{k_h}H^\top) \\ \star & -2 \end{bmatrix} \prec 0, \tag{B.5}$$

with some matrix $M = M^\top \succ 0$. Positive definiteness of M follows from the fact that $\mathcal{A}_2$ is Hurwitz. For the considered plants $P(s) = P_0(s)(1 + W(s)\Delta(s))$ with $\Delta(s) = \{-1, 0, 1\}$, the frequency-domain inequality (4.16) is satisfied with $k = 10$ and $\lambda = 1.05$. The numerical values for M obtained by solving the LMI in (B.5) with $k = 10$, $\lambda = 1.05$ for each plant are provided next. For the sake of presentation, the matrix M is partitioned as

$$M = \begin{bmatrix} M_{11} & M_{12} \\ M_{12}^\top & M_{22} \end{bmatrix}.$$

Nominal case $P(s) = P_0(s)$:

$$M_{11} = \begin{bmatrix} 3271.2854 & 82193.7721 & 534807.0409 \\ 82193.7721 & 3004106.9455 & 19530867.0090 \\ 534807.0409 & 19530867.0090 & 178395426.3242 \end{bmatrix},$$

$$M_{12} = \begin{bmatrix} 1574208.5725 & 1204195.6982 & 193.6223 \\ 55267119.4324 & 42759474.2704 & 5322.5116 \\ 532538659.7802 & 395523311.3937 & 42307.3639 \end{bmatrix},$$

$$M_{22} = \begin{bmatrix} 2026477013.5053 & 1493823228.4453 & 160464.7104 \\ 1493823228.4453 & 2160096399.0418 & 111424.8123 \\ 160464.7104 & 111424.8123 & 22.7148 \end{bmatrix}.$$

Perturbed case $P(s) = P_0(s)(1 - W(s))$:

$$M_{11} = \begin{bmatrix} 4880.7755 & 251826.0241 & 4191785.8755 & 39016697.2764 \\ 251826.0241 & 16510934.0492 & 302309543.5827 & 2743984637.6464 \\ 4191785.8755 & 302309543.5827 & 7587169458.3818 & 72761822581.433 \\ 39016697.2764 & 2743984637.6464 & 72761822581.433 & 881540530984.12 \end{bmatrix},$$

$$M_{12} = \begin{bmatrix} 199224663.4201 & 472787522.7299 & 328062752.3567 & 194.3089 \\ 13092392632.769 & 29888238457.862 & 20575239560.836 & 9294.7132 \\ 338418179289.5 & 767746791900.84 & 517746603541.71 & 186354.9208 \\ 4599831505793.6 & 9963315985129.3 & 7358773017871 & 2476973.7529 \end{bmatrix},$$

$$M_{22} = \begin{bmatrix} 27545089834524 & 65932522976148 & 46503156198819 & 16452428.8539 \\ 65932522976148 & 191562522048800 & 135108155437422 & 46019570.0211 \\ 46503156198819 & 135108155437422 & 183015500663288 & 31124546.6947 \\ 16452428.8539 & 46019570.0211 & 31124546.6947 & 23.2877 \end{bmatrix}.$$

Perturbed case $P(s) = P_0(s)(1 + W(s))$:

$$M_{11} = \begin{bmatrix} 5236.2926 & 279488.5565 & 4949323.5059 & 40714645.3085 \\ 279488.5565 & 17085147.7713 & 324839323.4501 & 2753567067.9138 \\ 4949323.5059 & 324839323.4501 & 7594793298.9262 & 70205901470.309 \\ 40714645.3085 & 2753567067.9138 & 70205901470.309 & 760739074665.45 \end{bmatrix},$$

$$M_{12} = \begin{bmatrix} 94681293.0386 & 511850879.0378 & 346410474.6424 & 222.5935 \\ 12829886381.139 & 31995174956.989 & 21948710840.095 & 10894.7828 \\ 331294005217.72 & 823668242739.55 & 564716797770.18 & 216726.8692 \\ 4049693004175.1 & 10108792675510 & 7305193912176.1 & 2238633.8050 \end{bmatrix},$$

$$M_{22} = \begin{bmatrix} 24473169128149 & 65339497795751 & 47724442948185 & 15035711.6402 \\ 65339497795751 & 194151848611488 & 149313749273044 & 49491927.7282 \\ 47724442948185 & 149313749273044 & 182828306824733 & 30773635.4780 \\ 15035711.6402 & 49491927.7282 & 30773635.4780 & 26.9618 \end{bmatrix}.$$

A numerical solution to the LMIs in (4.10) for each plant is then given by

$$P = M + \lambda H^\top H + \lambda k(C_u^\top H + H^\top C_u),$$

with $\tau_1 = \lambda k = 10.5$, $\tau_2 = \lambda k \omega_h (k-1)/k_h = 330.75$, $\tau_3 = 0$, and the resulting quadratic (ISS-)Lyapunov function is given by $V(x) = x^\top P x$.

B.3 Additional Details for Chap. 6

B.3.1 Describing Function for HIGS with an Additional Pole

Consider the hybrid integrator in (6.1) with $f_h(x_h, z) = -\alpha x_h + \omega_h z$ representing HIGS with an additional pole $\alpha \in \mathbb{R}$. With this choice, the "extended" hybrid integrator shows a certain resemblance with a first-order reset element (FORE) as considered in, e.g., [3, 5]. When including an additional pole $\alpha > 0$, it can be easily deduced from the proof of Theorem 2.3 in Chap. 2 that $\mathcal{H}$ in (6.1) is incrementally uniformly exponentially stable, which is thus a stronger property than what is derived for $\alpha = 0$, for which only incremental stability was shown. Also, amplitude invariance can be shown in a similar manner as was done in Theorem 2.4. For $\alpha < 0$, however, these properties are harder to verify due to the resulting unstable dynamics in "integrator mode". Still, for inputs that cross zero in finite time such as sinusoids, it can be shown that there can exist only one steady-state response which is also amplitude invariant. Here, the fact that the input crosses zero in finite time is explicitly used. By the grace of these properties it is possible to conduct a describing function analysis (see Chap. 2).

Consider $z(t) = \sin(\omega t)$ as an input to (6.1). Then, with $x_h(0) = 0$, the corresponding unique steady-state response is given by

$$u(t) = \begin{cases} \frac{\omega_h}{\omega^2+\alpha^2}\left(\omega e^{-\alpha t} - \omega\cos(\omega t) + \alpha\sin(\omega t)\right), & \text{for } 0 \le t < \frac{\gamma(\omega)}{\omega}, \\ k_h \sin(\omega t), & \text{for } \frac{\gamma(\omega)}{\omega} \le t < \frac{\pi}{\omega}, \\ \frac{\omega_h}{\omega^2+\alpha^2}\left(-\omega e^{-\alpha(t-\frac{\pi}{\omega})} - \omega\cos(\omega t) + \alpha\sin(\omega t)\right), & \text{for } \frac{\pi}{\omega} \le t < \frac{\pi+\gamma(\omega)}{\omega}, \\ k_h \sin(\omega t), & \text{for } \frac{\pi+\gamma(\omega)}{\omega} \le t < \frac{2\pi}{\omega}, \end{cases}$$

where $t = \gamma(\omega)/\omega$ is the switching instance, which follows from the intersection

$$\frac{\omega_h}{\omega^2+\alpha^2}\left(\omega e^{-\alpha\frac{\gamma(\omega)}{\omega}} - \omega\cos(\gamma(\omega)) + \alpha\sin(\gamma(\omega))\right) = k_h \sin(\gamma(\omega)). \tag{B.6}$$

While for $\alpha = 0$, the closed-form solution $\gamma(\omega) = 2\arctan(k_h\omega/\omega_h)$ can be found, for $\alpha \neq 0$ (B.6) does not admit a closed-form solution. Nevertheless, a numerical solution can be found efficiently via, e.g., standard zero-finding algorithms. The steady-state response to a sinusoid for different ratios of α/ω_h is shown in Fig. B.1.

The Fourier coefficients of the first harmonic in u are given by

$$a_1(\omega) = \frac{2\omega}{\pi}\left(\frac{\omega_h}{\omega^2+\alpha^2}\right)\left(\frac{\omega^2}{\omega^2+\alpha^2}\left(\frac{\alpha}{\omega} - e^{-\alpha\frac{\gamma}{\omega}}\left(\frac{\alpha}{\omega}\cos(\gamma) - \sin(\gamma)\right)\right) - \frac{2\gamma+\sin(2\gamma)}{4}\right) + \left(\frac{\alpha\omega_h}{\omega^2+\alpha^2} - k_h\right)\frac{1-\cos(2\gamma)}{2\pi},$$

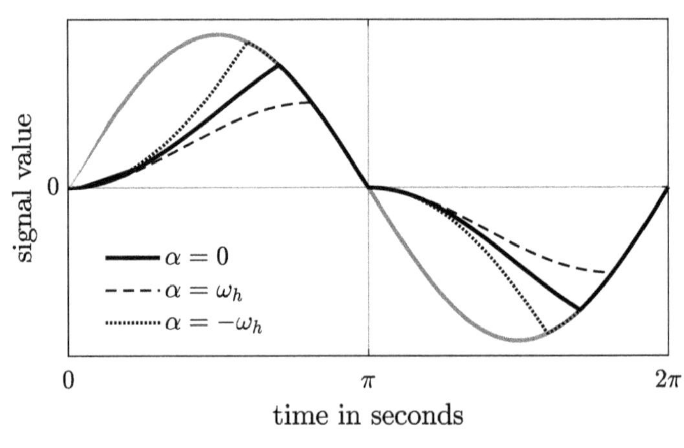

Fig. B.1 Response of $\mathcal{H}$ in (6.1) with $f_h(x_h, z) = -\alpha x_h + \omega_h z$ to a sinusoidal input for $k_h = 1$, $\omega_h = \pi$ rad/s, and different ratios of α/ω_h

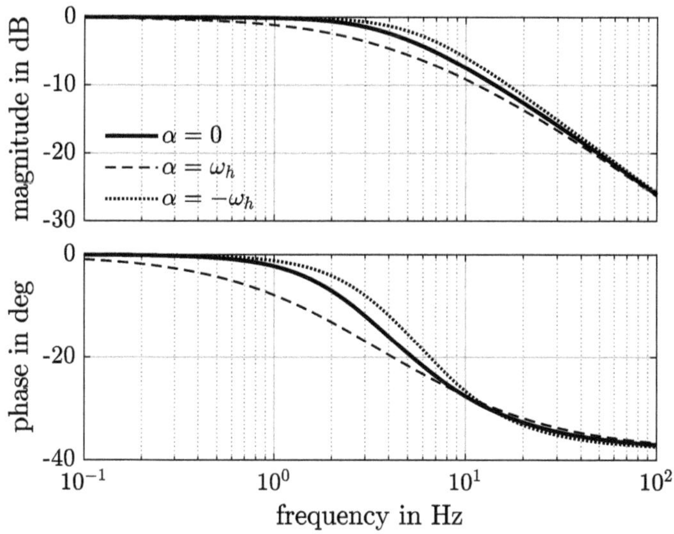

Fig. B.2 Describing function $\mathcal{D}(j\omega)$ of $\mathcal{H}$ in (6.1) with $f_h(x_h, z) = -\alpha x_h + \omega_h z$ for $k_h = 1$, $\omega_h = \pi$ rad/s, and different ratios of α/ω_h

and

$$
\begin{aligned}
b_1(\omega) = &\frac{2\omega}{\pi}\left(\frac{\omega_h}{\omega^2+\alpha^2}\right)\left(\frac{\omega^2}{\omega^2+\alpha^2}\left(1-e^{-\alpha\frac{\gamma}{\omega}}\left(\cos(\gamma)+\frac{\alpha}{\omega}\sin(\gamma)\right)\right)-\frac{\sin^2(\gamma)}{2}\right)\\
&+k_h+\left(\frac{\alpha\omega_h}{\omega^2+\alpha^2}-k_h\right)\frac{2\gamma-\sin(2\gamma)}{2\pi},
\end{aligned}
$$

and the describing function is given by $\mathcal{D}(j\omega) = a_1(\omega) + jb_1(\omega)$. Note that for $\alpha = 0$ one recovers the Fourier coefficients for HIGS as given in (2.17). The corresponding describing function characteristics for different ratios of α/ω_h are shown in Fig. B.2. It can be seen that the choice $\alpha > 0$ (stable pole) results in more damping, whereas the opposite is seen for the choice $\alpha < 0$ (unstable pole). For ratios satisfying $|\alpha/\omega_h| < 1$ the discrepancy with the case $\alpha = 0$ appears to be small. Note that in all cases, $\lim_{\omega\to\infty} \mathcal{D}(j\omega) = \frac{\omega_h}{j\omega}(1 + 4j/\pi)$, which resembles the describing function of the classical Clegg integrator, see also Chap. 2 and [2, 3].

References

[1] Andersen, E., Roos, C., Terlaky, T.: On implementing a primal-dual interior-point method for conic quadratic programming. Math. Program. **95**, 249–277 (2003)

[2] Clegg, J.C.: A nonlinear integrator for servomechanisms. Trans. Am. Inst. Electric. Eng. Part II Appl. Ind. **77**(1), 41–42 (1958)

[3] Horowitz, I., Rosenbaum, P.: Non-linear design for cost of feedback reduction in systems with large parameter uncertainty. Int. J. Control **21**(6), 977–1001 (1975)
[4] Löfberg, J.: Yalmip: a toolbox for modeling and optimization in MATLAB®. In: 2004 IEEE International Conference on Robotics and Automation, pp. 284–289 (2004)
[5] Zaccarian, L., Nešić, D., Teel, A.: First order reset elements and the clegg integrator revisited. In: Proceedings of the 2005, American Control Conference, 2005, vol. 1, pp. 563–568 (2005)

Index

S. van den Eijnden et al., *Hybrid Integrator-Gain Systems*, Lecture Notes in Control and Information Sciences 499, https://doi.org/10.1007/978-3-032-18101-5

GPSR Compliance
The European Union's (EU) General Product Safety Regulation (GPSR) is a set of rules that requires consumer products to be safe and our obligations to ensure this.

If you have any concerns about our products, you can contact us on

ProductSafety@springernature.com

In case Publisher is established outside the EU, the EU authorized representative is:

Springer Nature Customer Service Center GmbH
Europaplatz 3
69115 Heidelberg, Germany

www.ingramcontent.com/pod-product-compliance
Ingram Content Group UK Ltd.
Pitfield, Milton Keynes, MK11 3LW, UK
UKHW020934180926
475692UK00003B/29
9783032181008